Project Management
in Construction

ABOUT THE AUTHOR

Sidney M. Levy is an independent construction industry consultant in Baltimore, Maryland with more than 40 years of experience in the profession. He is the author of numerous books on construction methods and operations, including *Construction Databook, Design-Build Project Delivery*, and *Construction Superintendent's Operations Manual*. Mr. Levy was awarded the British Chartered Institute of Building Silver Medal for *Project Management in Construction*, Third Edition, in the category of Managing Construction.

Project Management in Construction

Sidney M. Levy

Fifth Edition

McGraw-Hill

New York   Chicago   San Francisco   Lisbon   London   Madrid
Mexico City   Milan   New Delhi   San Juan   Seoul
Singapore   Sydney   Toronto

Library of Congress Cataloging-in-Publication Data

Levy, Sidney M.
　　Project management in construction / Sidney M. Levy.—5th ed.
　　　　p.　　cm.
　　Includes index.
　　ISBN-13: 978-0-07-146417-8
　　ISBN-10: 0-07-146417-4 (alk. paper)
　　1. Construction industry—Management.　2. Project management.
　　3. Building—Superintendence.　I. Title.
　　HD9715.A2L44　　2006
　　624.068′4—dc22

　　　　　　　　　　　　　　　　　　　　　　　　　　2006015584

1 2 3 4 5 6 7 8 9 0　DOC/DOC　0 1 3 2 1 0 9 8 7 6

ISBN-13: 978-0-07-146417-8
ISBN-10: 0-07-146417-4

The sponsoring editor for this book was Larry S. Hager, the editing supervisor was David E. Fogarty, and the production supervisor was Pamela A. Pelton. It was set in Century Schoolbook by International Typesetting and Composition. The art director for the cover was Margaret Webster-Shapiro.

Printed and bound by RR Donnelley.

This book was printed on acid-free paper.

McGraw-Hill books are available at special quantity discounts to use as premiums and sales promotions, or for use in corporate training programs. For more information, please write to the Director of Special Sales, McGraw-Hill Professional, Two Penn Plaza, New York, NY 10121-2298. Or contact your local bookstore.

Contents

Preface

The importance of the project manager's role in the construction process cannot be overstated. From the two-dimensional plans and specifications that mark the first step of the building process to the ribbon cutting ceremony at project completion, the project manager is a key player.

The path from Point A to Point B is rarely without its detours and roadblocks and the project manager is charged with the responsibility of traversing those detours and tearing down the impasses.

A project manager must be technically competent and possess the management skills necessary to effectively control the teams of subcontractors, vendors, and field personnel required to provide the smooth flow of tradespeople and materials needed to get the job done. A little knowledge of accounting procedures, legal matters, and state and federal regulations is also best in order to effectively deal with the many forces that bear on the construction process.

Part-time instructor, father confessor, disciplinarian, and mediator, one thing is for sure, a project manager's daily routine is never the same. Only change is constant.

In the end, project management is an exercise in control: control over quality, schedule, and costs—each one a full-time job, yet all falling under the aegis of the project manager. *Project Management in Construction*, now in its Fifth Edition, examines some of the basic tenets of managing a construction project and explores new technologies that will impact how we do business in the future.

I have been associated with the construction industry for more than 40 years, progressing from summer jobs in high school to time clerk, labor foreman, assistant superintendent, project manager, senior vice-president of a major New England builder/developer, and now owner of my own consulting business in Baltimore, Maryland. These experiences have given me a broader view of the construction process, and this book is an attempt to share some of my experiences with you, the reader.

To quote an anonymous but savvy old-timer, "Smart men learn from experience. Wise men learn from the experience of others."

It is my hope that *Project Management in Construction, Fifth Edition* will provide some new insights, keeping that old adage in mind.

SIDNEY M. LEVY

An Introduction to the Construction Industry

The construction industry, like so many basic American industries, is being transformed to meet the new demands of the twenty-first century. Project delivery concepts are changing—design-bid-build, in both the private and public sectors is now recognized as less than efficient when it comes to both time and money, often promoting litigation and restricting innovation. Relationships between owners, design consultants, and contractors are changing as well, given how the emphasis on design-build has reinforced the positive effects of team collaboration. Technological advances have produced 3-D and 4-D building information modeling, resulting in the potential to construct a building, step-by-step, virtually within a computer before a shovel is ever placed in the ground.

These new building delivery systems and technologies place more responsibility on the project managers who, as before, are charged with "getting it built." The construction industry in the United States exceeded $1.1 trillion in 2005 and remains a vital part of the economy, providing jobs for more than 12 million people, along with untold millions in revenue to those industries dependent upon construction activities. We are a unique business, incorporating everything from small residential remodeling contractors to giant multinational constructors.

The construction industry can also be characterized as being highly fragmented, and although there are approximately 700,000 contractors in the United States, slightly more than 400,000 of them have less than four employees, while only one percent have more than 100. A highly competitive business, it is one in which profit margins are slim. Statistics compiled by the Construction Finance Management Association in

Princeton, New Jersey, reveal that over the past several years, net profit, after taxes, for industrial and nonresidential contractors ranged from 1.2 percent to 1.5 percent, a figure considerably less than the interest accrued in a passbook savings account.

Like with so many other businesses and institutions, the new century holds untold opportunities and challenges, as well as plenty of detours for the unwary. The paperless workplace predicted by computer gurus decades ago is now gradually unfolding. In mid-2005, General Motors announced they would build two automotive plants utilizing 3-D building information modeling techniques to define the project's geometry and produce fully integrated and coordinated structural, mechanical, and electrical details—without generating one piece of paper.

Critical Issues Facing Contractors in this New Millennium

As the first decade of the new century unfolds, some demanding issues in the construction industry have become apparent, while others remain more subtle. Both institutional perception and resource changes are taking place, affecting all facets of the industry, such as the following, which will act as the headings of upcoming sections in this chapter:

- How our industry is perceived
- Information technology
- Human resources—the changing workforce
- Productivity
- Quality control
- Project delivery systems
- The organization
- Construction technology
- Safety

How our industry is perceived

In 2004, the Construction Management Association of America (CMAA), in collaboration with FMI (the management consulting firm headquartered in Denver), conducted a survey to determine how the industry viewed the ethical practices of its peers. The survey was directed toward owners, architects, construction managers, general contractors, and subcontractors. Their responses were not heartening.

Eighty-four percent of respondents said they encountered situations that they considered unethical in their business dealings, while 61 percent said the industry was "tainted" by unethical acts. Thirty-four percent, meanwhile, claimed they had experienced unethical acts on several occasions. Of course, the construction industry alone can't be singled out for ethical lapses. Recently, the media has feasted on ethical and criminal acts committed by executives at Enron, WorldCom, and Tyco. Nevertheless, we in the construction business should revisit the way we do business in order to display for all to see just how the overwhelming majority of our firms operate: ethically, conscientiously, and with a strict work discipline. No portion of the industry was spared criticism by the CMAA—not owners, designers, or contractors.

Owners were blamed for authorizing work and then trying not to pay for it. They were accused of passing off some of their responsibilities to others and playing games with payments, as well as still shopping for prices once all the bids were received. And what about those reverse auctions on the Internet where some disreputable owners trolled fictitious low bids in the hope of hooking some desperate contractor?

Contractors, on the other hand, were accused of overbilling, front-end loading, bid shopping, and playing change-order games. Architects and engineers were chided for doing whatever was necessary to make their clients (owners) happy, often at the expense of the contractor, and for knowingly issuing drawings and other bid documents that were defective and deficient. Changing the perception of our industry is a difficult task, but the responsibility to do so rests with each of us—owner, design consultant, general contractor, and specialty contractor. We must launch conversations about ethical practices, and not ignore those that are unethical, just as we must make it policy to say "No" and walk away from situations that could compromise our integrity. Thus, in the end it is *our* responsibility to remove those few rotten apples from the barrel.

Information technology

The ability to communicate more rapidly and more accurately has transformed both the design and construction segments of the industry. Wireless mobility has freed the project manager and their field supervisors from their copper umbilical cord so they can now instantaneously transmit and receive verbal and written directives at the touch of a button from their office, from the field, or traveling in between. Architects and engineers have already advanced computer-assisted design in 3-D and 4-D modeling to the point where it promises to produce a seamless flow of design that will reduce systems interference issues to zero. Even further advances integrating design and construction are on the way.

By tapping into the global market, architects and engineers in this country can outsource work at the end of each workday to the other side of the world where the sun is just beginning to rise, thereby transforming each task into a round-the-clock effort.

Human resources—the changing workforce

The growing shortage of skilled workers and experienced managers that began to appear in the 1980s has reached dramatically high levels in today's marketplace, and remains one of the major challenges facing the industry. Signing bonuses for managers, once relegated to professional athletes, are now prevalent in many areas of the country, and reveal the desperation of some contractors to attract productive employees.

Company benefits over and above those included in collective bargaining agreements are now offered to select employees to keep them from jumping ship to their competitors. A major challenge of this twenty-first century will be to recruit and train new trade and manager entrants to the construction industry, a task that is vital to the interests of the country.

Union membership rolls have been decreased from a high of about 30 percent of U.S. workers in 1948 to slightly more than 8 percent in 2005, and with the coming retirement of many experienced tradesmen, the pool of skilled workers will be reduced even more. To attract people to this industry, we must work hard to change the public's perception of construction as the industry of the four Ds: dull, dirty, demanding, and dangerous. An aging workforce, the absence of apparent technological advances, and the lure of more attractive vocations have all contributed to this image.

The demographics of our population sounded the warning bell several decades ago, but we failed to recognize them. By the year 2010, it's estimated that the number of 55- to 64-year-old males will outnumber the 18- to 24-year-old group by at least 1.5 million. A survey of the age of field managers in 2004 revealed that 80 percent were over the age of 36, and nearly half were older than 45. Thus, we must fill these ranks as today's older managers approach retirement age.

The undocumented immigrant problem

The influx of immigrant workers that began in the late 1980s has accelerated, and today more and more jobsites require supervisors who are bilingual. The Pew Center in Washington, D.C., estimated that from 2000 to 2004, the undocumented migrant population increased by 10.3 million. About 20 to 25 percent of the entire construction workforce in this country falls within that category. This presents a twofold problem for construction managers: a logistical one—the need to fill the labor pool with unskilled workers, and a legal one—attempting not to run afoul of the law in doing so.

The Bureau of Labor Statistics says that the construction industry will have to add 100,000 workers each year through 2012 to keep up with demand. Thus, these pools of construction day laborers are responding to a demand, particularly in the residential construction field where low-skill jobs such as painting, roofing, and landscaping have sharply increased due to the housing boom, and a situation in which these types of jobs are rejected by the local workforce.

At the other end of the labor force, high-paid union workers are declining in number as union membership continues its downward spiral from a peak in the 1950s when it represented 35 percent of all workers in this country. Since then, the number of private-sector union workers has continued to fall. This is not too surprising in the construction industry when analyzing the cost of union labor, where many "burdened" hourly rates in 2005 were in the $75.00 to $95.00 range, compared to some open-shop contractors who had labor rates of $30.00 to $75.00 per hour.

Productivity

Opinions vary as to whether productivity in the construction industry has remained flat over the past decade or increased marginally. A 2003 *Construction Industry Productivity Survey Report*, prepared by FMI, revealed that 47 percent of respondents, primarily general contractors, and secondarily, specialty contractors, indicated that productivity had remained the same or actually decreased in some areas. Of note to project managers, 81 percent of those answering the survey believed they could save 5 percent or more on field labor costs through better management practices.

Companies were found to invest less in productivity improvement than in safety programs; however, this could be misleading since increased safety at the jobsite should, as a by-product, increase productivity by retaining core work groups.

Those reporting in the FMI survey produced a list of five items that impact productivity, as well as the four greatest external challenges to improving productivity:

Five items impacting productivity

- Lack of planning skills at the management level
- Lack of communication skills at the management level
- Poor communication between project manager and their field management team

- Lack of technical training at the craft level
- Cultural resistance to change (This can be interpreted as "I've always done it this way and it worked—so why change?")

Four greatest external challenges to improve productivity

- Poor quality of plans and specifications
- Poor coordination by owners, general contractors, and/or construction managers
- Slow responses from other members of the team: architects, engineers, customers, general contractors, and/or construction managers
- Lack of available and qualified craft personnel

Quality control

"Do it right, and do it right the first time" is a concept that will take on more importance in this and future decades. The shortage of skilled workers and experienced managers should increase the pressure placed on project managers from owners demanding, among other things, less call backs and less rework—in effect, higher quality.

In this competitive age, if your firm can't produce a quality product and produce it both quickly and at a competitive price, owners will look elsewhere. Organizations such as The International Organization for Standardization in Switzerland has developed two generic standards for the worldwide construction industry in order to drive the construction industry quality control engine. ISO Standard 9000 applies to quality management systems, while ISO deals with environmental management systems—but we don't have to look that far to see what else is being done to embrace quality.

In Providence, Rhode Island, Gilbane Construction, an industry leader in many areas, established a new department in 2005—Client Satisfaction—to raise client service to a new high by focusing on their needs and developing in-house programs that address them. Internal audits as a project begins, coupled with questionnaires to the owner and their design consultants, results in a regional and corporate level review to insure that the client's needs are being fully met. Gilbane's Director of Client Satisfaction says that the company strives to exceed the client's expectations and not accept that the job is "just OK."

Project delivery systems

Although guaranteed maximum price (GMP) type contracts and construction manager (CM) contracts continue to dominate project delivery systems, the search goes on for a better approach. As always, the cost of design and construction is a key concern of all parties; meeting schedules, often tightly compressed, keeps the pressure on. Avoidance of disputes between owners, design consultants, and contractors continues to be a prime objective of all parties, and a method to promote more accurate and error-free design documents is another goal pursued by the industry.

For the moment, design-build seems to address speed of delivery when including both design and construction time, lessening disputes due to the collaborative process it creates, and even results in lower overall costs in many cases because it generates fewer change orders. Design-build has been growing rapidly and some experts project that 45 percent of all projects will be design-build by the end of 2010. When coupled with the latest electronic advances in building information modeling, it would appear that design-build is the project delivery system to beat.

The organization

Of the reported 656,448 general contractors operating in the United States, the overwhelming majority remain small businesses with modest annual volumes, operating in a limited geographic area.

McGraw-Hill's *Engineering News Record Magazine* in their 2004 edition of the Top 400 Contractors in the United States reveals the wide gap between #1 Fluor Corporation with $13 billion in sales and the #400 firm with $112 million in sales. That leaves about 656,000 other contractors with a significantly smaller annual sales volume. The demographics of the general contracting business validates this small business concept, as shown next (Table 1.1):

TABLE 1.1   Number of General Contractors by Revenue and Employees

Size of business by annual revenue	Number of establishments
Less than $25,000	29,127
$100,000 to $249,999	166,948
$250,000 to $499,999	118,463
$500,000 to $4.9 million	312,788
$5 million to $9.9 million	16,021
Over $10 million	13,101
Total	656,448

Size of establishment by number of employees*	Number of establishments
1 to 4 employees	409,256
5 to 9 employees	123,389
10 to 19 employees	67,093
50 to 99 employees	10,958

*Does not include companies with employees in excess of 99; therefore, total does not equal 656,448.

SOURCE: Department of Commerce—U.S. Census

Construction technology

In the 1970s, spurred on by their concerns over an aging population and the potential lack of skilled workers in their construction industry, the Japanese began to develop a whole series of construction robots including excavation and real-time compaction robots, as well as rebar bending, concrete placement, and floor-finishing robots. Second- and third-generation models emerged in the mid-1990s and it appeared that the robots' commercial production was just around the corner when the Japanese economy spiraled downward at the end of that decade putting most of these expensive programs on hold. Nevertheless, they showed what could be done.

Global Positioning System (GPS) satellites were used back then with increased frequency by survey parties, and now GPS is being investigated not only as a way to verify line and grade, but to guide unmanned excavating machines to produce those lines and grade. Combined with 3-D modeling, it is now possible to create a three-dimensional site plan on a remote console and, similar to a video game, grip two joysticks and actually direct an unmanned bulldozer to work the site. The equipment manufacturer, John Deere, is currently equipping 5 percent of its dozers with GPS units and by 2015 predicts that 20 percent of their new bulldozers will be equipped with automatic grade controls. This may be the first of a wave of automated construction machinery to be produced by U.S. manufacturers. After all, the Japanese have shown us it can be done!

Safety

Once again, the scarcity of skilled workers places added importance on maintaining a safe working environment, not only to polish the industry's image, but to retain the integrity of productive work teams.

More owners, aware of safety from both a moral and cost standpoint, are requiring that contractors provide them with a history of safe working conditions as part of the project bid criteria. Although construction accounted for approximately 6.6 percent of the total workforce in the United States in 2000, it had the dubious distinction of accounting for 19.5 percent of all workplace fatalities. As worker compensation insurance rates continue to remain a significant factor in the calculation of a company's overhead—and therefore its competitiveness—builders need to be more aware of the cost implications of a poor safety record as well as a good one.

The changing marketplace

Tomorrow's managers will have to become more astute and selective in defining their markets in the face of stiffer competition, and more general contractors are looking to specialization, or niche marketing, to do so.

Niche marketing will become more important as each company seeks to exploit its experience and expertise in the field of its choice and thus narrow the field of competition. Marketing or sales development, long ignored by many contractors who depended upon word-of-mouth in a defined geographic marketing area, has now become not only essential, but sophisticated, as witnessed by the proliferation of contractor Internet web sites. When the longest economic expansion in the nation's history ended dramatically in mid-2001, contractors needed to reassess their market strategy to assure that they had the necessary tools available to survive and thrive in the coming years.

The changing role of the general contractor.   The character and role of the general contractor has changed dramatically over the past 50 years or so. The time when the general contractor employed crews of laborers, carpenters, masons, and operating engineers, and owned substantial numbers of excavating equipment ("iron"), performing significant amounts of work with their own forces, is largely over. Construction projects became more sophisticated in design, and as competition intensified, the reliance on specialty contractors, *subcontractors*, soon became the order of the day. As early as 1991, subcontractors accounted for 75 percent of all construction company establishments, and that number has been growing ever since. Change is nothing new to contractors. When activity in new construction lessens, contractors pursue renovation, rehabilitation, and interior tenant fit-up work. When private sector work decreases, the contractor looks to the public sector for projects. But in the coming years, along with the usual concerns over costs and getting the job done on time, other factors will occupy the thoughts and actions of progressive contractors. The electronic age is here to stay and those contractors that fail to embrace the advantages it has to offer will find themselves at a disadvantage. Old ideas must give way to the new, presenting yet further challenges that must be surmounted, just as the problems of the past were.

The project manager's role.   With all of the changes taking place in the industry, the project manager's role remains constant: control over both the work process and the costs associated with that work. Management of a construction project can be divided into four components:

- **Construction engineering**   The proper technique of assembling materials, components, equipment and systems, and the selection and utilization of the best construction technology to do so.

- **Management of the construction process**   Establishing the most effective way to implement the construction process, including proper scheduling and the coordination and control of the flow of labor, materials, and equipment to the jobsite.

- **Human resources management** Since labor productivity and a harmonious working environment are essential elements of a successful project, control over human resources becomes important, more so than ever these days where shortages of both skilled workers and experienced managers exist.

- **Financial management** Construction is a high-risk business with historically low profit margins. Control over costs, cash flow, and adequate project funding is critical to the success of any business endeavor, and construction is certainly no exception.

All of these key management functions, to some degree or other, will fall upon that most visible member of the construction team, the project manager, who must not forget the seven criteria essential to the successful completion of a project.

- The project was completed on time.

- The complete project cost remained within budget.

- The quality levels expected were achieved.

- The project was completed with no unresolved disputes and no outstanding claims.

- The contractor maintained a professional relationship with the designers—the architect and engineers.

- The contractor maintained a mutually beneficial relationship with all subcontractors and vendors.

- The contractor-client relationship was a good one.

The project manager's role in the construction process may vary from company to company depending, primarily, upon the annual revenue and sophistication of the individual firm and the availability of support staff. Some companies assign the project manager, estimating buy-out responsibility along with their management duties, while other companies assign these tasks to specific departments and staff.

But the one project management responsibility that remains constant is the effective orchestration, guidance, and control of the construction process from beginning to end. Project management means managing the construction project—and that is what this book is all about.

2

The Start of the Construction Process

We're called *contractors* in no small part because contracts are such an integral part of our industry. Every day we deal in contracts with owners both private and public, with subcontractors and vendors, and on occasion with architects and engineers. Although these contracts often exceed millions of dollars and we sign our names to them agreeing to abide by all of their terms and conditions, how often do we *actually read them and fully understand their contents?*

The predominant form of standard construction contracts are those published by three organizations: the American Institute of Architects (AIA), the Associated General Contractors of America (AGC), and the Construction Management Association of America (CMAA). The Engineer's Joint Contract Documents Committee also publishes a series of standard contract forms used primarily by the engineering profession. Although frequently modified by owners, the basic contents of these standard contract forms should become familiar to all whose responsibility it is to abide by their provisions.

In the later part of 1997, the American Institute of Architects issued their newly revised contracts, copyrighted 1997, which superseded their 1987 editions. Many owner–general contractor contracts written as late as 1998 still used the 1987 edition, and many contracts currently in effect utilize the 1987 copyright. In 2004, AIA issued 12 new contract documents, six of which pertain to design-build, and in 2005, four forms dealing primarily with architectural services were introduced.

Other less frequently used contracts are also under review, such as turnkey and joint venture contracts. The design-build contract and the design-build process deserve their own space, thus a later chapter in this book will be devoted entirely to them. Many of the AGC contract documents were 1999 issues, and CMAA has 2005 updates for their construction manager contracts. Even before a standard form or modified contract is executed, at times an owner, general contractor, or subcontractor will agree to proceed with a limited amount of work based upon the issuance of a letter of intent with the assumption that a fully executed contract will follow shortly thereafter.

The Letter of Intent

A letter of intent is generally a temporary document authorizing the commencement of construction in a limited fashion. Limits are placed upon the scope of work to be performed, the dollar value of the work to be performed (often expressed as a "not to exceed" amount), the time frame in which the work is to be completed, and restrictions on subcontracts or purchase orders to be awarded in order to comply with the provisions of that letter of intent. When a full scope contract is ultimately awarded, that which was included in the letter of intent will be incorporated and any payments made under the letter of intent will usually be credited to the contract sum.

A number of reasons exist for using a letter of intent, including

- An owner may wish to start demolition of a recently vacated office space while negotiating with a new tenant desirous of a quick move in.
- Having received a verbal loan approval from their lender, an owner may wish to commence a limited amount of construction work on a new project while awaiting full written approval of that loan.
- When a project is "fast tracked," an owner may wish to commit to a certain portion of work while the final budget and/or design is being completed. The letter of intent can be used to purchase some long-lead items such as reinforcing steel or structural steel shop drawings that would jumpstart the project.

A letter of intent must be specific in nature:

- It should clearly define the scope of work to be performed. If plans and specifications define that scope of work, these documents ought to be referenced. If plans and specifications are not available, an all-inclusive narrative should define the exact nature of the required work.
- It should include either a lump sum to complete the work, or a "cost not to exceed" amount, including the contractor's fee.
- It should include payment terms.

- It should contain a date when the work included in the letter of intent can commence, and in some cases, when the letter of intent expires.

- It should contain a statement stipulating that the scope of the work and its associated costs will be credited to scope/cost included in any formal construction contract, if subsequently issued.

- It should include a termination clause setting a time limit on the work, or an event that triggers termination, such as the issuance of a formal contract. A termination clause "for convenience" is often included, allowing either party to terminate work upon written notification, and stating the method by which a settlement of costs up to that point will be established.

- The letter of intent is to be signed and dated by all concerned parties.

A typical letter of intent between an owner and contractor might be worded as follows:

> Pursuant to the issuance of a formal contract for construction, the undersigned (owner) hereby authorizes the (contractor) to proceed with the tree removal in the areas designated on Drawing L-5, prepared by ABC Engineers, and dated July 8, 2006. All debris including tree stumps will be removed from the jobsite. Prior to commencement of the work, all erosion control measures will be installed by the Contractor according to Drawing L-8, issued by ABC Engineers, and dated July 1, 2006.
>
> Maintenance of soil erosion measures will be required from the date of installation until this letter of intent is terminated on or about September 15, 2006.
>
> All of the above work is to be performed at cost plus the 15-percent contractor's overhead and profit fee. Daily work tickets and copies of all subcontract agreements and purchase orders will be presented by (contractor) to (owner's representative) as substantiation for all costs."

Signed: *Contractor*   Signed: *Owner*

Defining costs in the letter of intent

When any formal agreement between an owner and contractor includes reimbursement of costs, the definition of what constitutes "costs" in the initial document will greatly reduce any disagreements when subsequent invoices are presented for payment. This subject will be dealt with more fully when discussing cost plus contracts later in this chapter, and must also be considered when issuing a letter of intent.

Scope of work, tasks, and reimbursables included in letters of intent can include such items as shop drawing preparation, and cancellation charges for any materials/equipment ordered if a further construction contract is not forthcoming or if the letter of intent is terminated prematurely. Reimbursable expenses may extend to in-house costs incurred by the general contractor for estimating, accounting, and even interim project management and superintendent salaries. The owner can be presented with a list of reimbursable costs appended to the letter of intent as an exhibit to avoid future misunderstandings.

When a formal construction contract is issued, the segregated costs associated with the work performed under the letter of intent are generally applied against the costs for the total project. It is important to accumulate and segregate all reimbursable costs as they are incurred by assigning a separate cost code to all labor, material, equipment, and subcontract commitments. This will permit easy retrieval of all related costs when reimbursement is requested from the owner.

Subcontractor commitments via the letter of intent

While operating under the terms and conditions of the letter of intent, the general contractor may have to make certain commitments to subcontractors and vendors, and any purchase orders or subcontract agreements issued should contain the same restrictive provisions as the agreement between the general contractor and owner.

For example, if the owner's letter of intent contains provisions for the preparation of reinforcing steel drawings, placing an order for some nonstock sizes and even partial fabrication, the same restriction(s) placed upon the general contractor should be transferred to the reinforcing bar vendor.

The letter of intent termination clause

A typical termination provision in a letter of intent would be similar to the following:

> Upon receipt of a written directive to cease the work covered under the terms and conditions of this letter of intent, the Contractor shall immediately stop all work. All costs for work-in-place as of that date will cease. Cancellation costs for work-in-progress will be honored upon a receipt of a detailed explanation for all such costs documented by purchase orders or other commitments and related Stop Work Orders.

Any such termination clauses should be included in all vendor purchase orders or subcontract agreements. It is important that the Project Manager notify all vendors and subcontractors promptly, both verbally and in writing, upon receipt of any termination notice issued by the owner.

It is risky business to proceed with any phase of construction work without some form of written authorization. In this day and age, corporate personnel are very mobile, moving from one job to another, and that familiar owner representative who initially authorized the work may be gone from the company, along with all traces of any prior verbal commitment.

Requesting a letter of intent when the occasion arises is not only the proper business approach but is also a way of preserving a good relationship, since misunderstandings will be lessened or eliminated entirely.

Prevalent Types of Construction Contracts

Construction contracts today generally fall into one of the following categories:

- Cost of the Work Plus a Fee
- Cost of the Work Plus a Fee with a Guaranteed Maximum Price (GMP)
- Stipulated or Lump-Sum
- Construction Management
- Design-Build

Less frequently used contracts between owner and general contractor include

- Turnkey
- Joint Venture
- Build-Operate-Transfer (BOT) and its several variations

Each form of contract has its own caveats, and a closer look at each type may be helpful to further understand their unique provisions.

Cost of the work plus a fee

What could be simpler—a contract where the contractor bills the owner for all work-related costs plus their fee for overhead and profit? Well, the cost-plus contract requires a great deal of thought, preparation, and administration in order to work successfully.

This form of contract demands an atmosphere of trust between owner and contractor and means keeping communication channels open in order to convey the status of work and their costs as they accumulate. First of all, a definition of what constitutes "cost" is often a point of interpretation between owner and contractor, and needs to be fully defined upfront. Secondarily, this form of contract lends itself to situations where a complete set of drawings and/or specifications have not yet been developed, so defining the scope of work is crucial.

What constitutes "cost."   One has only to look at the American Institute of Architects' Document A111,1997 edition, *Cost of the Work Plus a Fee with a Negotiated Guaranteed Maximum Price*, to provide a starting point for a fully developed list of costs—both reimbursable and nonreimbursable:

Costs to be reimbursed

1. Labor costs—basic hourly rates, premium rates, and labor "burdens." Many owners question the high hourly rates of some collective bargaining agreements when all benefits are included, and if the type of trades to be employed on the project is known, a breakdown of each trade's billable rate attached as an exhibit may prove useful (see Figure 2.1).

2. Wages—the salaries of the contractor's supervisory and administrative personnel when stationed at the site *with the owner's approval*. If any nonfield-based personnel costs are to be reimbursed (for instance, estimating and accounting), they should be listed here.

3. Taxes—insurance, employer contributions, assessments

4. Subcontract costs

5. Costs of materials and equipment incorporated in the completed project

6. Costs of other materials and equipment, temporary facilities, and related items fully consumed in the performance of the work

7. Rental costs for temporary facilities, machinery, equipment and hand tools *not customarily owned by construction workers* whether rented from the contractor or others.

8. Cost of removal of debris from the site

9. Costs of document reproduction, faxes and telephone calls, postage, parcel post, and reasonable petty-cash disbursements

10. Travel expenses by the contractor while discharging duties connected with the work

11. Cost of materials and equipment suitably stored offsite, *if approved in advance by the owner*

12. Portion of insurance and bond premiums

13. Sales and use taxes

14. Fees—assessments for building permits and other related permits

15. Fees for laboratory tests

16. Royalties—license fees for use of a particular design, process, or product

WAGE RATE SCHEDULE

STRAIGHT TIME

	IRONWORKERS	
	FOREMAN	JOURNEYMAN
PENSION FUND	4.25	4.25
WELFARE FUND	4.00	4.00
ANNUITY FUND	5.00	5.00
ED & BUILDING FUND	0.85	0.85
INDUSTRIAL FUND	0.03	0.03
PENSION SUPPLEMENT FUND	0.65	0.65
D.C. LABOR MANAGEMENT	0.15	0.15
LOCAL # 7 APPRENTICE	0.23	0.23
JAC BUILDING FUND	0.50	0.50
TOTAL BENEFITS	15.66	15.66
WAGE RATE	39.65	35.55
EMPLOYERS FICA	2.46	2.20
EMPLOYERS MEDEX	0.57	0.52
EMPLOYERS FUTA	0.32	0.28
EMPLOYERS SUTA	2.34	2.24
LIABILITY INSURANCE	4.44	4.14
WORKMAN'S COMP	12.00	11.20
TRAVEL SMALL TOOLS	2.85	2.65
TOTAL	80.29	74.45
TOTAL	**80.29**	**74.45**

DOUBLE TIME

	FOREMAN	JOURNEYMAN
WAGE RATE	39.65	35.55
ANNUITY FUND	6.00	5.30
EMPLOYERS FICA	2.46	1.62
EMPLOYERS MEDEX	0.57	0.38
EMPLOYERS FUTA	0.32	0.21
EMPLOYERS SUTA	2.34	2.24
TOTAL	51.34	45.30
DOUBLE TIME	131.63	119.75
TOTAL	**131.63**	**119.75**

RATES ARE SUBJECT TO CHANGE PER COLLECTIVE BARGAINNING AGREEMENTS.

PLEASE NOTE THE ABOVE RATES DO NOT INCLUDE 15% OVERHEAD & PROFIT MARK UP.

FIGURE 2.1 Wage rate schedule.

17. Data processing costs related to the work

18. Deposits lost for causes other than the contractor's negligence

19. Legal—mediation and arbitration costs, including attorneys' fees arising out of disputes with the owner *with the Owner's prior written approval.*

20. Expenses incurred by contractor for temporary living allowances

21. Cost to correct or repair damaged work provided that such work was not damaged due to negligence or was nonconforming

Costs not to be reimbursed

1. Salaries and other compensation of the contractor's personnel stationed at the contractor's principal office or offices, *except as specifically provided for in the contract*

2. Expenses of the contractor's principal office

3. Overhead and general expenses, *except as provided in the contract*

4. The contractor's capital expenses

5. Rental costs of machinery and equipment except as specifically spelled out

6. Costs due to the negligence of the contractor

7. Any costs *not specifically included in costs to be reimbursed* (This transfers the responsibility onto the contractor to include a *comprehensive* list of reimbursable costs since they will not be able to later claim they "forgot" to include some miscellaneous costs which are *always* reimbursed.)

8. Costs, other than approved change orders, that would cause the GMP price to be exceeded

Any and all changes to these standard "costs," both additions and subtractions, need to be clearly spelled out in the agreement. The project manager must alert their superintendent and their accounting department to properly identify all applicable costs by project number and by proper cost coding. When requisitions are prepared, documentation of all reimbursable costs are to be attached to that request for payment.

Cost-plus contract pitfalls.   The following are a few pitfalls to avoid when administering a cost-plus contract without a GMP:

- The scope of the work included in the agreement must be clearly defined and if the scope is increased, any associated increase in cost must be presented quickly to the owner in writing either as a fixed amount or an estimated cost.

- A statement of "costs to be reimbursed" and those not to be reimbursed must accompany any cost-plus agreement.

- When a project manager is assigned to the project and time spent administering the project is a reimbursable cost, establish a method of accounting for his/her time in the field and in the office when working on the project. It will be helpful to keep a simple one- or two-sentence log of activities associated with these hours so that, if called upon in the future, the project manager can recall the exact nature of those activities.

- Prepare weekly reports tracking actual costs versus the estimate. If no estimate was presented initially, prepare weekly costs to apprise the owner of those week-to-week costs. Note: Some vendors and subcontractors are notoriously late in submitting invoices, so always add a caveat to the reported costs such as "costs received to date" or "additional costs may accrue during final accounting."

- Attempt to convert the cost-plus contract to a lump sum or GMP as quickly as possible—if the nature of the work so dictates.

Remember the five critical elements in a cost-plus-fee contract:

- Define the scope of work as precisely as possible
- Identify all scope changes as soon as they occur.
- Establish approximate or firm costs for these changes.
- Notify the owner of the changes and related costs as soon as they are identified—and *do it in writing*.
- Hope for the best!

The stipulated or lump-sum contract

A stipulated or lump-sum contract is most frequently used in competitive bid work in either the public or private sector where a complete set of plans and specifications have been prepared by the owner's design consultants. Contractors are expected to estimate the cost of the work contained in a specific set of bid documents—no more, no less. Any deviation of the scope of work as interpreted from these bid documents, except if amended later by other contract provisions, will result in a change of scope where any adjusted costs will be dealt with by change orders.

The problem of defining scope.   Although defining the scope may appear to be rather straightforward, it is not. The "intent" of the plans and specifications can often be interpreted in many ways by each participant in the construction process. Since the architect is generally designated by the contract as the "interpreter" of the plans and specifications created, the final decision on what constitutes the obligation of all parties to the contract rests with that authority—unless challenged and resolved by either negotiation, arbitration, or litigation. The use of requests for information (RFIs) during the bidding process is one way of clearing up any discrepancies, errors, and omissions in the plans and specifications, and all bidders should afterward receive the architect's response.

Allowances and alternates in the stipulated-sum contract.   Allowances are frequently included in a contract when the exact nature of the work or quality levels desired were not clearly defined when the contract was formulated. As the owner and architect provide definition, the contractor must establish the actual cost, compare it with the allowance amount, and adjust the contract sum accordingly. Unless the contract specifies otherwise, the following costs are to be included in developing the total cost of the allowance in question:

- Cost of materials and equipment
- Cost of unloading and handling at the site, plus installation costs
- Overhead and profit (included in the *contract* but not in the allowance)

Alternates, meanwhile, can sometimes be looked upon as the owner's wish list. They are generally items that may be added to the contract scope at a predetermined cost. One problem relating to alternates is the time frame in which they are to be accepted or rejected, and many contracts do not include a date or time frame within the budget for acceptance or deletion. Take a simple example: an alternate for a recessed entry mat in a vestibule as opposed to the contract provision for a smooth concrete surface upon which the owner can place a mat. The decision to accept or reject the alternate must be made prior to the placement of the concrete slab. Acceptance after the slab is poured will necessitate chopping out the slab at considerably more cost.

When alternates are included in the contract and no date of acceptance or rejection is specified, the project manager must prepare a list and advise the owner of the latest date for which the alternate can be accepted at the cost included in the contract.

Lump-sum contract pitfalls.   The following are a few pitfalls to avoid when administering a lump-sum or stipulated-sum contract:

- A thorough review of the bid documents is essential to uncover any ambiguities, errors, and omissions in the plans and specifications.

During the bidding process, these problems can be presented to the architect who will issue a ruling in the form of a response to the contractor's RFI or via a Q&A follow-up to questions posed by several bidders. It is important that all parties are notified of these queries, and the project manager should verify that the design consultant's directive has been sent to all bidders.

- Although it may be precarious to qualify a bid on a public project without risking a rejection of the bid, in private work the contractor *can* qualify their bid and list their qualifications if they have concerns about the intent of the documents. However, some contractors are reluctant to do so for fear that their bid will be rejected. Generally, if the bid for a private project is competitive, qualified or not, the owner will probably not reject the bid but will arrange an interview to discuss these questionable issues. In public bidding, the agency may consider minor deviations in the contractor's bid if they are deemed "in the public interest" to do so.

- When awarding contracts to subcontractors, avoid including the words "to include plans and specifications." Relying on this phrase as the sole basis for determining their scope of work can be dangerous. Instead, include a scope letter defining the scope of work more specifically. Remember the subcontractor's interpretation of the plans and specs may be considerably different from yours. (More on this subject in Chapter 6.)

- Prior to negotiating a contract with a subcontractor or vendor, review the estimate thoroughly. If any items were inadvertently omitted, attempt to incorporate them into the subcontractor agreement or vendor purchase order. Remember, this is the time when your negotiating power is greatest.

- Obtain the subcontractor's/vendor's agreement that they have received and reviewed *all appropriate documents*—plans, specifications, addendums, and so on—that will be referenced in the subcontract agreement or purchase order.

The Cost-plus-a-fee with a GMP contract

The cost-plus contract with a guaranteed maximum price (GMP or Gmax) is frequently used because it allows the owner to gain the protection of the maximum cost of the construction while retaining the potential for cost savings. It is basically a cost-plus-a-fee contract with a cap on it, and many of the caveats that apply to the cost-plus-a-fee contract also apply to this contract. The GMP contract is often used for "fast tracked" projects when incomplete or sketchy construction documents are all that is available at the time of contract preparation. Usually,

70- to 80-percent complete design documents are sufficient to negotiate a GMP contract, and this very process can result in numerous misunderstandings between contractor and owner.

The importance of the contract qualification statement. Because the GMP contract is most often awarded before the plans and specifications are 100-percent complete, the question of what the contractor should assume for the 20 to 30 percent of the remaining design can be answered in a comprehensive qualification statement attached to the contract as an exhibit (Figure 2.2). In this exhibit, the contractor will list specific items that have been included in the scope of work anticipated (but not as yet in the final design) as the drawings are completed.

Other contractors get more specific and include a scope sheet for each division of work, where they not only list every item included in their guaranteed maximum price but the cost of that item, in order to establish a quality level. Figure 2.3 shows a portion of one such exhibit, this one dealing with electrical work.

Fees and savings. The contractor's fee is usually prenegotiated, based upon a percentage of cost. Any changes that increase the scope of work will be allowed a percentage increase in overhead and profit. When minor deductions in scope occur, the contractor should not be expected to include a credit for fee reduction since the amount of deleted work will probably not have materially affected the overall management of the project. When major portions of work are deleted, a credit for overhead and profit may be warranted, however.

The GMP contract contains a "savings" clause specifying that any savings will be shared by the owner and the contractor in varying percentages. Some owners prefer to have the contractor receive a greater portion of the savings, theoretically creating more incentive for the builder to search for potential savings. Thus, a 50/50 savings basis provides the owner with an incentive to review and accept value engineering suggestions proposed by the contractor.

Cost certification provisions. A standard feature of the GMP contract is a requirement for a cost certification audit when the project has been completed. The owner has the option to audit the contractor's books to determine the extent and nature of any savings, or to verify that all costs charged to the project are proper. If this audit is to be conducted by an independent accounting firm, provisions for the audit should be included in the contract either as a contractor-excluded cost or as a separate line item setting aside a specific sum for the audit.

In anticipation of an owner audit, the project manager must monitor, identify, and isolate all costs as they are incurred, and as they relate to

<u>**Exhibit F**</u>

<u>**Schedule of Assumptions and Qualifications**</u>

Pricing Qualifications

The Guaranteed Maximum Price provided for in Section 6.1 of the Agreement is subject to the following assumptions and qualifications. As used below, "we" and "our" refers to the Contractor. If an item is described as "included," then the cost of completing that item is included in the Guaranteed Maximum Price.

Division 1 - General Conditions/General Requirements
Inclusions/Exclusions

1
    Underpinning is said to be not required by the structural engineer and is therefore not
2   included
3   Asbestos, lead, or hazardous materials testing, removal and remediation is not included.
4   Utility and service company fees and charges for connections or meters are not included.
5   Temporary electric consumption cost during construction is not included.
6   A payment and performance bond is not included in the base, an alternate has been provided in the schedule of alternates.
7   Builders risk insurance is assumed provided by Owner and is not included
    Costs associated with the Health Department inspections are not
8   included
9   Construction site security, other than separation of work zones by construction fences is not included.
10 With timing going to be critical, we assume that the Owner will assist in obtaining permits with ISD / BFD / BW&S/ BRA as may be required.
11 As discussed we have included removing of the topping slab, waterproofing the existing glass block and pouring a new topping slab at the elevation.
    We have assumed Civil drawings to be 1"=10' not 1"=20' as
12 shown.
13 We have assumed MEP drawings to be 1/8"=1' not 1/4"=1' as shown.

Allowances included within the General Conditions

1   Police details, fire watch and street permits: $15,000 is included
2   Winter weather protection will most likely be required for the waterproofing, and concrete sidewalks In addition, protection will likely be required for the facade restoration after the completion of the sidewalks.
    $25,000 is included.
    An allowance of $25,000 is included to refeed, relocate and coordinate all existing
3   MEP'S.

FIGURE 2.2   Contract qualification statement.

the various line items in the estimate and requisition. This will save countless hours and reduce owner frustration at the end of the project if any job related costs are called into question. Most payroll reporting systems require a project manager to apportion their daily activities to specific projects or cost codes, and it is a good idea for the project manager, in a separate log, to enter a brief one- or two-sentence entry describing their time devoted to the GMP project.

Division 2 – Site work / Demo
Inclusions/Exclusions

1   Below grade obstructions, rock, and/ or removals of these are not included
2   The furnishing and installing of the new gas line including the meter is by the gas company.
     has included the excavation and backfill of this line as shown.
3   Relocation, repairs, or replacement of uncharted utilities.
4   Interior building demolition along with the removal of the existing floor slab in basement is currently
     underway and as such is not included. We have included approximately 110 lf of wall demo as outlined
5   Removal of any other topping slab at 1st floor level other than the area shown On JB.D.1:1
6   The work associated with the connection of existing roof drains to   was   said to not be required
     and as such, we have not included.
7   The City of   will require flowable fill in the utility trenches and as such this is included
8   The current demo operation will included the removal of the existing walls into the former street
     vault areas
9   As discussed, we have assumed a balanced site.   Unit prices for exporting and importing fill are as
     listed in the alternates.
10 Paving roadways as shown on   is excluded, we have included patching of the roadway at new utilities.
11 Since there is no geotech report at this time we have excluded all dewatering.
12 We have assumed the existing soils are adequate for base material and will remain in place.
13 Our pricing is based on filling, grading and compacting the existing soils at the rebuilt sidewalks
     as necessary to meet the proposed grades. We have assumed the existing soil base meets the
     City of   gravel requirements and that new base material is not required.
     In the base we have not included the work associated with drawings showing the extended
14 sidewalk.

Allowances included within the Site work / Demo
     As the full scope of the demo required for the remove of the existing concrete topping
1   slab
     and unknown supports down to the assumed structural slab is included
     $10,000.
2   We believe that there will be a need for saw cutting of the existing column for new beam pocket: $2,500.
3   As we also do not have an elevation, we have included an allowance of $5,000 to support the
     existing sewer as we install the adjoining water line.
4   We have included an allowance of $2,500 for misc. demolition.

Division 3 - Concrete
Inclusions/Exclusions
     We have not included any work scope associated with existing concrete foundation
1   walls.
2   As agreed, cosmetic patching is excluded in tenant space. An allowance for cosmetic patching in
     public space is included.
3   Structural concrete patching in tenant space other then what is shown on   is excluded.

FIGURE 2.2   (*Continued.*)

Awarding subcontract agreements.   Prior to the issuance of a subcontract agreement, some GMP contracts require the project manager to submit all subcontract proposals to the owner for their review, together with a recommendation for any awards. In fact, some contracts include a provision requiring the owner to approve a subcontract agreement, in writing, before it is issued. If that's the case, the owner must be made aware of the time restraint necessary for their review and comment. Thus, the project manager should then forward each subcontractor recap to the owner with a cover letter stating the time frame in which approval is required.

—	4" Conduits from Switchboard to Pull Box	160.00	lf	8.00 /lf	1,280
---	2" Conduits from Tel/Data Room to Pull Box	200.00	lf	6.00 /lf	1,200
---	Marche Electric	1.00	ls	38,400.00 /ls	38,400
---	3" Conduits for HVAC Feeders	300.00	lf	8.00 /lf	2,400
---	225 Amp Feeders	300.00	lf	40.00 /lf	12,000
---	400 Amp Feeders	50.00	lf	50.00 /lf	2,500
---	3" Conduits for HVAC Feeders	250.00	lf	8.00 /lf	2,000
---	600 Amp Feeders	100.00	lf	70.00 /lf	7,000
	Secondary Service				**74,780**
16000.05	**Distribution Equipment**				
—	225 Amp Panel	2.00	Ea	2,750.00 /ea	5,500
—	225 Amp Panel	3.00	Ea	2,750.00 /ea	8,250
—	400 Amp Panel	1.00	Ea	3,500.00 /ea	3,500
—	600 Amp Panel	1.00	Ea	550.00 /ea	550
	Distribution Equipment				**17,800**
16000.06	**Branch Circuit Wiring**				
—	Duplex Receptacles	34.00	Ea	35.00 /ea	1,190
—	Duplex Receptacles - Wiring	850.00	lf	3.25 /lf	2,763
—	Duplex Receptacles - Homeruns	255.00	lf	5.50 /lf	1,403
—	Duplex Receptacles - WP	2.00	Ea	55.00 /ea	110
—	Duplex Receptacles - Wiring	50.00	lf	3.25 /lf	163
—	Duplex Receptacles - Homeruns	15.00	lf	5.50 /lf	83
	Branch Circuit Wiring				**5,710**
16000.07	**B.C.W. Equipment**				
—	Equipment Terminations; allowance	1.00	Ea	1,500.00 /ea	1,500
—	Propeller Steam Unit Heaters	3.00	Ea	450.00 /ea	1,350
—	Elevator Power	1.00	Ea	1,500.00 /ea	1,500
—	Sump Pump	2.00	Ea	350.00 /ea	700
—	Sewage Ejector	2.00	Ea	850.00 /ea	1,700
—	Water Booster	0.00	Ea		
—	1000 GPM Fire Pump/Jockey Pump System	0.00	Ea		
—	Compressor for Dry Piping System	2.00	Ea	750.00 /ea	1,500
—	Awning	1.00	Ea	550.00 /ea	550
—	Electric Unit Heaters	1.00	Ea	450.00 /ea	450
—	Split System AHU's	0.00	Ea		
	B.C.W. Equipment				**9,250**
16000.08 Lighting					
—	Wall Mounted Batting Tunnel Fixture – W1	0.00	Ea		
—	1x4 Surface Mounted Fixture - FS1	23.00	Ea	140.00 /ea	3,220
—	2x4 Surface Mounted Fixture - FS2	12.00	Ea	140.00 /ea	1,680
—	Exit Signs	24.00	Ea	225.00 /ea	5,400
—	Fixtures – Wiring	2875.00	lf	3.50 /lf	10,063
—	Fixtures – Homeruns	865.00	lf	5.25 /lf	4,541
—	Lighting Control	17136.00	Sf	0.60 /sf	10,282
—	Fixture Labor	96.00	hrs	80.00 /hrs	7,680
—	Fixture Wiring	1600.00	lf	3.25 /lf	5,200
—	Fixture Wiring - Homeruns	480.00	lf	5.50 /lf	2,640
—	2x4 Surface Mounted Fixture - FS3	27.00	Ea	140.00 /ea	3,780
—	Exterior Wall Sconce - Type M INSTALL ONLY	38.00	Ea	275.00 /ea	10,450
—	Emergency Battery Unit	2.00	Ea	350.00 /ea	700
—	Batting tunnel Lighting- Allowance 10 fixtures	10.00	allw	200.00 /allw	2,000
	Lighting				**67,635**
16000.11	**Fire Alarm**				
—	Empty fire alarm conduits	17,136.00	Sf	0.40 /sf	6,854
—	**Fire Alarm**				**6,854**
16000.12	**Communication**				
—	Security (Empty Conduit)	17,136.00	Sf	0.30 /sf	5,141
—	Tel/Data (Empty Conduit)	17,136.00	Sf	0.30 /sf	5,141
	Communication				**10,282**
	Electrical			13.89/sf	222,965
	17,136.00 sf				

FIGURE 2.3   A portion of an exhibit concerning electrical work.

Only qualified subcontractors should be invited to submit bids, and if any nonqualified subcontractors submit bids unsolicited, the proposal should be forwarded to the owner. If an unrealistic low bid is received from an unsolicited subcontractor who has a past record of poor performance or poor quality work, the bid should be forwarded to the owner with a comment stating that this was an unsolicited bid and is not to be considered. The reasons for the bid rejection should be included. In the absence of such a procedure, an owner may learn of a particular low bid, and not being made aware of it, may be of the opinion that they are not being afforded the most competitive pricing.

To be safe, the project manager should forward all unsolicited bids to the owner—if in fact the contract calls for the owner to review these bids—with comments and recommendations by the general contractor. In most cases, the owner will follow the general contractor's advice.

On the other hand, if a contractor does not prequalify subcontractors and receives an unrealistically low bid from one of them, forwarding it to an owner with a comment that the bid should be rejected will likely prompt the client to respond, "Well why did you allow an unqualified subcontractor to submit a bid on my job in the first place?"

Change orders and value engineering.   There may be a tendency to fold extra work into the GMP without the issuance of a change order, the reasoning being that substantial savings have accrued to absorb these costs. However, those illusive "savings" may quickly disappear and reduce or totally eliminate any surplus, a portion of which should have been returned to the contractor. Any change in scope requested by the owner or the architect/engineer should be cause for issuing a change order.

Value engineering.   One of the many functions of a project manager administering a GMP contract will be to look for savings throughout the life of the project. All subcontractors and vendors should be requested to review their work with an eye toward developing possible cost savings. As these suggestions are received, the project manager must review and analyze them to determine whether a savings in one trade may result in an increase in another trade, in effect resulting in no savings at all.

This process of "value engineering" has received its share of criticisms, and in several cases, rightly so. For example, the substitution of two small rooftop exhaust fans for one larger unit may result in a savings in the mechanical portion of the work, but will it add costs to electrical circuitry and an additional framed rooftop opening? All such "value engineering" suggestions should be routed through the architect and engineer for comment before being formally submitted to the owner. The more adept a contractor becomes at developing meaningful cost savings or value engineering suggestions, the easier it will be to build a solid reputation as an effective administrator of GMP contracts.

GMP contract pitfalls.   The following are some pitfalls to avoid when administering a GMP contract:

- If the GMP was based upon less-than-complete plans and specifications, carefully review all future design development drawings to ascertain that the scope of work emerging as a "for construction" set is what was anticipated by the contractor initially, complete with inclusions and exclusions If it appears that this scope has deviated from the original concept, alert the owner immediately and identify the changes and their respective costs.

- When receiving value engineering cost savings, review them carefully with all related suppliers and subcontractors to verify that the actual savings being suggested are, in fact, true savings and that no hidden costs lurk somewhere down the line.

- Resist the owner's request to submit "actual costs" to date versus "estimated costs" early in the project. Any early "savings" may dramatically change as later costs develop. Respond to any such requests by saying that it is too early to project true costs at this stage of the work.

- Do not allow scope increases or decreases to occur without increasing or decreasing the guaranteed maximum price, and do this as quickly as possible.

- Instruct field supervisors to identify all materials and equipment, and attach signed receiving tickets with the project name, number, and where the item was used. This will make it much easier to verify costs if and when a detailed audit is required.

Construction manager contracts

Construction manager (CM) contracts have gained much popularity in recent years as a project delivery system for several reasons:

- The CM concept allows an owner to engage a construction professional to work with their design team as the project progresses through the design development stage, thereby utilizing the contractor's knowledge of costs, constructability issues, and local market conditions.

- It essentially provides an owner with an arm's-length management team, allowing them to avoid a third-party relationship with a builder.

The Construction Management Association of America (CMAA) defines CM work as follows:

- A project delivery system comprised of a program of management services
- Defined in scope by the specific needs of the project and the owner
- Applied to a construction project from conception to completion, to control time, cost, and quality
- Performed as a professional service under contract to the owner by a construction manager
- Selected on the basis of experience and qualifications of the CM firm
- Compensated on the basis of a negotiated fee for the scope of services rendered

The CM concept can be subdivided into the following:

- *CM—For Fee (also referred to as Agency):* A process whereby the required construction management services performed by the CM are reimbursed via a fee based upon a percentage of costs. The final, ultimate cost of the project is not guaranteed—it is what it is.
- *CM—At Risk:* The CM in this case also provides all required construction management services, including reimbursable expenses and a fee based upon a percentage of costs; however, the CM guarantees the final cost of the project. Some critics of this form of CM contract charge that the CM now serves two masters, the owner and their own company since decisions may be made that favor the CM's bottom line instead of the client's interests. Any CM with such a reputation will soon be out of business, making this the biggest counterargument to the two-masters proposition.

Construction management services during preconstruction and construction. The CM will provide professional staff to the owner prior to or during the design phase—including estimating, scheduling, purchasing, and project management. The purpose being to develop the owner's construction program with the design development team in order to insure it meets the client's schedule and budget restraints. The CM may provide these services as a lump-sum proposal or cost-plus-a-fee, with or without a GPM. The owner may or may not elect to award the second phase, the construction phase of the project once the preconstruction portion is concluded.

Regarding construction services, the CM provides the staff and related facilities to administer and manage the construction project, acting as the owner's agent. During construction, the CM will be awarded either an "agency" or an "at-risk" contract.

Of course, many owners engage CMs to perform both functions—preconstruction and construction services—so one contract may be issued incorporating both phases. Some contracts have an escape clause, however, so if the owner is not satisfied with the CM's performance during the preconstruction phase, the construction portion of the contract can be voided, allowing the owner to seek the services of another CM.

The preconstruction phase of a CM contract.   One major advantage of the CM process is that an owner can obtain the services of a team of construction professionals to act on their behalf during the preparation of the project's design. The CM's staff of experienced professionals, having day-to-day contact with subcontractors, local labor pools, equipment manufacturers, and material suppliers (as well as a detailed database of construction component costs) can provide invaluable assistance in determining the most cost-effective design commensurate with the owner's budget and project delivery dates. A clear understanding of the services required by the owner is essential in establishing a comprehensive list of reimbursable expenses assigned to these activities.

CMAA's owner and construction manager contract.   CMAA Document A-1—the Owner and Construction Manager contract (2005 edition) includes detailed information regarding the basic services associated with the predesign, design, and construction services to be provided by a construction manager. In general, this includes the following:

- *Predesign phase:* Tasks include working with the owner to develop a management plan, selecting an architect, developing a master schedule, developing a budget and preliminary estimate, and fostering a management information system.

- *Design phase:* Work to do in this phase includes monitoring the designer's compliance with the owner's program, reviewing the design documents, making revisions to the master schedule, preparing estimates as required, performing value engineering, advising on contract awards, prequalifying bidders, conducting pre- and postbid conferences, and developing MIS programs for scheduling, cost reporting, and cash flow.

- *Construction phase:* Providing project management is part of this phase, as well as conducting cost administration and analysis procedures. Monitoring safety, schedules, costs, changes in scope, and closeout and occupancy procedures are other tasks that must be handled.

Some contracts do not prohibit construction managers from performing certain work tasks with their own forces if they are experienced in these tasks and can demonstrate that their involvement will be cost-effective. In that event, the CM is allowed to include a certain percentage for administrative costs and profit just as though that portion of the work had been subcontracted to another firm.

CM fees.   The fee charged for construction manager services varies depending upon whether preconstruction *and* construction services are required, and whether the CM will be a "for fee" or "at risk" contract. Fees in either case are significantly lower than those charged by general contractors performing lump-sum or GMP work because the CM is reimbursed for most field-incurred expenses; therefore, most of the fee will go to the contractor's "bottom line."

These "reimbursable expenses" are specifically listed in the CM's proposal, and their "cost" includes an applied percentage for overhead and profit. These reimbursables are referred to as "reimbursables with a multiple"—in other words, the owner will pay the CM for specified costs multiplied by a factor of 1.5 or 2.0, or whatever is agreed upon. Therefore, the cost of a superintendent's weekly salary, including fringe benefits, may possibly be $1500. At a multiple of 1.5, the owner will agree to pay the CM $2250 for the "cost" of this superintendent's services; at a multiple of 2, the super's cost is billed at $3000 per week.

Exclusive of preconstruction services and reimbursables, a "for fee" CM will be in the range of 2 to 4 percent whereas the fee for a CM "at risk" could be as high as 8 percent. Preconstruction services are often quoted on a lump-sum basis in the proposal that includes specific duties and responsibilities during that phase and a time frame for carrying out those duties and responsibilities.

CM contract pitfalls.   One of the pitfalls to avoid in administering a CM contract is failure to include a complete and inclusive list of reimbursables for field-related expenses (which is why a CM will administer a project from the field, transferring all office-related project functions to the field, thereby removing those costs from their central office overhead).

A typical CM list of standard reimbursables includes the following:

- Project Field Office Set-Up:
  - Office complex, trailers, security fencing
  - Office equipment, duplicating machine, computers, miscellaneous (staplers, hole punches, and so on)
  - Utility connections for telephone, data, water/sewer (if applicable), electricity
  - Signage
  - Printers and scanners

- Office furniture—desks, chairs, filing cabinets, conference tables
- All IT-related costs, Internet connections, wireless costs, and so on
- Project Field-Office Expenses:
  - Supplies for office equipment, periodic maintenance
  - Copy machine supplies
  - Fax machine supplies
  - Utilities—power, telephone, sanitary, water
  - Postage, package deliveries, overnight delivery service
  - Wireless charges
  - Reproductions
  - Automobile and trucking expenses; repairs
  - Travel expenses
  - Security
  - Office maintenance and cleaning services
- Site-Related Expenses
  - Engineering (if requested)
  - Initial survey, interim lay-outs, final survey
  - Testing
  - Shop drawings—receipt, review, and transmission to A/E, and then their return to the appropriate subcontractor/vendor
  - As-built drawings (either preparation, or the review of drawings prepared by others)
  - Safety/first aid
  - Photographs
  - Project Maintenance (if requested)
    - Site security—fencing; including maintenance, lighting
    - Erosion control
    - Access roads—installation and maintenance
    - Fire extinguishers and maintenance of same
    - Personal safety equipment
    - Portable toilets—delivery and periodic maintenance
    - Clean-up/dust control
    - Dumpster services
    - Trash chutes
    - Pest control

The joint venture agreement

In today's marketplace, large national and international building firms are venturing into new geographic areas seeking work to increase or maintain their already substantial annual volume. These incursions are often the result of their national account clients building in a new geographic area, or construction giants who are just looking to increase their market share. Expansions into local markets may offer opportunities to established local contractors. Having little to no experience with local subcontractors, vendors, and labor markets, large contractors may seek out local partners to work with them in what is known as a joint venture. This can benefit both firms if the joint venture agreement is properly prepared, allowing shared responsibilities and shared profits to enrich both parties. A typical joint venture agreement (contract) is set forth in Figure 2.4 for those interested in pursuing this type of work.

Turnkey contracts

Turnkey contracts are often associated with the process engineering industry in the design and construction of petrochemical and chemical plants where the owner is also buying the design and engineering expertise of the contractor. Although there are several variations on the turnkey contract concept, the most universally accepted definition is a project whose costs will not be paid until the contractor completes the project and "turns the keys" over to the owner. In a turnkey project, the cost of funds required by the contractor to pay for all expenses incurred over the life of the project will be included in their total project costs. Monthly requisitions are not submitted to the owner for payment, but once the project has been completed and accepted, the contractor receives payment in full.

Build-operate-transfer

The build-operate-transfer (BOT) concept gained popularity in the last three decades of the twentieth century, but more so in Europe and Asia than in the United States. It is a process whereby the builder/developer provides architectural, engineering, construction, and financial services to construct a project which they will not only *build,* but also *operate* for a specific number of years before *transferring* the title to the owner. These projects are applicable where a revenue-producing facility will be constructed since the concept of BOT is based upon the builder/developer receiving sufficient revenue to cover design, construction, financing, and operational costs while also generating a profit. BOT-type undertakings typically appear in the public sector and consist of such revenue-producing projects as toll roads, bridges, and tunnels. Although dozens of these BOT projects, and their variations, have been built in Asia, Great Britain, and Europe, the Dulles Toll Road in Virginia is one of only a few of America's ventures into this sophisticated project delivery system.

JOINT VENTURE AGREEMENT

THIS AGREEMENT made and entered into this _____ day of _____, 19____
by and between

WITNESSETH:

WHEREAS, _____
_____(hereinafter Owner) has
advertised for bids for the construction of _____
_____
_____

which bids are to be submitted on or about _____
and

WHEREAS, the parties hereto have agreed to form a Joint Venture to submit a joint bid for and, if possible, to obtain a contract with Owner for such construction.

WHEREAS, the parties desire to enter into a joint Venture agreement in order to fix and define between themselves their respective interest and liabilities in connection with the submission of such bids and the performance of such construction contract in the event that it is awarded to them.

NOW, THEREFORE, in consideration of the mutual promises and agreements herein set forth, the parties hereby agree to constitute themselves as joint venturers for the purpose of submitting a joint bid to Owner for the performance of the construction contract hereinbefore described, and for the further purpose of performing and completing such construction contract in the event that it is awarded to them on such joint bid, and the parties hereby agree that such bid shall be filed and such construction contract, if awarded to them, shall be performed and completed by them as a joint venture subject to the following terms and conditions:

1. NAME AND SCOPE OF JOINT VENTURE

1.1 The bid shall be filed and the construction contract, if awarded to the parties hereto, shall be entered into in the names of the parties as joint venturers, and the obligations of the parties under such bid and construction contract shall be joint and several. Such construction contract, if awarded to the parties hereto based on their joint bid, shall be carried out and performed by them in the name of "_____
_____ , a Joint Venture"
(hereinafter Joint Venture) all money, equipment, materials, supplies and other property acquired by the Joint Venture shall be held jointly in that name.

FIGURE 2.4 A typical joint venture contract.

Contracts with government agencies

Local, state, and federal public agencies have contract forms that sometimes borrow heavily from those of the American Institute of Architects (AIA) or Associated General Contractors of America (AGC). They will also include pages of various local, state, and federal laws and ordinances,

1.2 No payment shall be made by the Joint Venture to any of the parties hereto in reimbursement of expenses incurred by such parties in connection with the preparation of the bid for and securing the award of the construction contract, unless otherwise agreed in writing.

1.3 It is the intent of the parties hereto that the joint bid contemplated and provided for herein shall be satisfactory and acceptable to both parties. If the parties are unable to agree upon a joint bid this Joint Venture shall terminate upon written notice by the dissenting party to the other party or parties, which written notice shall be delivered to the other party or parties prior to the time of the bid, and, in such event no party shall have any liability to any other party or parties.

1.4 This Agreement shall not be interpreted or construed so as to extend beyond the submission of such joint bid and the performance of such construction contract, nor to create any permanent partnership or permanent Joint Venture between the parties and shall not limit any of the parties in their right to carry on their individual businesses for their own benefit, including other work for the Owner.

2. PROPORTIONATE SHARES

2.1 Except as otherwise provided in Paragraphs 4.2 and 4.3 hereof, the interest of the parties in any profits and assets and their respective shares in any losses and liabilities that may result from the filing of such joint bid and/or the performance of such construction contract, shall be as follows:

_____	_____ %
_____	_____ %
_____	_____ %
_____	_____ %

with such percentages being referred to hereinafter as the respective party's Proportionate Share.

2.2 It is the intention of this Agreement, and the parties hereby agree that in the event of any losses arising out of, or resulting from the performance of said construction contract, each party hereto shall assume and pay its Proportionate Share of such losses. If for any reason any one of the parties hereto sustains any liabilities or is required to pay any losses arising out of or directly connected with the performance of such construction contract, or the execution of any surety bonds or indemnity agreements in connection therewith, which are in excess of its Proportionate Share in the losses of the Joint Venture, the other party or parties shall reimburse such party in such amount or amounts as the losses paid and liabilities assumed by such party exceed its Proportionate Share in the total losses of the Joint Venture, so that each member of the Joint Venture will then have paid its Proportionate Share of such losses; and to that end each of the parties hereto agrees to indemnify the other party or parties against and to hold it harmless from any and all losses of said Joint Venture that are in excess of such party's Proportionate Share. Provided, however, that the provisions of this sub-paragraph shall be limited to losses that are directly connected with or arise out of the performance of said construction contract and the execution of any bonds or indemnity agreements in connection therewith, and shall not relate to or include any speculative, prospective, incidental or indirect consequential losses that may be sustained or suffered by any of the parties hereto.

2.3 The parties shall, from time to time, execute such applications for bonds and indemnity agreements, and other documents and papers as may be necessary in connection with the submission of said joint bid for and the performance of such construction contract. Provided, however, that the liability of each of the parties hereto under any agreements to indemnify a surety company or surety companies shall be equal to the Proportionate Share of each of said parties in the Joint Venture.

FIGURE 2.4 (_Continued._)

as well as executive orders that must be complied with. Under the canopy of Equal Employment Opportunity (EEO), or provisions for fulfilling requirements for Disadvantaged Business Enterprises (DBE), Minority Business Enterprises (MBE), Women-Owned Business Enterprises (WBE), and the Americans with Disabilities Act (ADA), these supplemental or special conditions are often in print so small as

2.4 If any of the parties hereto is a subsidiary of another corporation, the performance by such subsidiary of the obligations assumed hereunder shall be guaranteed by the parent corporation of any such subsidiary.

3. MANAGEMENT OF JOINT VENTURE

3.1 Authority to act for and bind the parties to this Joint Venture in connection with all or any part of the performance of said construction contract shall be vested in the Management Committee, which may, from time to time delegate all or any part of such authority to one of the parties and/or to any individual or individuals upon unanimous consent of the parties. Neither the Management Committee nor any party hereto shall have the authority to act for or bind any other party except in connection with the performance of said construction contract. Except as provided in Paragraphs 4.2 and 4.3 each party shall have a voice in the Management Committee equal to its Proportionate Share. Except as otherwise noted herein the Management Committee may act upon consent of the party or parties having a Proportionate Share or Shares totalling more than fifty percent (50%). The parties hereby designate the following indviduals to represent them on the Management Committee:

PARTY   REPRESENTATIVE

_____   _____
                                         Representative Alternate

_____   _____
                                         Representative Alternate

_____   _____
                                         Representative Alternate

Such designations may be changed by any party at any time upon written notice to the other parties by the Chief Executive of such party.

3.2 _____
is hereby designated as the Managing Party, subject, however, to the superior authority and control of the Management Committee. The Managing Party, shall appoint the Project Manager through whom the Managing Party shall have direct charge over and supervision of all operational matters necessary to and connected with the performance of said contract, except as otherwise provided herein.

3.3 Any delegation of authority to any individual or party may be revoked by majority vote of all the parties; provided, however, that if the authority of the individual serving as Project Manager is revoked, the Managing Party shall have the right and obligation to appoint another individual to serve in that capacity who is acceptable to the parties hereto.

3.4 Decisions (or, at their option, the establishment of Guidelines to be followed by the Managing Party) regarding the following matters shall be made upon the unanimous vote of the Management Committee:

a.) Significant financial matters such as borrowing, debt guarantees, lease commitments and the investment policy with respect to Joint Venture funds.

b.) The sale of Joint Venture assets; including the terms of such sale and the agent therefor, if any.

c.) Settlements, in excess of a threshold amount set by the Management Committee, of claims or litigation by or against the Joint Venture.

d.) Transactions between any joint venturer and the Joint Venture, including agreements, if any, concerning rates of payment or reimbursement for employees, equipment, temporary or permanent materials, or management, data processing and/or other services.

FIGURE 2.4   (*Continued.*)

to defy readability, but must nevertheless be read and understood. Some of the more frequently encountered acts and their provisions are set forth in the following sections.

The Davis-Bacon Act.   The Davis-Bacon Act requires the payment of "prevailing" wage rates established for specific geographic areas of the

3.5 The Managing Party shall be designated as the "tax matters partner" (as said term is used in sections 6221 through 6232 of the Internal Revenue Code) for the Joint Venture.

3.6 Management Committee meetings shall be held as needed, but in no event less frequently than once every three months. Job progress reports, a recently updated construction schedule, and the most recent copies of the financial reports described in Paragraph 5.2 shall be presented and reviewed by the parties at such meetings. Any other matters of interest to the Joint Venture may be investigated at, or as a result of, such meetings. Any party may, upon written notice fifteen (15) days in advance of same, call a Special Meeting of the Management Committee.

3.7 Notwithstanding the provisions of Paragraph 3.8, if any dispute between the parties affects or threatens the orderly or timely progress of the Work, the Joint Venture shall proceed with the Work as directed by the Managing Party in writing, whose decision with respect to matters affecting the prosecution of the Work shall be final and binding unless an objecting party provides written notice of its objections within twenty (20) days after receipt of the Managing Party's written directive. In no event shall any dispute be permitted to delay the progress of the Work.

3.8 In the event of any dispute, including those which have been the subject of a formal objection pursuant to Paragraph 3.7, the parties shall exhaust every effort to settle or dispose of same. If, after the Chief Executive Officers of all of the parties have met on no less than two separate occasions in an attempt to settle or dispose of such dispute, then such dispute shall be settled by arbitration under the American Arbitration Association Construction Industry Rules, and judgement upon the award rendered by the arbitrator (s) may be entered in any court having jurisdiction, and the arbitration decision shall be final and binding on the Joint Venture and on all parties. The venue of such an arbitration shall be _____ .

4. CAPITAL CONTRIBUTIONS AND DEFAULT

4.1 The Management Committee shall from time to time determine the amount of working capital required to carry out and perform said construction contract, and each party shall contribute its Proportionate Share of such working capital whenever requested to do so. Such contributions shall be made within ten (10) days after request therefore.

4.2 If any party fails or is unable to provide its proportionate share of the funds required by the Joint Venture, the interest of said party in the return of investment and profits of this Joint Venture shall be decreased to the proportion that the amount actually provided by it bears to the total amount of the funds provided by all parties, and the interest of any party which may have contributed more than its proportionate share of such funds shall be increased in the same proportion. Nothing contained herein shall increase or decrease the proportionate liability of the parties hereto for losses suffered or sustained by the Joint Venture. The amount unpaid, plus simple interest which shall be charged at the rate per year of 3% above the prime rate of interest charged by the Morgan Guaranty & Trust Company of New York (but not exceeding the maximum allowed by law), shall continue to be a charge against the defaulting party until repayment. It is understood that the subsequent payment of working capital in arrears by any party hereto, which has failed or refused or was unable to contribute its appropriate share of the working capital and funds, shall not cure the default of such party, except by the express written consent of the other parties hereto not so in default.

Reduction in a defaulting party's share of the profits and increases in the share of the other parties shall be calculated as of the time of each default in contributions and as of the time of excess contributions by other parties. The profit shares as so adjusted may be further adjusted to reflect any subsequent default or excess contributions.

FIGURE 2.4 (*Continued.*)

country by the Department of Labor, for laborers and tradesmen hired to work on federally funded construction projects valued in excess of $2000. Davis-Bacon also applies where federal funds have been provided for local and state projects. A prevailing wage rate schedule is included in the bid documents for the contractor's use in preparing their estimate and, if awarded a contract, these rates are to be employed as a minimum wage for all workers hired for the project.

The defaulting party shall have no representative on the Management Committee and shall have no right to participate in the affairs of the Joint Venture until either (1) all of the defaulted contributions and default interest have been paid to the Joint Venture or (2) distributions to the non-defaulting parties have included repayment of all of the excess contributions and payment of all default interest.

4.3 If any party hereto shall dissolve; become bankrupt, or shall file a voluntary petition in bankruptcy, the remaining party or parties shall do all things necessary to wind up the affairs of this Joint Venture, including the completion of said construction contract, the collection of all monies and property due to the Joint Venture, the payment of all debts and liabilities of the Joint Venture, and the distribution of its assets. Such dissolved or bankrupt party shall have no further voice in the performance of said construction contract or in the management of the Joint Venture, nor shall any trustee, legal representative, or successor of any type. The participation of such dissolved or bankrupt party, or its representative, in the profits of the Joint Venture shall be limited to that Net Proportion (Contributions of defaulting party minus default interest charges) which the contributions of such party to the working capital of the Joint Venture bears to the total contributions to such working capital made by all of the parties, but such dissolved or bankrupt party and its representatives shall be charged with and shall be liable for its full share, as fixed in Section 2 hereof, of any and all losses that may be suffered by the Joint Venture under said construction contract, or any additions or supplements thereto or modifications thereof.

5. BANKING AND ACCOUNTING

5.1 All contributions of working capital made by the parties hereto, and all other funds received by the Joint Venture in connection with the performance of said construction contract, shall be deposited in such bank or banks as the parties may designate in separate bank account(s) bearing the name of this Joint Venture. Withdrawals of such funds may be made in such form and by such persons as the parties may from time to time direct. All persons authorized to draw against the funds of the Joint Venture shall be bonded in such company or companies and in such amounts as the parties shall determine.

5.2 Separate books of account of the transactions of the Joint Venture shall be kept in accordance with generally accepted accounting principles. Such books, and all records of the Joint Venture, shall be available for inspection by any party at any reasonable time. Periodic audits shall be made of such books at such times and by such persons as the parties may direct, with a certified audit performed annually (unless otherwise agreed, in writing, by the parties hereto) and copies of the audit reports shall be furnished to each party. Monthly financial statements and cost reports shall be prepared, with contract profit reported on a Percentage-of-Completion method. No less frequently than every three months forecasts of cash flow, final contract revenue, cost and profit and reports setting forth the status of change requests, shall be prepared and copies furnished to each party. Upon completion of the construction contract, a final audit shall be made and copies of such audit report shall be furnished to each of the parties.

6. MISCELLANEOUS

6.1 The parties may determine from time to time during the course of this Agreement that some of the assets held and acquired by the Joint Venture may be divided among or paid to the parties in accordance with their Proportionate Shares. Upon the completion of the construction contract, the assets held and acquired by the Joint Venture shall be divided between the parties and the profits or losses accrued in the performance of said contract shall be divided between or paid by the parties, as the case may be, in accordance with the terms of this Agreement, and this Agreement shall then

FIGURE 2.4   (*Continued.*)

This "prevailing wage rate" is generally close to union labor rates and includes not only the hourly wage rate for each trade but also a provision for fringe benefit levels, which if not met by the contractor, must be paid to the worker "in kind." For example, if a wage rate of $15 per hour is required and fringe benefits worth another $5 per hour are also mandatory, a contractor must account for $5 in fringes, or add any shortfall to the worker's weekly paycheck. The only workers who can be paid terminate; provided, however, that if claims of any nature or legal action of any type are brought against the Joint Venture or any of the parties at any time after such distribution by any third party or parties not signatory to this Agreement and such claims and/or legal action relate to or arise out of this Agreement, the performance of the Construction Contract and/or the work product thereof, this Agreement shall be considered to have remained in full force and effect and the rights and obligations of the parties hereto with respect to such matters shall be determined by this Agreement, the passage of time notwithstanding.

6.2 The interests and rights of each party in this Joint Venture shall not be transferable or assignable, except that any party may assign its share in any money to be received by it from the Joint Venture for the purpose of obtaining a loan or loans from any bank or other lending agency.

6.3 The scope and limits of insurance which shall be obtained by the Managing Party on behalf of the Joint Venture and, as appropriate, the joint venturers, individually, shall be as mutually agreed by the parties. Said insurance program shall not necessarily be limited by the minimum requirements set forth in the contract with the Owner and shall clearly define what liabilities, if any, are to be insured against by each joint venturer.

6.4 This Agreement shall be construed and governed by the laws of _____

6.5 The following Attachments are attached hereto and made a part hereof:

    A. Policy Statement on Business Conduct

    IN WITNESS WHEREOF, the parties have caused this Joint Venture Agreement to be executed by their duly authorized officers or agents on the date first written above.

ATTEST:

_____   _____

ATTEST:

_____   _____

ATTEST:

_____   _____

FIGURE 2.4 *(Continued.)*

less than the prevailing wage are those classified as apprentices and trainees registered in state-approved apprenticeship or training programs.

The contract work hours and safety standards Act. The contract work hours and safety standards Act (CWHSSA) requires the contractor to pay time and a half for overtime hours (over 40 in any workweek).

The Copeland Act. The Copeland Act (Anti-Kickback Act) makes it a crime for anyone to require a labor or mechanic employed on a federal of federally assisted project to kickback any part of their wages. Some disreputable contractors, required to pay Davis-Bacon wage rates, wages that were higher than those workers earned previously on private projects, required these employees to kickback a portion of their weekly wages. This law makes the practice a crime.

Certified payroll requirements. Minimum hourly rates for skilled and unskilled labor should be established and included in the contract documents, where a certification of compliance is required (Figure 2.5) with the submission of weekly payroll costs on special forms provided in the bid documents (Figure 2.6).

These certifications, following the provisions of the Davis-Bacon act must be strictly followed since falsification is a violation of federal law. For the project manager embarking on their first public works project, it is critical that they read all of the bid documents, particularly the general, supplementary, and special conditions section of the specifications.

General and supplementary provisions. In the general and supplementary conditions section of the specifications, there are stipulations dealing with subcontractors and methods by which equal opportunity requirements are to be met. Some sections of the specifications may dictate how a site logistics plan is to be established, while other items include the various closeout procedures and requirements that must be addressed.

Special conditions. The special conditions portion of the contract specifications in many public works projects include limitations on fees to be applied to change-order work, and if present, these provisions should be included in any subcontract agreements.

A typical requirement:

> On work performed by a general contractor with their own forces, their allowance for overhead and profit will likely be as follows:

For amounts up to and including $5000	15% overhead and profit
For amounts between $5001 and $25,000	10% overhead and profit
For amounts exceeding $25,000	5% overhead and profit

> For work performed by subcontractors, the general contractor is allowed to include 5-percent overhead and profit.
>
> Subcontractors are permitted the same overhead and profit as that just outlined for the general contractor performing work with their own forces.
>
> Total subcontractor fees, including those of their subcontractors, cannot exceed the amounts stipulated in the preceding section.

Requirements for payment for the offsite storage of materials and equipment are frequently spelled out in this section of the specifications. If materials or equipment are stored in a location any distance from the

Wage and Hour Division   Budget Bureau No. 44-R1093

STATEMENT OF COMPLIANCE

Date _____

I, _____ , _____ do hereby state:
    (Name of signatory party)   (Title)

(1) That I pay or supervise the payment of the persons employed by _____ on the _____:
                                               (Contractor or Subcontractor)

that during the payroll period commencing on the ____ day of _____, 19__ and ending the ___ day of _____, 19__, all persons employed on said project have been paid the full weekly wages earned that no rebates have been or will be made either directly or indirectly to or on behalf of said _____ from the full weekly wages earned by any person and that no deductions have been made either
    (Contractor or Subcontractor)

directly or indirectly from the full wages earned by any person, other than permissible deductions as defined in Regulations, Part 3 (29 CFR Subtitle A), issued by the Secretary of Labor under the Copeland Act, as Amended (48 Stat. 948, 63 Stat. 108, 72 Stat. 967; 76 Stat. 357; 40 U.S.C. 276c), and described below:

_____

_____

_____

(2) That any payrolls otherwise under this contract required to be submitted for the above period are correct and complete; that the wage rates for laborers or mechanics contained therein are not less than the applicable wage rates contained in any wage determination incorporated into the contract; that the classifications set forth therin for each laborer or mechanic conform with the work he performed.

(3) That any apprentices employed in the above period are duly registered in a bona fide apprenticeship program registered with a State apprenticeship agency recognized by the Bureau of Apprenticeship and Training, United States Department of Labor, or if no such recognized agency exists in a State, are registered with the Bureau of Apprenticeship and Training, United State Department of Labor.

(4) That:
    (a) WHERE FRINGE BENEFITS ARE PAID TO APPROVED PLANS, FUNDS, OR PROGRAMS
        ☐ In addition to the basic hourly wage rates paid to each laborer or mechanic listed in the above referenced payroll
           payments of fringe benefits as listed in the contract have been or will be made to appropriated programs for the benefit of such employees, except as noted in Section 4(c) below.

    (b) WHERE FRINGE BENEFITS ARE PAID IN CASH
        ☐ Each Laborer or mechanic listed in the above referenced payroll has been paid as indicated on the payroll, an
           amount not less than the sum of the applicable basic hourly wage rate plus the amount of the required fringe benefits as listed in the contract, except as noted in section 4(c) below.

    (c) EXCEPTIONS

EXCEPTIONS (CRAFT)	
Remarks	
Name and Title	Signature

The wilful falsification of any of the above statments may subject the contractor or subcontractor to civil or criminal prosecution. See section 1001 of title 18 and section 231 of title 31 of the United States code.

Form WH-348 (1/68) Purchase this form directly from the Supt. of Documents

FIGURE 2.5 Certification of wage compliance.

jobsite, there may be provisions in the general, supplemental, or special conditions requiring the contractor to reimburse the inspector for any costs incurred to travel to that area and to inspect the item being requisitioned.

Subcontractors need to be made aware of these provisions at the time of contract negotiations to avoid misunderstandings at a later date. The project manager should never assume that the subcontractor has read and

U.S. DEPARTMENT OF LABOR
WAGE AND HOUR DIVISION

PAYROLL

(For Contractor's Optional Use; See Instruction, Form WH-347 Inst.)

NAME OF CONTRACTOR ____ OR SUBCONTRACTOR ____   ADDRESS

PAYROLL NO.   FOR WEEK ENDING   PROJECT AND LOCATION   PROJECT OR CONTRACT NO.

(1) NAME, ADDRESS AND SOCIAL SECURITY NUMBER OF EMPLOYEE	(2) # OF EXEMP- TIONS	(3) WORK CLASSIFICATION	OT OR ST	(4) DAY AND DATE														(5) TOTAL HOURS	(6) RATE OF PAY	(7) GROSS AMOUNT EARNED	(8) DEDUCTIONS				(9) NET WAGES PAID FOR WEEK
				HOURS WORKED EACH DAY																	FICA	WITH HOLDING TAX	OTHER	TOTAL DEDUCTIONS	

FORM WH-347 (1/68)

FIGURE 2.6 Typical certified payroll reporting form.

understood all of these miscellaneous provisions. It has been the author's experience that very few subcontractors thoroughly read any of these special, general, and supplementary conditions, and when brought to their attention during the project they seemed upset that they were expected to abide by such provisions.

The notice to proceed in a Public Works project.   This document, generally in the form of a letter, transmitted to the contractor by the government agency, is the official notification of the starting date of the contract—the date from which the contract time will be charged. In some cases, there may be two Notices to Proceed: the first one issued for mobilization of the contractor's field office (the contract timeclock generally does not start with this notification), and the second one stipulating when the contractor is to commence construction (this one *does* start the contract clock).

Public Works provisions that can affect subcontractor negotiations.   Prompt payment provisions are being included in an increasing number of public works projects. The "pay-when-paid" clause in many general contractor-subcontractor agreements will therefore be void when these prompt payment clauses are included in government contracts. The prompt payment provisions typically state that the general contractor is obligated to pay their subcontractors within 30 days of receipt of payment. These subcontractors, in turn, must pay their subcontractors within 30 days of receipt of their payment. If payment is to be delayed, the contractor or subcontractor must respond in writing stating the reason for the delayed payment (for example, subcontractor is in default of the contract, or must replace defective work, or has failed to pay the vendor).

Change-order clauses in government contracts.   Some contracts may contain clauses which state that no changes other than those for project *enrichment* or extra work ordered by the owner's representative or architect will be approved. The term *enrichment* can have one meaning for the owner, but a different one for the contractor.

What about items added when the local building official or fire marshal walks through the job on an inspection tour? If additional exit lights are required or added emergency lights are requested, is this a legitimate change order falling within the concept of *enrichment?* A contractor could argue strongly that this is a compliance with code requirement, and without these extra cost items the project is worthless since it will not receive a certificate of occupancy. That surely qualifies these added cost items as *enrichments.*

AIA Document A201, 1997 edition, specifically mentions that the contractor is *not* required to ascertain that the contract documents are in accordance with building codes (more on this subject in Chapter 3).

Administering contracts with public agencies can be a demanding task. The project manager should be thoroughly familiar with all phases of the contract because they could probably be enforced to the letter by an overzealous inspector.

Watch for owner-inserted changes to the standard contract. Rare is that contract for private sector work that's not amended by an owner's lawyer. Some of the amendments to the base contract or the general conditions attachment can be deadly if not recognized by the project manager before the contract is executed. Traps await those who do not take the time to read and understand all of the terms and conditions in a contract, as well as the content of all of the exhibits and addenda attached to it. The following is a sample of some of the more onerous provisions, and the sections of the contract and general conditions where such amended articles might be found:

Article 3 of the General Conditions pertaining to the Contractor

- Regarding shop drawings—"The contractor shall submit complete and accurate submittal data at the first submission. If the submittal is returned requiring resubmittal, only one such additional submittal will be reviewed at the owner's cost. Any additional submittals will be reviewed at the contractor's cost."

- Schedules—"If any of the work is not on schedule, the contractor shall immediately advise the owner in writing of proposed action to bring the work back on schedule. In such an event, the owner will require the contractor to work such additional time over regular hours (including Saturdays, Sundays, and holidays) at no additional cost to the owner, in order to bring the work back on schedule."

- Contractor responsibility for details not shown on the drawings— "The contractor has constructed several projects of this type and has knowledge of the construction and finished product." This is a trap yet to be sprung. If some minor portions of work are omitted from the contract documents, the contractor may be required to provide them at no cost since they have acknowledged having constructed several similar projects.

- "If the drawings and specifications conflict, then the greater quantity and/or quality shall apply."

Article 7 of the General Conditions—Changes in the Work

- "The owner at all times shall have the right to participate directly in the negotiations of change-order requests with subcontractors and material suppliers."

- "If the owner and contractor are unable to agree on the amount of any cost or credit to the owner resulting from a change in the work, the contractor shall promptly proceed with, and diligently prosecute, such change in work and the cost or credit to the owner shall be determined on the basis of reasonable expenditures and savings." This means that you cannot refuse to perform extra work if the cost of this work is in dispute prior to starting the work. The preferable method in such a case is to invoke the Construction Change Directive procedures, as outlined in the standard AIA contract.

Article 9 of the General Conditions—Payments and Completion

- "Unless otherwise agreed to in writing by the owner, the project shall not be considered substantially complete if the items in the punch list would reasonably require more than two weeks to complete."

Article 13 of the General Conditions—Miscellaneous Provisions.   The following provision type is applicable primarily to renovation and rehab work.

- "The contractor shall review the structural capability of the structure prior to allowing installation of temporary lifting devices or staging equipment or the temporary off-loading and storage of materials. Costs associated with the architect's review or redesign of structure to accept the temporary construction loading shall be borne by the contractor."
- Contingency accounts may include limits on what can be considered a contingency item, such as outlined in the following:
- "Any estimating errors or other errors in the contractor's bid are not cause for reimbursement from the contingency account."
- "Any funds remaining in the contingency account will be cause for a deduct change order and cannot apply to the project's overall savings."
- "It is understood that the amount of the contingency reserve is the maximum sum that the contractor will seek or is entitled to recover from the owner for costs that fall within the categories and definition of contingency costs."

The potential for restrictive clauses in the contract documents, sometimes in obscure places, makes it important for a project manager to carefully review the owner's agreement, highlight the important provisions, and be ready to deal with them if the occasion arises.

Chapter

3

The General Conditions to the Construction Contract

Most construction contracts are supplemented by other documents such as general, supplementary, and special conditions. These supplementary contract obligations may be incorporated into the project specifications manual or in stand-alone documents. The most widely used contract supplement is the American Institute of Architects AIA Document A201—General Conditions of the Contract For Construction issued in its present form in December 1997.

Too often, project managers consider these documents "boilerplate" and don't give them the attention they deserve. But nestled amongst it all are many provisions that a contractor should approach cautiously, while others offer a general contractor considerable protection.

AIA A201—General Conditions of the Contract for Construction

The previous edition of AIA A201—General Conditions of the Contract for Construction was dated 1987 and sometimes is appended to contracts even though it has been superseded by the 1997 issue. A project manager should read this document from cover to cover at least once, and re-read selected sections from time to time to either support a position or defend against one. Let's look at some of the provisions of this document and how they affect the administration of a project.

Article 1: General provisions—the contract documents

Along with defining the components of the contract documents, Section 1.5.2 in Article 1 requires the contractor to stipulate that they have visited the site and are somewhat familiar with local conditions under which the work is to be performed. This is not a statement to be taken lightly, and quite often a contractor's failure to visit the site and observe conditions that could impact their contract obligations will result in the denial of future claims. For example, if construction has just gotten underway and a site condition not indicated on the drawings is encountered, which would have been apparent upon visiting the site, it may be very difficult to initiate a claim for extra work based upon the notion that its condition wasn't noted in the contract documents.

Let's say an abandoned well was found in the area of a proposed footing, and the well cap or well cover was clearly visible. The architect could invoke the provisions of Article 1 as the reason for disallowing a contractor's claim for additional costs for a structural fill required to raise the subgrade under the footing or lower the elevation of that footing to achieve proper bearing. The abandoned well would have not only been visible, but a prudent contractor should have noted its location and even removed the well cap/cover to determine its depth. Notifying the architect during the bid process would have clarified this matter.

Article 2: Owner

A provision in Article 2 directs the owner to designate in writing a representative with the authority to bind the owner to matters requiring an owner's approval or authorization. This appointment of an owner's representative can speed up the communication process between the contractor and owner when field conditions occur that require a prompt owner/architect decision. This section of the General Conditions also stipulates that the owner is obliged to present reasonable evidence that financial arrangements have been made to satisfy the requirements of the construction contract. The contractor may obtain a copy of this financial commitment by writing to the owner and requesting the same.

Another proviso deals with reproducibles. Unless the contract stipulates to the contrary, the owner shall furnish the contractor, *free of charge, sufficient copies of plans and specs* that are *reasonably* necessary for the execution of the work. (Does this include sufficient copies for each subcontractor as well? A strong case could be made for several sets for major subcontractors, and one set for minor subcontractors.) The owner, per this article, has the right to subcontract portions of the work by giving written notice to the general contractor. (*Author's note:* But suppose the general contractor is a union contractor and the owner hires, say, a nonunion electrical contractor to install their data communications work? Although the owner has the right to do so, a prudent general contractor should respond by requesting that they be held harmless from any labor disputes that arise out of these arrangements.)

Article 3: Contractor

Article 3 is a key component and should be read and comprehended fully. This is true in part because it deals with shop drawings and requires the contractor to take field dimensions of any existing conditions related to the work; thus, any errors, omissions, or inconsistencies discovered should be reported promptly to the architect. In subparagraph 3.12.4 and 3.12.5, this article states that shop drawings are not *contract* drawings. (This voids a contractor's argument that approval of a shop drawing is proof that the item has been accepted by the architect. If it has been accepted, but is deemed of lesser quality than the specified item, the architect may request, and in fact is entitled to, a credit.)

Contractors would be wise to read subparagraphs 3.12.4 through 3.12.10 relating to shop drawing submissions, review, and disposition in their entirety.

With respect to the review of contract documents for errors and omissions, it is recognized that such a review is being performed by a contractor and not a licensed design professional, thus the contractor is *not required* to ascertain that the plans and specifications are in accordance with laws, statutes, ordinances, building codes, and rules and regulations. Paragraphs 3.2.2 and 3.2.3 absolve the contractor for damages resulting from errors, inconsistencies, or omissions in the contract documents, or for differences between field measurements and conditions and the contract documents *unless the contractor recognized such an error, inconsistency, omission, or difference and failed to report it to the architect.*

Article 3 assigns the contractor responsibility and control over construction means, methods, techniques, and sequences unless the contract documents dictate otherwise. In another section, it restates that the contractor has no responsibility to ascertain that the contract documents are in accordance with applicable laws, statutes, ordinances, building codes, and rules and regulations. (*Project managers take note:* This will diffuse arguments about responsibility for costs of those last-minute additions to the scope of work when building officials, during one of their many walk-through inspections, require more exit lights, fire alarm devices, and so on.) The contractor is obligated in this section of the General Conditions to submit a construction schedule for the architect's *information.* (Previous issues of this A201 document required submission of a schedule for the architect's *approval.*)

The question of "costs" to be included in an allowance is frequently raised by the contractor and owner who often don't agree on what is to be included or excluded. For example, does an allowance include the contractor's overhead and profit or not? Article 3.8 answers this question and others concerning the reconciliation of an allowance item. Unless otherwise stated in the contract, an allowance includes

- The cost to the contractor of all materials, labor, and equipment for the item delivered to the site, including taxes and trade discounts
- The contractor's cost to unload and distribute equipment and materials and all related labor and installation costs, which shall be included in the contract sum
- The contractor's overhead and profit for the allowance, which should be included in *the contract sum and not in the allowance.*

If the allowance, when reconciled, is more or less than stated in the contract, the contract sum is to be adjusted by a change order and shall reflect the difference between the actual cost and the allowance. If the actual cost of the allowance exceeds its contract value, the contractor's overhead and profit can be added to the coverage amount when a change order is issued. Lastly, Article 3 states that the contractor shall not be required to provide professional services which constitute the practice of architecture or engineering unless specifically called for in the contract.

Article 4: Administration of the contract

Article 4 repeats, once more, the contractor's right to control the construction means, methods, techniques, sequences, or procedures since these are "solely the contractor's rights and responsibilities." The architect is charged with the duty to review shop drawings with reasonable promptness. (The exact time frame for shop drawing review is frequently spelled out in other parts of the contract documents—for instance, special, supplementary conditions—but the word "reasonable" does restrict the review time frame to some degree.)

The architect is charged with the authority to interpret and decide matters concerning performance or requirements of the contract documents or interpretations and decisions consistent with the intent of, or reasonably inferable from, the contract documents. *(Author's note*: There appears to be a lack of checks and balances in this arrangement. The person who prepared the plans and specs is given authority to interpret these documents and rule on their intent. This is like asking the fox to watch the proverbial chicken coop!)

Article 4.2.7 stipulates that architect approval of a specific item does not indicate approval of an assembly to which the item is a component. This provision could apply to a value engineering proposal presented by the contractor where one component of an assembly is substituted and approved, but its substitution invalidates the assembly. For example, the substitution of the type or gauge of roof coping or flashing may be approved by the architect but will not meet the manufacturer's recommendations, hence a roofing bond will not be issued by the manufacturer of their agent.

Article 4 includes procedures for filing a claim and sets a time limit of 21 days after the occurrence of the event as the time frame within which the claim must be filed. The subject of claims for concealed conditions or unknown conditions cited in subparagraph 4.3.4 make it an important section to read and understand. The sentence referring to "differing conditions" may be helpful for those contractors filing a claim for unsuitable soils or excessive rock when actual conditions encountered differ *materially* from those set forth in the geotechnical report accompanying the bid documents. There is a fine legal line between "differing conditions" and "materially different conditions," just ask your company attorney.

When the contract labels all sitework as "unclassified," this in effect means that whatever deleterious material the contractor uncovers must be removed and replaced with acceptable fill, at no cost to the owner. However, if such a claim is based upon conditions uncovered that vary significantly from those anticipated after a reasonable review of the geotechnical report, invoking the "differing conditions" provisions of subparagraph 4.3.4 may allow the contractor to recover some of these extra costs.

Another important part of this article is subparagraph 4.3.10 which disallows any claims for consequential damages. This theoretically denies the contractor the right to use the Eichleay formula, which is a method to establish a contractor's claim to recoup unabsorbed or underabsorbed corporate overhead costs in delay claims. Dispute resolution procedures are included in Article 4 and mediation is deemed the first step in that endeavor. If mediation fails to resolve the claim, arbitration is the next step in the process, thus several paragraphs in this section outline the steps to be taken to invoke mediation and arbitration.

Subparagraph 4.5.4 is entitled "Limitation on Consolidation or Joinder," which may be of interest to general contractors if they have reached the arbitration stage in dispute resolution matters. The term *joinder* means to join with. If, for example, both general contractor and subcontractor have a claim against the owner, they cannot join together in one arbitration proceeding, but must pursue their claims individually—the subcontractor should file for arbitration against the general contractor, and the general contractor should file for arbitration against the owner, if such is the case.

Article 5: Subcontractors

The general contractor is directed in Article 5 to submit in writing a list of proposed subcontractors to the architect who should promptly review and reply as to whether the owner has objections to any of the subcontractors presented. The contractor is cautioned not to proceed to contract with a subcontractor which the owner/architect has objected to; however, if the rejected subcontractor could have been deemed reasonably capable of performing the work, the contract sum will be increased or decreased by the substitution of an owner-acceptable subcontractor. When a general contractor is considering invoking this provision, the GC should request a written statement from the owner containing their reasons for the rejection of any subcontractor, and if the general contractor disagrees with the owner's assessment, they must respond to the owner in a letter stating their views, all of which should be properly and completely documented.

The author of this book had just such an experience as this when he was requested to obtain bids on low voltage alarm systems for a senior living community project. The owner rejected all subcontractors' bids and requested that a bid be solicited by a low-voltage systems supplier they had employed on several recent projects in other parts of the country. This "preferred" supplier failed to submit their bid in a timely fashion, failed to submit a proper schedule of values for comparison with the other bidders, and failed to deliver materials and equipment in a timely fashion. It was apparent that this vendor felt that their "preferred" position with the owner would cover a multitude of sins, until the author sent a letter to the owner advising them of serious delays in the offing unless *their* vendor met the current project's demands. Within a day or two the much needed equipment arrived at the jobsite.

Provisions in this article require the contractor to bind all subcontractors to the contractor by the terms of the contract documents and assume all obligations and responsibilities which the contractor maintains toward the owner. This is a standard "pass-through" provision which is actually beneficial to the general contractor. According to Article 5.4 the general contractor must include in their subcontract agreement a provision allowing the assignment of a subcontract to the owner under specific conditions spelled out in detail in the subparagraph dealing with the Contingent Assignment of Subcontracts. This allows the owner some degree of protection in case a general contractor defaults on the project. Rather than renegotiate with existing subcontractors to complete the work, possibly at much higher costs, the owner merely "takes over" all active subcontract agreements.

Article 6: Construction by owner or by separate contractors

Article 6 affords the owner the right to subcontract certain portions of the work; however, as stated previously, if the general contractor anticipates that a conflict could arise due to collective bargaining agreements, the owner must be advised of these potential problems in writing.

Article 7: Changes in the work

Article 7 deals with change orders and the construction change directive (CCD). A change order, according to Article 7, is based upon an agreement between owner, contractor, and architect, but a CCD can be issued by an owner requesting a change in the contract scope when there is no agreement on cost.

If the CCD issued by the architect affects the contract sum, an adjustment in the contract sum will be based upon one of the following:

- A mutual acceptance of a lump sum, properly itemized and documented.
- Unit prices contained in the contract (if, in fact, there are any in the contract)
- The cost—to be determined in a manner agreed upon by all parties as well as a mutually acceptable fixed amount or percentage for the contractor's fee and overhead
- A time and material cost approach, which should include
  - The cost of labor and fringe benefits
  - The cost of materials, supplies, and equipment, including transportation costs
  - The rental costs of equipment whether rented or company-owned
  - The costs of bonds, insurance premiums, and any related fees
  - Additional costs of supervision and field office personnel directly attributable to the work

Subparagraph 7.3.6 of this article contains a full explanation of the CCD process. Subparagraph 7.3.8 is also of importance to contractors and states that all amounts in the CCD *not in dispute* can be included in the current application for payment. This is an important consideration because it means that an interim billing can be included in a current application for payment for change-order work completed but for which no formal, fully executed change order has been previously received. Change-order work that can stretch out for months can only be billed when it is fully complete. *Don't forget to consider potential changes in contract time when formulating the CCD. Will the contract time remain unchanged, or will it decrease or increase?*

Article 8: Time

The definition of "contract time" is included in Article 8. Unless stated otherwise, the Contract Time is the period of time, including authorized adjustments (change orders), that is required to attain Substantial Completion of the project. The term "day," unless otherwise stated, is meant to be a calendar day. Methods by which delays and extensions of time are to be handled are included in this section of the general conditions.

Article 9: Payments and completion

In Article 9, the contractor is directed to submit a Schedule of Values for approval by the architect prior to the submission of the first Application for Payment. If there are no objections from the architect, this Schedule of Values will be incorporated in the contractor's monthly requisition requests and serves as the basis for determining the percent of completion for each line item in that requisition.

This article stipulates that payment is allowed for the onsite storage of materials and equipment, but any request for offsite storage payments must be made to the architect in writing, and is conditional upon meeting the following terms and conditions:

- A presentation of a procedure by the general contractor to insure that the title to the materials/equipment will pass to the owner. This can be accomplished by submitting a bill of sale, which will automatically transfer the title to the owner once payment is received.
- Evidence is presented that the cost of insurance, storage, and the subsequent transportation to the site will be paid by the contractor.
- An insurance certificate is furnished which documents that coverage will remain in effect during storage and transportation to the site.

Article 9 includes conditions that permit the architect to withhold certification for payment on the monthly requisition...

- If defective work has not been repaired/replaced
- If third-party claims have been filed
- If the contractor fails to make payments to the subcontractors or pay for labor, materials, and equipment incorporated into the building
- If reasonable evidence exists that the work cannot be completed for the unpaid balance of the contract sum
- If there's damage to the owner or another contractor
- If reasonable evidence exists that the project will not be completed within the contract time
- If the contractor consistently fails to perform the work in accordance with the contract documents

Article 9 of the general conditions document directs the contractor to pay each subcontractor upon receipt of the owner's payment the amount which each subcontractor is entitled to receive. This, in effect, is confirmation of the standard "pay when paid" clause. This article states that subcontractors may, upon written request to the architect, obtain information regarding percentage of completion paid by the owner to the contractor for their portion of the work.

Article 9 also deals with the subject of Substantial Completion, both defining this stage of completion and payments due the contractor when this phase of construction has been achieved. Partial occupancy and/or use of the project is described in Article 9.8, along with the owner's and contractor's responsibilities when phased occupancy is being considered. Last but not least, final payment procedures are set forth, stating that no payment will be made until the contractor does the following:

- Provides an affidavit stating that all labor, materials, and equipment incorporated into the building have been paid for

- Produces a certificate as evidence that insurance will remain in effect for 30 days after final payment and will not be canceled prior to that date

- Provides a written statement that the contractor is aware of no reason that current insurance coverage will not be renewable to cover the period set forth in the contract documents

- Shows consent of surety to final payment (if a bond was required)

- Offers up a release from the general contractor to the owner with respect to any claims, liens, or other encumbrances arising out of the contract—for instance, a final waiver of lien

Article 10: Protection of persons and property

Safety precautions and contractor procedures to protect persons and property during construction are called out in Article 10. This article requires the owner to advise the contractor of the absence or presence of any hazardous materials likely to be encountered on the site. If hazardous materials are discovered on the site, the contractor is to cease work, notify the owner, and await further instructions from the same.

Article 11: Insurance and bonds

In conjunction with specific limits of insurance contained in the bid documents, Article 11 provides more detail about insurance coverage. Unless otherwise stated in the contract, the owner will purchase and maintain builder's risk "all-risk" insurance. When partial occupancy occurs, it shall not commence until the insurance company or companies have consented to partial occupancy, or use, by endorsement or otherwise.

Article 12: Uncovering and correction of work

If a portion of the work to be inspected by the architect/engineer has been covered or enclosed contrary to the architect/engineer instructions, Article 12 requires the contractor to uncover the work if requested by the architect/engineer, and also recover both at the contractor's expense. If the architect/engineer did not earlier specifically request to inspect the work before being covered, and then decides to do so, the costs to uncover and replace would be cause for a change order to the owner. If when uncovered, the work is found to comply with the contract documents, all related costs will be borne by the owner, but if defective work is exposed, all costs to remove, repair, and replace will become the contractor's responsibility.

Of importance to the contractor are the contents of Article 12.2.2 which deal with corrective work after substantial completion has been attained. According to the provisions of this article, if during the one-year warranty period the owner fails to notify the contractor of work to be corrected (thereby not affording the contractor an opportunity to do so), the owner waives all future rights to require the contractor to perform that warranty/guarantee work.

Article 13: Tests and inspections

Procedures for inspections and testing are provided in Article 13. Also included in Article 13, seemingly misplaced, is a statement concerning the payment of interest on late remittances to the contractor. Article 13.6.1 stipulates that, in the absence of an agreement to the contrary, interest will accrue on payments to the contractor that remain unpaid from the date payment is due. The interest rate will be the prevailing one at the place where the project is located.

Article 14: Termination or suspension of the contract

The contractor may terminate the contract for the reasons set forth in Article 14. Conditions under which the owner may terminate the contract "for cause" are also included in this article. Article 14 allows the owner to suspend or terminate the contract "for convenience," the circumstances of which are set forth in subparagraphs 14.3 and 14.4.

The 1987 edition of AIA A201

The current 1997 issue of the A201 document contains 154 substantive changes from the 1987 edition. If any contracts currently administered include the 1987 general conditions provisions, the project manager should carefully review this older document and take note of its predecessor provisions.

AIA Document A201CMa—General Conditions for the Construction Manager Contract

Although many of the provisions contained in AIA A201 are similar to those included in the construction manager (CM) version, there are enough fundamental differences to warrant a project manager administering a construction manager contract to read this document from cover to cover at least once. Article 3 of the CM version of the General Conditions defines the *Contractor* as the person or entity that performs the construction administered by the construction manager. The *contractor* is to "carefully study and compare the contract documents with each other," and they are not liable to the owner, construction manager, or architect for "damage resulting from errors, inconsistencies, or omissions" unless the contractor recognized these deficiencies and "knowingly failed to report it."

This article requires the contractor to take field dimensions, as required, and compare them with those indicated on the contract drawings. The contractor is also directed to report any errors, omissions, and inconsistencies to the construction manager and architect promptly. A provision in Paragraph 3.7.3 states that the contractor's responsibility does not extend to verification that the contract documents comply with applicable building codes, laws, ordinances, and other appropriate rules and regulations, similar to the AIA provision.

The topic of "Allowances" is discussed in this article and defines the elements of cost allowed for incorporation in the allowance item. Provisions for the submission of the contractor's construction schedule are included in this article and although it is to be submitted for the owner's and architect's *information*, it must be submitted to the construction manager for *approval*. This subtle difference between "information" and "approval" in both the AIA and CM versions can mean quite a bit when a contractor is assembling a claim for delays. An "approved" schedule becomes a baseline schedule, and deviations from that baseline schedule can form the basis for requests for compensable costs associated with the delays. When the schedule is submitted for *informational purposes*, it serves that purpose only—for information, not acceptance.

Article 4 of A201/Cma establishes the basic responsibilities of the construction manager:

- The CM will determine, in general, if the work is being installed in accordance with the contract documents and will inspect for defects and deficiencies in the work.

- The CM will provide coordination of activities required for the work of the contractors under their supervision.

- The CM will review and certify all requests for payment.

- Although the architect will have the right to reject work that does not conform to the contract documents, no such action will take place until the construction manager has been notified. Subject to the review by the architect, the construction manager also has the responsibility to reject nonconforming work.

- The CM will receive all shop drawings, and review and approve them as consistent with the contract requirements. The CM will also coordinate them with information received from other contractors and pass them on to the architect.

- The CM will prepare change orders and construction change directives for presentation to the architect and owner.

Article 4.7.4 is entitled "Continuing Contract Performance" and states that pending final resolution of a claim, the contractor shall proceed with the performance of the contract. All too often, a contractor, out of extreme frustration in their attempt to resolve a dispute, will tell the owner, architect, or CM: "I've had it. I'm going to stop the job until this claim (or change order, or dispute) is settled." Unfortunately, if that threat is carried out, the contractor will be declared in default of the contract and their bargaining power will be significantly decreased.

Article 4 establishes arbitration as the first step in the dispute resolution process. Article 7, "Changes in the Work," follows more or less the same procedures as those in the AIA General Conditions documents and includes similar provisions for the preparation of a construction change directive (CCD) in the event that the contractor and construction manager cannot reach an agreement on a lump sum amount for the work. Article 8, entitled "Time," includes discussions regarding delays in the work, and contains a specific provision in subparagraph 8.3.3 that allows the contractor to recover damages under other provisions of the contract documents.

The Construction Management Association of America publishes their own version of construction manager General Conditions in CMAA Document A-3-2005 edition. These general conditions are meant to be appended to CMAA's Document A-1—Standard Form of Contract between Owner and Construction Manager. This 37-page document has many similarities with the AIA General Conditions.

The Associated General Contractor's Version of General Conditions between Owner and Contractor-AGC Document No.200

The Associated General Contractors of America, Inc. (AGC) developed a series of construction contracts and their AGC Document No.200 combines a Standard Form of Agreement and General Conditions between

Owner and Contractor in one document. Contractors using the AGC contract document will find some provisions in the general conditions portion similar to the AIA A201 document, while other provisions may have the same effect but are worded in a slightly different manner. With respect to uncovering errors, omissions, and inconsistencies in the contract documents, AGC requires the contractor to promptly advise the owner of any defects in the plans and specifications, but recognizes the fact that the contractor is not acting in the capacity of a licensed design professional. In other words, the contractor should not be held accountable for building code or building regulation violations.

The matter of warranties is treated quite differently from that in the AIA version of the General Conditions. Although the contractor is bound by a one-year warranty, any extended warranties required by the contract will be assigned to the owner after this standard one-year warranty period has expired. The owner then assumes the responsibility to notify the vendor or subcontractor of warranty issues, and the contractor is bound only to provide *reasonable* assistance in enforcing the provisions of these extended warranties. Subparagraph 3.10.3 of AGC Document No.200 requires the contractor to designate a Safety Representative whose duty it will be to enforce safety rules and regulations on the site.

Article 10.2, entitled "Mutual Waiver of Consequential Damages," may be looked upon as a double-edged sword. Recovery of consequential damages, most often those created by delays, cannot be pursued by the contractor, and by the same token, the owner cannot threaten the contractor with consequential damages if they are seeking recovery of costs caused by contractor-generated construction delays.

The dispute resolution menu

Appended to AGC Document No.200, as Exhibit No.1, is a Dispute Resolution Menu, a thoughtfully prepared series of steps to be taken to resolve disputes. The owner and contractor can elect to resolve any potential disputes in one of five ways, as indicated on this checklist-type form which has become an integral part of construction contracts.

- Using a Dispute Resolution Board (DRB), which consists of one member selected by the owner, one selected by the contractor, and a third by two owner- and contractor-selected members. The DRB will meet periodically to track the construction process and make advisory recommendations along the way to avoid or settle any potential disputes or claims.

- Using advisory arbitration, which will be conducted in accordance with the Construction Industry Rules of the American Arbitration Association.

- Holding a mini-trial in which top management from the owner's and contractor's organization submit their individual positions to a mutually selected individual who will make a nonbinding recommendation to the parties (a process similar to mediation proceedings).

- Invoking binding arbitration pursuant to the Construction Industry Rules of the American Arbitration Association.

- Litigation

The Engineers Joint Contract Documents Committee General Conditions

In 1990, a joint committee composed of the National Society of Professional Engineers, the Consulting Engineers Council, the American Society of Civil Engineers and the Construction Specifications Institute prepared several contract documents for use by engineers. In cases where roadwork, infrastructure, or other civil engineering projects are concerned, engineers are the designers, and it was felt that these designers should have their own contract and General Conditions documents.

The Standard General Conditions of the Construction Contract, prepared by the joint committee, includes a rather definitive index that requires a full 12 pages of text. This document contains several unique features. Article 2, entitled "Preliminary Matters" requires that a preconstruction conference be arranged within 20 days after contract signing. The purpose of the conference is to establish a working understanding of the project and discuss schedules and procedures for handling shop drawings and other submittals.

Article 3, "Contract Documents—Intent, Amending, Reuse," states that although it is the intent of the contract documents to describe a "functionally complete project," the "intended result will be furnished and performed whether or not specifically called for." This phrase puts the contractor on notice that the "intent" of the documents is of equal importance to the scope defined by the plans and specifications. Although several articles prohibit the design engineer from dictating means, methods, techniques, sequences, or procedures of construction, the contractor is prohibited from working overtime on Saturdays, Sundays, or legal holidays without the written consent of the engineer. Throughout this document, the term *engineer* is substituted where the word architect would normally be used. Thus, all communications between the owner and contractor must pass through the engineer. Procedures for change orders, the rejection of work, and the approval of progress payments are similar to the AIA General Conditions document, but Article 16, "Dispute Resolution," is rather unique. The Engineers Joint Committee requires that a dispute resolution method and procedure be spelled out in a separate document designated "Exhibit GC—A Dispute Resolution Agreement." If no such agreement has been created, the owner and contractor may use whatever remedy is at their disposal, as long as it does not conflict with any contract language.

A Word to the Wise

A project manager should take the time to read all of the contract documents relating to their project, including any general, supplementary, and special conditions. These important contract documents define various situations that may arise during the life of a project, and specify how they are to be handled. These documents elaborate on each party's duties, rights, and obligations, the full understanding of which is necessary to intelligently and professionally manage the construction project. And as the warning printed on the front page of AIA A201 emphasizes: *"This document has important legal consequences."*

Bonds and Insurance

The discussion of bonds is an interesting one because the process of achieving bond-ability is really a study of the inner workings and health of a contractor. With new developments in project insurance today, this requires a rethinking of existing practices.

A simple definition of both bonds and insurance should clear up some misconceptions.

- **Bonds** These provide *guarantees* of project performance and completion to a third party, typically an owner.

- **Insurance** A *loss-sharing mechanism* that guards the policy holder from damages that may be incurred in the future.

This Risky Business

Everyone in the industry recognizes the inherent risks in the construction business—it is an open-air manufacturing facility producing a one-off product; subject not only to the vagaries of the weather, but potential labor shortages, disputes, and material shortages. Contracts containing a fixed price for some projects may extend two years or more, while razor-thin projects can be wiped out faster that you can say "Help!"

A recent study conducted by Ken Roper and Michael McLin of FMI Corporation supports the tenuous nature of the contracting business:

- Dun & Bradstreet reported that more than 10,000 contractors fail annually.

- The surety industry has lost $3.4 billion between 1998 and 2003.

- Most company failures are due to poor operational execution.
- Many general contractors are just one difficult project away from bankruptcy.

The nearly 30-percent business failure rate among general contractors and subcontractors leaves a residual of unfinished projects and losses to their owners in the billions. That's why more and more owners are requiring contractors to provide bonds—primarily payment and performance bonds—for their new projects. The federal government has required contractors to be bonded since 1893. In 1935, the Miller Act forced contractors to provide payment and performance bonds on all public works contracts valued at more than $100,000, and payment bonds for all contracts exceeding $25,000. All 50 states, plus the District of Columbia and Puerto Rico, have enacted similar laws requiring bonds.

Why Contractors Fail

A 2005 report issued by the Surety Association of America (SAA), which was prepared after their review of 86 claims, identified the five top factors contributing to contractor failure.

1. Unrealistic growth or change in the scope of business—37 percent of cases reviewed
   - Change in the type of work performed
   - Change in the location of work performed
   - Significant increase in the size of individual projects
   - Rapid expansion
2. Performance issues—36 percent of cases reviewed
   - Inexperience with new type of work
   - Personnel do not have adequate training or experience
   - Insufficient personnel
3. Character/management/personal issues—29 percent of cases reviewed
   - Contractor retires, company is sold, changes in leadership or focus
   - No ownership or management transition plan exists to ensure continuity of operations in the event of a principle's death or disability
   - Key staff leaves the company
   - Staff inadequately trained on company policy and operations
4. Accounting issues/financial management problems—29 percent of cases reviewed
   - Inadequate cost and project management systems
   - Estimating and procurement problems
   - Lack of adequate insurance
   - Improper accounting practices [not adhering to the Audit Guide for Construction Contractors (AICPA)]

5. Other factors (which also seem to point to management problems)
   - Economic downturn and/or high inflation
   - Weather delays
   - Poor site conditions and/or building plans
   - Labor difficulties (lack of skilled labor)
   - Material and equipment shortages
   - Owner's inability to pay
   - Onerous contract terms

The SAA developed a profile of the qualities associated with a solid contractor organization. They include

- Formal and on-the-job training for all levels of employees
- Logical, incentive-based compensation programs
- Tenure for proven field supervisors, and internal promotion wherever possible
- Depth at all levels of the organization
- Succession planning
- An up-to-date, distributed organization chart
- A culture of loyalty, ownership, and urgency
- Visionary, inspirational leadership
- Low turnover

Finance:

- Solid management of cash flow and overhead
- Profit-focused company
- Timely payment of bills
- Management of debt and retainage
- Reasonable growth without over extending resources

Marketing:

- Superior estimating skills and systems to manage costs
- Satisfied customers
- A well-defined market niche and a 12–36 month growth plan
- A company culture where everyone is a great salesperson

Project Control:

- Closely managed projects with early warning systems to catch potential problems

- Litigation avoidance
- Productive field managers trained to improve processes

  Planning:

- Disaster preparedness
- A continuity plan with adequate life insurance
- A shareholders agreement detailing buy-sell understanding for multiple shareholders
- Only qualified and interested family members in management
- A detailed business plan
- Strengths, weaknesses, opportunities, and threats

  Where does your company fit in this profile?

Bonds and the Bonding Process

Bonds are, in effect, a three-party arrangement between a surety (bonding company), a principal (contractor), and an oblige (owner). The most frequently employed bond forms are the payment and performance bond and the bid bond. With respect to a payment bond and performance bond, the surety obligates itself to pay the oblige if the principal does not meet the performance criteria (of a performance bond) or payment requirements (of a payment bond) of the bond.

The bid bond provides financial assurance that a contractor intends to enter into a contract with the owner at the price in their submitted bid. If the contractor fails to accept the offer of an award, they forfeit the value of the bond and the owner is free to use those proceeds toward the contract sum of the next lowest bidder.

The terminology of bonds

The following terms are related to bonds:

Calling the bond.   Notification to the bonding company by the owner (oblige) that the contractor (principal) has failed to live up to the terms of the contract for construction, and that the owner requests that the bonding company (surety) provide sufficient funds to cover the contractor's unsatisfied contract commitments.

Consent of surety.   On the successful completion of the construction project, when all bills have been paid and all provisions of the contract fulfilled, the terms and conditions of the payment and performance bond will be considered to have been met. The contractor will then request surety to "sign-off" on the bond, getting *consent of the surety* that the bond requirements have been fulfilled, before releasing the bond.

Dual obligee.   When two parties have a financial interest in the project, such as an owner and the lending institution that provided the funds, the bonding company will have a financial obligation to both parties— a dual obligation—if the contractor fails.

Guarantor.   The underwriter or surety company, as opposed to the insurance company representing surety.

Obligee.   The project owner and others, if there is a dual oblige.

Penal sum.   The amount of the bond; generally, the amount of the construction contract.

Premium.   The cost of the bond.

Principal.   The entity requesting the bond (for example, the contractor, subcontractor, architect, engineer, and so on).

The letter of credit

Not quite a substitute for a bond, a letter of credit can be used in those instances where a bond is not available to a contractor or subcontractor who may have reached their bonding limits. A contractor's inability to furnish a bond does not necessarily indicate lack of financial strength; contractors have bonding limits and a bonding company may wish to have a contractor, near or at their limit, complete a current bonded project before being issued a new one.

A bank-issued letter of credit (LOC) is a cash guarantee whereby the bank would freeze a certain portion of the contractor's, or subcontractor's, liquid assets in an amount equal to the value of the LOC. If commitments under which an LOC is issued are not met, the owner would then "call" the LOC and receive its proceeds.

There are three basic types of LOCs.

- **Conditional** Requires some burden of proof from an owner that the contractor has failed to perform in some respect
- **Stand-by** Deals with payment of a specific sum within a particular period of time
- **Transactional** Applies to a specific transaction

The bonding process

High bonding limits are prized by contractors since, in the eyes of the industry, it conveys strength—both in performance and in financial matters. The initial procedure for obtaining bonding is, in itself, a process referred to as the three Cs: character, capital, and capacity, as explained next:

- **Character** The reputation that the contractor has in the community; essentially, a track record of the successful completion of projects, and the absence of excessive litigation
- **Capital** Strong financial statements and evidence of good accounting practices
- **Capacity** Strong management, a history of producing profitable projects, a history of acceptable estimating, and the adoption of cost control procedures

Prequalifying for a bond

A bonding company will require the contractor to furnish documents that provide an in-depth look at their organization. Among the information included is

- An organization chart of key employees, noting their responsibilities and including their resumes
- A business plan outlining the type and size of work sought, prospects for that work (a sales development plan), the geographic area in which the company plans to work, and a statement of growth and profit goals
- Current work in progress, a history of completed projects (with name, address, phone/e-mail of owner), including contract sum, completion date, gross profit earned, and current backlog of work
- A continuity of business plan, outlining how business will continue upon death or retirement of the present owner; life insurance policies on key personnel required
- Evidence of a bank line of credit
- Letters of recommendations from owners, architects/engineers, subcontractors, and major suppliers

Financial statements must accompany the application and should include the following:

- Fiscal year-end statements for at least the past three years, along with the latest audited and certified statement
- A balance sheet showing assets, liabilities, and net worth

- Income statement showing the gross profit on contracts, operating profit, and net profit before and after provision for taxes
- A statement of cash flow
- Accounts receivable and payable schedules
- A schedule of general and administrative expenses
- Explanatory footnotes—qualifications made by the accountant
- A management letter conveying the CPA's findings, observations, and recommendations

A Grant Thornton study in 2005 compared the change in importance of the various criteria involved in obtaining bonding between the period 1996 and 2005. It's interesting to see how bonding companies have shifted their emphasis on some basic issues, while others remain essentially the same. See Table 4.1.

Insurance

Standard contract insurance requirements are generally limited to Commercial General Liability (CGL) and Contractors Professional Liability insurance (CPL). The CGL policy offers third-party coverage to the contractor, arising out of its operations and premises that may be either owned by, or under the control of, the contractor. It provides bodily injury and property damage liability coverage.

TABLE 4.1 The Shift in Bond Issue Importance

Criteria	2005	1996
Strength of balance sheet	98%	90%
Financial statement presentation	97%	91%
Equity	92%	75%
Debt	81%	68%
History of successful projects	81%	68%
Consistent profitability	78%	63%
Experience in type of project	78%	63%
Use of CPAs with industry knowledge	75%	68%
Claims history	67%	Not available
Reputation of firm and principals	66%	65%
Financial statement disclosure	65%	70%
Accounting practices	63%	59%
Experience in geographic area	59%	43%
Size of over/under-billings	55%	36%
Overhead expenses	47%	35%
Contract volume	36%	33%
Succession planning	30%	Not available
Safety record	10%	20%

Contractors Professional Liability insurance (CPL) provides payment on behalf of the contractor for damages resulting from bodily injury and/or property damage caused by the insured and arising out of ownership and maintenance or use of the premises where construction operations are taking place. Umbrella liability coverage provides coverage in excess of that provided by the underlying liability insurance policy.

Builder's risk

Generally excluded from a general contractor's basic insurance requirements, unless otherwise required, builder's risk—also known as *course of construction* insurance—provides coverage for loss or damage to the structure incurred during construction.

Two basic types of builder's risk insurance exist:

- **All risk** Covers all risks except those expressly excluded
- **Named peril** Covers only certain risks identified in the policy

Workers' compensation insurance

Contractors are very familiar with workers' compensation insurance, which is required by the state in which they're operating, but are project managers conversant with how a poor accident record affects these insurance premiums? A bad accident record requires three years of good experience before workers' compensation insurance rates are lowered.

Premiums are established using the following formula:

$$\text{WCIP} = \text{EMR} \times \text{Manual rate} \times \text{Payroll units}$$

WCIP stands for Workers' Compensation Insurance Premiums.

EMR is Experience Modification Risk, the multiplier determined by the previous work experience of the contractor, which is used to forecast future benefit payments to employees who have filed claims.

Manual rate is the rate structure assigned to each type of work performed. Various trade crafts are classified into "families" based on their potential exposure to injury. Each "family" is assigned a four-digit number corresponding to their premium rate, which takes into account the worker accident claims experienced for that particular family of trades.

Payroll units is a number determined by dividing the contractor's annual direct labor costs by 100. A poor accident record can increase worker compensation insurance premiums to the point where the company's overhead must be increased to compensate for these costs, making them less than competitive on hard-bid projects.

Subcontractor default insurance

When it becomes necessary to have subcontractors bonded, but a particular subcontractor cannot obtain a bond, the general contractor can inquire about subcontractor default insurance. This coverage includes reimbursement of costs incurred to complete the unfinished portion of the subcontractor's work, in case of default. These types of policies generally involve a deductible, a copayment percentage, and an upper limit of liability. The insurance company will assess the general contractor's method of prequalifying and managing the subcontractor by reviewing their management procedures and policies. The cost of the policy is determined by project size and geographic location, among other criteria.

Controlled insurance programs

A controlled insurance program, more commonly referred to as CIP or OCIP (owner-controlled insurance program), has been around since the mid-1960s, but has gained popularity the last 10 years as a way to control insurance costs while still maintaining the desired coverage.

The program.   When initiated by a general contractor, the CIP is basically a wrap-around process, allowing the contractor to use their greater purchasing power to buy all of their CGL and CPL insurance for the project. Each subcontractor thus provides a credit for the cost of project insurance they would have included in their bid.

The OCIP works the same way, except that not only the subcontractors but also the general contractor provide the owner with a credit equal to the price tag of their combined insurance costs, excepting auto-liability coverage. The differences between conventional insurance coverage and the CIP approach are rather straightforward, as shown in Table 4.2.

A typical exhibit to a subcontract agreement advising of CIP and requiring the subcontractor to provide that insurance not included in the controlled program is shown in Figure 4.1.

TABLE 4.2   The Differences Between Conventional Insurance Coverage and the CIP Approach

Type of Coverage	Conventional Approach	The CIP Approach
Workers' compensation	Each contractor and subcontractor	Held for all parties*
CGL	Owner, gen. contractor, subcontractor	Held for all parties
Builder's risk	Owner or contractor	Held for all parties
Auto liability	Contractor, subcontractors	Same as conventional

*Some states offer workers' compensation directly to contractors and subcontractors.

EXHIBIT B: Subcontractor, Sub-Subcontractor Provided Insurance

This project is an Owner Controlled Insurance Program (OCIP) Project. Therefore, much of the insurance necessary for the project is provided by the OCIP. There are, however, additional insurances requirements for "off site" liability exposures that are required in the OCIP manual. There are also additional insurance requirements under this Exhibit B.

Each Subcontractor of every tier shall purchase and maintain the following insurance (s) during the term of this Project:

Automobile Liability insurance for not less then:
$1,000,000. Bodily injury/Property Damage Combined Single Limit.

This Insurance must apply to all owned, leased, non-owned or hired vehicles to be used in the performance of the Work, and the policy shall include an Additional Insured Endorsement naming the Owner, its directors, officers, representatives, agents and employees, Construction Manager, and Developer as an Additional Insured with respect to their operations at the Project Site.

Coverage is primary and non-contributory with respects to Owner, its directors, officers, representatives, agents and employees, Construction Manager, and Developer.

NOTE: Automobiles are defined in accordance with the 1986 ISO insuring agreement. This definition includes, but is not limited to, a land motor vehicle, trailer or semi-trailer designed for travel on public roads, whether licensed or not (including any machinery or apparatus attached thereto).

Workers Compensation, including Employers' Liability Insurance with minimum limits of:
   (a) Workers' Compensation- Statutory Limits with Other States Endorsements
   (b) Employers Liability
        $1,000,000. Bodily Injury with Accident – Each Accident
        $1,000,000. Bodily Injury with Disease- Policy Limit
        $1,000,000. Bodily Injury with Disease- Each Employee

To protect Subcontractor and Sub-subcontractor from and against all workers compensation claims arising from performance of work outside the project site under the contract.

FIGURE 4.1   An exhibit to a subcontract agreement advising of CIP and requiring the sub-contractor to provide that insurance.

OCIP and CIP programs are generally not cost-effective for small projects since the effect of combining premiums for all general and sub-contractors may not result in sufficient savings to offset the added costs to administer the program. But on large projects, significant savings can accrue, and by combining all of the insurance premiums for all of the companies working onsite, the insured can present one large account for insurance companies to bid on, rather than a number of small accounts.

However, everything comes with a price. Unless the general contractor or owner has a strong safety program in place, backed up by a full-time Safety Supervisor, an accident prevention program that isn't effectively controlled could drive insurance premiums through the roof.

General Liability Insurance for contract operations not physically occurring within the Project Site. Five (5) Year Completed Operations Coverage Extension with a limit of liability not less than:

$1,000,000 Per Occurrence
$1,000,000 Personal Injury and Advertising Injury
$2,000,000 General Aggregate (On a Per Project Basis)
$2,000,000 Products/Completed Operations Aggregate

The Coverage to be written on the ISO standardized CG 00 01 (10/01) or substitute form providing equivalent coverage. Such insurance shall cover liability arising from premises, operations, independent contractors, products-completed operations, personal and advertising injury, and liability assumed under an insured contract (including the tort liability of another assumed in a business contract). Such policy shall include an Additional Insured Endorsement naming the Owner, its directors, officers, agents, represents and employees, Construction Manager, and Developer as additional insured's.

IMPORTANT – The additional insured endorsement shall be maintained for a minimum period of at least (5) five years after end of the project completion. This endorsement wording should be equivalent either of the following ISO endorsements:

- CG 20 10 (11/85)
- Both CG 20 37 (10/01) and CG 20 10 (10/01)
- CG 20 26 (11/85)

Coverage is primary and non-contributory with respects to Owner, its directors, officers, representatives, agents and employees, Construction Manager, and Developer.

Wrap up exclusion to be removed five (5) years after completion of your work.

Excess Liability insurance written on an occurrence form for contract operations not physically occurring within the Project Site. Five (5) Year Completed Operations Coverage Extension with a limit of liability not less than:

Subcontractors of all tiers
$5,000,000 Per Occurrence
$5,000,000 Annual Aggregate

FIGURE 4.1   (*Continued.*)

Insurance terms

Interpreting the insurance lingo is sometimes confusing. Thus, the following glossary was created to help you decode some of the terms used.

Additional insured.   An entity other than a named insured who is to be protected under the terms of the policy. Usually these additional insured are added by "endorsement" or referred to in the policy as such.

Aggregate limit.   The maximum amount of coverage that an insurer will pay for all losses during a specific period of time (usually the coverage date) no matter how many claims are made.

Excess Liability insurance will include coverage for Automobile Liability for the, Subcontractor, Sub-subcontractors while on-site and off-site.

Such policy shall include an Additional Insured Endorsement naming the Owner, its directors, officers, agents, represents and employees, Construction Manager, and Developer as additional insured's.

Contractors Equipment insurance, for all construction tools and equipment whether owned, leased, rented, borrowed or used on Work at the Project Site is the responsibility of the Subcontractors and Sub-subcontractors, the Owner or Construction Manager shall not be responsible for any loss or damage to tools and equipment. The Contractor's Equipment insurance shall include a waiver of subrogation against the Owner, its designee(s), Construction Manager, Developer, other contractor(s) and subcontractor(s) of all tiers to the extent of any loss or damage. If the Subcontractor or Sub-subcontractor does not purchase such insurance, he will hold harmless the Owner, its designee(s), broker(s), Construction Manager, Developer, other contractor(s) and subcontractor(s) of all tiers for damage to their tools and equipment.

Waiver of Subrogation

Subcontractors at every tier shall require all policies of insurance that are in any way related to the Work and that are secured and maintained by the Subcontractor and all tiers of Sub-subcontractors to include clauses providing that each underwriter shall waive all of its rights of recovery, under subrogation or otherwise against the Owner, their designee(s), broker(s), Construction Manager, Developer, contractor(s) and all tiers of subcontractors.

Construction Manager shall require Subcontractors and all tiers to waive the rights of recovery (as aforesaid waiver by Construction Manager) as stated above.

Notice of Cancellation/Termination

Each Certificate of Insurance shall provide that the insurer must give the Construction Manager at least 30 days prior written notice of cancellation and termination of the subcontractors and sub subcontractors coverage there under.

Insurance Carrier

Insurance Carrier must be licensed to do business in the State of (or whichever state in which work is being performed) and have a Best Rating of A VII or better.

FIGURE 4.1 *(Continued.)*

All risk insurance. A policy against damage to the property written to insure all risks of loss or damages, as opposed to a policy insuring against "named perils"—that is, specific hazards against which the policy insures. However, all policies exclude insurance against some hazards, so this is actually a misnomer.

Annual aggregate limit. The maximum amount payable under an insurance policy for all losses occurring during a calendar or fiscal year.

Blanket insurance. Insurance that covers more than one type of property in one location, or one or more types of property in several locations.

<u>Contractors Pollution Liability Coverage (Asbestos Abatement/Removal Operations)</u>

Contactors engaged in testing for, monitor, clean up, removal, containing, detoxify, neutralize, transporting, handling, storing, treating, disposing of or processing as waste pollutants or in any way respond to, or access the effects of pollutants should provide Pollution Liability coverage in the amount of $5,000,000 per claim/$5,000,000 aggregate. No exclusion for Asbestos.

<u>Warranty Work</u>

Any Contractor or Subcontractor who has completed their work at the Project site and whose OCIP insurance has been terminated, who return to the site to perform warranty work must do so under their own insurance. Evidence of Insurance shall be provided to the Owner and Property Owner, Construction Manager and Developer before return to the site for such warranty work.

<u>Proof of Insurance</u>

Certificates of Insurance and Additional Insured Endorsement in form satisfactory to Owner and Hopkins and Construction Manager and Developer shall be delivered 60 days prior to the commencement of work. In addition, the Construction Manager shall make available a list of all subcontractor, work schedule and proofs of insurance of each prior to the first payment to the Construction Manager pursuant to the Agreement.

Not less than two weeks prior to the expiration, cancellation or termination of any such policy, the subcontractor shall supply a new and replacement. Certificate of Insurance and Additional Insured endorsement as proof of renewal of said original policy.

FIGURE 4.1 *(Continued.)*

Broad form property damage insurance. Covers damage to work installed by, or on behalf of, a contractor.

Business interruption insurance. A temporary shutdown of a company's activities due to physical damage to its property or another's property. Generally provides reimbursement for salaries, taxes, rents, and other continuing expenses during shutdown, and also includes loss of net profits which would have been earned during the period interruption.

Care, custody, and control. A frequent exclusion in liability insurance denying coverage of the insured's liability for damage to another's property while in the insured's care, custody, or control.

Coinsurance. A provision that obligates the insured to either purchase insurance to a specific percentage of the total value of the insured property or to bear a fraction of each loss in proportion to the deficiency in the amount of the insurance purchased.

Combined single limit. A single limit of liability coverage for bodily injury and/or property damage.

Commercial General Liability insurance (CGL).   A broad form of insurance coverage for the legal liability of an insured for bodily injury and property damage arising out of operations, products, and completed operations and independent contractors—but excluding coverage for liability arising out of the use of automobiles.

Comprehensive General Liability insurance.   Same as preceding entry. The name previously used for this insurance.

Endorsement.   Document attached to a policy which modifies the policy's original terms.

Excess limits.   Limits of liability that may be purchased in excess of the limits included in the basic policy.

Extended reporting period endorsement.   A device offered by insurers writing a claims-made form whereby the coverage is extended to future claims that may be reported, or that may occur after the expiration of the policy under which the claims had been made.

Fidelity bond.   Insurance that protects an insured employer against loss of money or other property due to the dishonesty of an employee.

Owners' and contractors' protection liability.   Coverage to protect an owner and contractor for any liability incurred as the result of independent contractors employed by the owner or contractor.

Products liability insurance.   Coverage protecting the insured for liability arising out of defects in products manufactured, sold, or distributed by the insured.

Professional liability.   Liability for injury, personal injury, death, and property damage arising out of the negligent act or omission of a professional (with respect to construction, this would include an architect, engineer, or attorney).

Single limit.   Maximum of the insurer's liability for all types of bodily injury, property damages, or personal injury claims arising out of one accident, regardless of the number of persons incurring an injury.

Retro date.   The inception date of the first policy written on a claims-made basis.

Split limit. Separate limits for bodily injury and property damage claims. These policies contain three separate limits: one for bodily injury to each insured person, one for bodily injury to two or more persons injured in the same accident, and one for property damage per accident.

Subrogation. The legal right of anyone who has paid an obligation owned by another to collect from the party originally owning the obligation. An insurer, after paying the insured for damages incurred, may attempt to recover damages from the third party that actually created the damages.

Subrogation waiver. A waiver by the named insured giving up any right of recovery from another party.

Tail. An insurance term referring to the lapse of time between the occurrence of an accident and the eventual resolution of any related claims.

Organizing the Project Team

After the flurry of activity that precedes the signing of a construction contract, time must be set aside to get organized. Telephone bids requiring written confirmation need to be pursued and all subcontractor material and equipment quotes, as well as e-mails and bidding instructions by the architect, must be sorted out and placed in the appropriate files. Once construction is underway, information storage and retrieval must be quick and easy.

Correspondence from the various design consultants (as well as from the owner) will be incoming, while responses to all of their questions and comments will be marked outgoing. Shop drawings and reports of various types will need to be logged in, sorted, distributed, logged out, and filed. This process must be made simple and uncluttered. Individual project manager idiosyncrasies must give way to standard office filing procedures to ensure that everyone in the office follows the same method of document storage. Far too often, a project manager has been required to temporarily or permanently take over a job in midstream from a recently departed project manager only to find that documentation was so poor, improperly filed, or even nonexistent that it was extremely difficult to pick up the threads of continuity.

Of course, there are other compelling reasons for proper job organization, too, but either way, the time to start that organization process is now. Integration is the "buzz word" of the construction industry today, and organizing items with this concept in mind is important as we begin to manage the project. Estimating, purchasing, accounting, and project management in today's construction industry must operate as an integrated unit where various types of software can communicate with each other. As we proceed through this chapter, the thought of "How can I link my project management systems together" is one to bear in mind.

Organizing the Job in the Office

A first step toward organizing the job is to review the contract with the owner to determine what provisions and modifications have been made that are unique to this project.

As you recall, Article 3 of the 1997 edition of AIA Document A201—General Conditions requires the contractor to "carefully study and compare the various drawings and other contract documents relative to the work . . . any errors, inconsistencies, or omissions discovered by the contractor shall be reported promptly to the architect as a request for information." Future potential disputes can often be avoided if this procedure is followed.

Changing CSI specification division numbering

In 2004, the Construction Specifications Institute (CSI) reissued their standard Specification Division numbering system, adding many more divisions beyond the previous Division 17 designation. Figure 5.1 shows a list of the new division numbers and titles that, if not already applied to the current in-house project, will most likely be used in future projects.

Addenda and bulletins

If during the bidding process the architect or engineer issued a number of addendums to the plans and/or specifications, it is now time to assemble, categorize, and identify them for future use in the administration of the project, and (just as importantly) during the negotiations with subcontractors and suppliers.

The difference between an *addendum* and a *bulletin* may not be clear to all parties reading this, so it deserves some explanation. The term *addendum* applies to changes in scope made *before* the issuance of a signed contract with the general contractor. Changes made to either the plans and/or specifications *after* the contract signing are referred to as *bulletins*.

It's interesting to note that not all architects and engineers abide by these standards. The author of this book was involved with a $42 million project several years ago, and the contract with the owner included Addendums 1 through 4. However, the architect, after reviewing the "contract" drawings, made several *changes* to those drawings and issued the "For Construction" drawings with Addendums 1 thru 5 in the title block. The author immediately contacted the architect and requested that the drawings be reissued reflecting Addendums 1 through 4, *plus* Bulletin 1.

Division Numbers and Titles

PROCUREMENT AND CONTRACTING REQUIREMENTS GROUP
Division 00   Procurement and Contracting Requirements

SPECIFICATIONS GROUP

GENERAL REQUIREMENTS SUBGROUP
Division 01   General Requirements

FACILITY CONSTRUCTION SUBGROUP
Division 02   Existing Conditions
Division 03   Concrete
Division 04   Masonry
Division 05   Metals
Division 06   Wood, Plastics, and Composites
Division 07   Thermal and Moisture Protection
Division 08   Openings
Division 09   Finishes
Division 10   Specialties
Division 11   Equipment
Division 12   Furnishings
Division 13   Special Construction
Division 14   Conveying Equipment
Division 15   Reserved
Division 16   Reserved
Division 17   Reserved
Division 18   Reserved
Division 19   Reserved

FACILITY SERVICES SUBGROUP
Division 20   Reserved
Division 21   Fire Suppression
Division 22   Plumbing
Division 23   Heating, Ventilating, and Air Conditioning
Division 24   Reserved
Division 25   Integrated Automation
Division 26   Electrical
Division 27   Communications
Division 28   Electronic Safety and Security
Division 29   Reserved

SITE AND INFRASTRUCTURE SUBGROUP
Division 30   Reserved
Division 31   Earthwork
Division 32   Exterior Improvements
Division 33   Utilities
Division 34   Transportation
Division 35   Waterway and Marine Construction
Division 36   Reserved
Division 37   Reserved
Division 38   Reserved
Division 39   Reserved

PROCESS EQUIPMENT SUBGROUP
Division 40   Process Integration
Division 41   Material Processing and Handling Equipment
Division 42   Process Heating, Cooling, and Drying Equipment
Division 43   Process Gas and Liquid Handling, Purification, and Storage Equipment
Division 44   Pollution Control Equipment
Division 45   Industry-Specific Manufacturing Equipment
Division 46   Reserved
Division 47   Reserved
Division 48   Electrical Power Generation
Division 49   Reserved

Div Numbers - 1

FIGURE 5.1   New division numbers and titles.

The architect did recognize the difference between the two terms and agreed that Addendum 5 should have been designated Bulletin 1, but since their firm had already printed 25 sets of drawings at a total cost of $6000, they were reluctant to discard them and reprint all new sets, but he assured the author that costs associated with Addendum 5 would be recognized as extra and reimbursable costs if the represented scope increases. In the spirit of cooperation, no reprinting was demanded, but this incorrect designation plagued the project for months to come as bids from vendors and subcontractors were solicited with drawings referencing Addendums 1 through 5. In fact, many of the bidders took issue with the architect, implying he was trying to "put one over on us." As a result of the mislabeling, too, special provisions had to be included in purchase orders and subcontract agreements referring to Addendum 5 as Bulletin 1. Necessary adjustments had to also be made to previously issued proposals as well.

Dealing with addenda and bulletins.   Addenda and bulletins represent changes to the initial set of bid documents. Albeit some of these changes will have occurred before contract signing and some *after* contract signing, but organizing either type of change is basically the same, so when addenda are discussed in the following paragraphs, understand that the same procedure would also apply to the posting of bulletins.

When addenda have been issued as changes to the specifications, they can be incorporated into the "contract" specifications books in one of several ways:

By cutting out the line item(s) changes and pasting them directly over the lines they supersede and then identifying the addenda from which they were extracted by writing the number in the margin. If these cutout portions are taped on one end only, they can be lifted for comparison between the original document and the revised one.

The changes can also be handwritten above or below the affected lines in the specifications—if they aren't too wordy, that is, when addenda add full pages or even full sections to the specifications book, however, remove the binding and put the specifications, with tabbed sections for addendums, in a large capacity three-ring binder.

A duplicate book should be prepared for the job superintendent and maintained in the field office.

Addenda as sketches.   Addenda may also take the form of sketches, details, or drawing clarifications printed on $8^1/_2 \times 11$ paper, and these types of addenda can be taped directly to the construction drawing where the changes occurred. If this is not done, it is possible that the job superintendent may forget that an area has been changed by addenda and might proceed with a detail that had been subsequently revised. Subcontractors referring to the super's set of plans and specifications should also be alerted to the changes.

It is helpful to maintain one book in the office and one in the field, each containing all the addendums *in sequential order*—both as printed pages and $8^{1}/_{2} \times 11$ sketches. In spite of the precautions taken to update "contract" specification manuals and drawings to reflect the changes contained in the addendums, at times this book will prove invaluable as a cross-check of what changes took place and which addendum directed that change. Changes to changes can be confusing to all concerned and when there are numerous addenda sketches issued by the A/E pertaining to the same drawing or drawings, the project manager should request a reissuance of the affected drawing or drawings.

The project manager must not only maintain a file of all $8^{1}/_{2} \times 11$ addendum type changes, but appropriate copies should be sent to all affected subcontractors and vendors prior to any final contract or purchase-order negotiations. In fact, the documents should be sent by transmittal, because all too often a subcontractor or vendor will state that they never received a particular addendum and thus did not include that scope of work in their bid, but the transmittal will counter that argument.

Project files

All the written materials coming into, and going out of, the construction office will end up in files, and it is important that they are easily retrievable. Although this sounds so simple as not to warrant discussion, how many times have you filed an important document and couldn't remember into which file it was placed?

Of course, letters to and from the architect, engineer, and owner will be filed in folders entitled "Correspondence with Architect," "Correspondence with Engineer," and "Correspondence with Owner," but all document filing is not so easily compartmentalized. Some project managers handle their filing under the theory of job security: "They can't fire me because I'm the only one who knows where the important documents are hidden."

But what if that project manager is on vacation and the office calls with an irate boss on the other end wanting to know where that (*expletive deleted*) letter to Client J. Importante is?

The central file.   A companywide filing system must be established, if it isn't already in place, so that all filing is uniform. This master list of file categories can be used by the project manager as a checklist to create specific files for the current job. At the same time, a more abbreviated list of file folders can be prepared for the field office to assist the superintendent in getting organized in the field. The main purpose of filing is to be able to retrieve a document that is needed, and usually rather quickly. When a project manager is on vacation, or sick, or tied up in a meeting out of the office, someone else should be able to retrieve a document if a standardized filing method is employed companywide and everyone is familiar with that system.

Some project managers prefer to keep a separate file of their own, or use three-ring binders in which to accumulate files. This is perfectly permissible as long as it doesn't replace the standardized company filing system.

The chronological file.   At times, the chronological (or "chrono") file can be a lifesaver. In addition to the central, subject-related files for a particular project, another dual-tracking filing system can be established, and all that's needed is to make another copy of each outgoing or incoming document, or both.

These duplicate documents should be filed *chronologically* instead of by subject matter since there may come a time when the project manager can remember *when* an important letter was sent or received, but because of its subject matter, it may have been filed in any number of subject files. If a chronological file is created, a quick flip through the time period when it was supposedly sent or received will retrieve the document in a hurry.

Organizing the Estimate

During the period when the project was competitively bid or negotiated, the estimate was probably modified several times and in many different ways. Numerous adds and deducts to various elements of the estimate invariably occur as late bids are received and analyzed, and adjustments are made.

Once a contract has been awarded, all additions, deductions, and corrections made to the estimate during the bid process must be rechecked by the project manager and allocated to the proper cost category. For instance, the apparent low bidder for electrical work may have taken exception to the inclusion of temporary power as part of their scope of work, and the estimating department added X dollars to cover these costs in the General Conditions portion of the estimate. However, a late-breaking competitive price from another electrical subcontractor included this work in their bid, so an adjustment to Division 16–Electrical must be made, as well as a deduct to the General Conditions item in Division 1 of the estimate.

All these last-minute add and deduct adjustments to the estimate should be made in an organized manner so that it will be relatively easy to determine at a later date how the final numbers were derived. When all of the line items have been adjusted to reflect the correct budget amounts, the buy-outs can begin.

This updated and revised estimate will also serve as the basis for a Schedule of Values, generally required by the architect as part of the monthly requisition process. And since the requisition format and approval of a Schedule of Values by the architect is required prior to submission of that first request for payment, it is best to submit both forms well in advance in case any changes are requested.

After all of the changes to the initial estimate have been made, it is a good idea to prepare a new, clean sheet which incorporates all of them. *Do not*, however, throw away any original estimates or original worksheets. They will prove invaluable if it becomes necessary to retrace all steps from the initial estimate to the revised one.

Don't throw anything away; even those documents after a contract is awarded—they may be needed in case of a future claim or dispute. Keep all of these scraps of paper, notes, and records of telephone calls. In some states, courts have ruled that if a general contractor can prove that they used a telephone bid from a subcontractor that resulted in their being selected as "apparent low bidder" based upon the incorporation of that bid, the subcontractor may not be allowed to withdraw their bid without incurring damages or penalties.

Investigating allowance and bid alternates

A review should be made of any specification sections having to do with allowances and/or alternates. First, allowance items should be segregated in the Schedule of Values and noted on monthly requisition forms as "allowance" items, with their corresponding value assigned. At some point during the construction process, these allowances will be reconciled and adjusted accordingly. Establish a separate Allowance Schedule (Figure 5.2) if there are numerous allowances included in the contract. This should be done for several reasons:

- It ensures that these items possibly buried in a contract exhibit are given the attention they require.

- They highlight the need for resolution and provide a basis for review at weekly project meetings.

- They alert the owner to the savings and potential overruns in the Allowance account.

- They keep the owner current on the monies spent to date for each "open" allowance item.

ALLOWANCES and T&M

PROJECT

DATE:

CSI	Origin	Name	Amount	Spent to Date	Remaining
01007	Contract	Police Details / Sidewalk Permits	$ 15,000	$ 15,000	$ -
01503	Contract	Weather Protection	$ 25,000	$ 5,307	$ 19,693
02051	Contract	Removal of existing topping slab	$ 10,000		$ 10,000
02052	Contract	Saw cutting of the existing column	$ 2,500		$ 2,500
02053	Contract	Misc. Demolition	$ 2,500	$ 1,046	$ 1,454
02054	COR #3	Relocate / Protect MEP's For Masonry Wall Removal @ S line	$ 3,000		$ 3,000
02055	COR #1	Interior Barricade	T&M	$ 4,976	$ 9,024
02056	COR #3	Surgical Demo and Finish removal @ B.O.	$ 3,600	$ 1,700	$ 1,900
02101	Contract	Support of the existing sewer	$ 5,000		$ 5,000
03101	Contract	Structural Concrete Patching	$ 10,000		$ 10,000
03102	Contract	Cosmetic Concrete Patching	$ 10,000		$ 10,000
03103	Contract	Structural Repairs to slab @ Brookline Ave.	$ 5,000		$ 5,000
04201	Contract	Repairs behind existing yellow bricks	$ 20,000	$ 20,000	$ -
04202	Contract	Repairs for existing Limestone	$ 15,000	$ 15,000	$ -
05101	COR #3	Steel Supports @ Counter / Bar A3.1	$ 3,500		$ 3,500
05102	COR #3	Steel Supports for Menu Board A3.1	$ 5,000		$ 5,000
06401	COR #3	Install and refinish bowling alley top	$ 12,240		$ 12,240
09251	Contract	Patching and repair of the wall at corridor 101	$ 750		$ 750
09252	COR #3	Restore Finishes @ Baseball Ops	$ 10,000		$ 10,000
09253	COR #3	Traffic coating @ Gate E and Concessions	$ 35,547		$ 35,547
09681	COR #3	Floor Prep @ Traffic Topping	$ 2,500		$ 2,500
15501	Contract	Refeed, relocate and coordinate all existing MEP's	$ 25,000		$ 25,000
16051	Contract	Elec. Demo outside of the basement	$ 17,500		$ 17,500
16052	Contract	Lighting in the batting tunnel	$ 2,000		$ 2,000
			$ 240,637	$ 63,029	$ 191,608

FIGURE 5.2 Establishing an allowance schedule.

While reviewing the section of the specifications relating to allowances, determine how they are to be handled: Do they include materials only or do they include labor and materials? Were the allowance items to include the contractor's overhead and profit or were they listed as costs without mark-ups? (If no mention is made relating to overhead and profit, refer back to A201—General Conditions, Article 3.8, which clearly states the composition of an allowance item, if not specified to the contrary in other contract documents.)

Are the allowance items to be bid on competitively, and are such bids forwarded to the architect for review and selection before an award is made? Do the allowance items have definable scope or will more information be required from the architect at some future date? For example, if the finish hardware for the project is an allowance item, how is the hardware schedule to be developed? Will it be prepared by the architect or by a preselected hardware supplier, or a combination of both?

Oftentimes, an architect will select a local hardware vendor to work with them to develop a schedule within the price range of the allowance. Although the selection of hardware may not appear to be a critical activity at the start of the project, it generally is one that requires prompt attention. If hollow metal doors and frames are required for the project, either stock or custom design, the process of shop drawing preparation, submission, approval, and actual delivery will take quite some time, and hollow metal vendors require an approved hardware schedule and possibly templates of the various hardware items before they start to prepare their shop drawings. When these frames are not on the job prior to the start of partition work, the drywall subcontractor or masonry subcontractor will surely be within their rights to request an "extra" to retrofit these frames into their framed work.

Prefinished, premachined wood doors, if required on the project, will require sending an approved hardware schedule to the manufacturer before any production schedules can be established since delivery of special veneer doors often require 12 to 14 weeks lead time. So it's important to recognize and prioritize the resolution of any allowance items that impact the construction schedule.

Alternates

Alternates present a somewhat different problem in that they need to be accepted or rejected within a time frame that does not impact the schedule. Too often, neither the bid documents nor the contract establishes a time limit on the selection of an alternate. Each alternate may require a different time frame for selection. For example, a selection of lobby quarry tile set in a mortar bed instead of surface applied carpet, must be made well in advance of the forming and placement of a concrete slab in that area.

Alternates dealing with the application of vinyl wall fabric in lieu of paint in certain areas of the building are obviously less urgent. Alternates involving the purchase and installation of equipment, must allow time to not only purchase the equipment but also to receive shop drawings, have them reviewed, and permit installation into the proper sequence of work.

The project manager, after reviewing both allowance and alternates, would be wise to send a letter to the architect listing each allowance item and alternate, and requesting that these selections be made in advance of the critical dates established for each item.

Shop drawings and the shop drawing log

A review of the specifications will establish the procedures for shop drawing submissions such as:

- Is there a special stamp required by the subcontractor and general contractor that must be used on each shop drawing submission?

- How many copies of shop drawings are required for submission, and how many sepias or other reproducibles are needed?

- How are samples to be handled?

- Are all types of shop drawings to be sent to the architect, or can structural drawings be mailed directly to the structural engineer with a copy of the transmittal only going to the architect?

All of these procedures should be discussed and clarified for presentation at that first subcontractor meeting. The shop drawing log (Figure 5.3) will become an essential document in tracking and processing shop drawings and should contain, at a minimum:

- The number assigned to each submission that can also be tracked if revised and resubmitted.

- The name of the firm submitting the shop drawing.

- A brief description of the drawing.

- The party to which it was sent (for instance, architect/engineer) and the date it was sent.

- A due date, as well as the date the drawing was returned to the project manager.

- The response time, which will allow the project manager to review lengthy review times with the A/E at a subsequent job meeting.

- The action taken—approved, approved as noted, rejected.

Figure 5.4 is a shop drawing log that groups submittals according to the Construction Specifications Institute's numbering system instead of listing them chronologically.

Number-Rev	Author Co	Description	To Company	Sent	Due	Rec'd	Days +/-	Action
		Drawings						
003-05120-0	Iron Works, Inc.	New angles and deck at Sidewalk	Architects,	12/28/2004	1/4/2005	1/6/2005	2	Approved as Noted
004-05120-0	Iron Works, Inc.	Storefront framing and stair submittal	Architects,	1/10/2005	1/17/2005		-6	For Review & Approval
001-07161-0	Restoration Preservation	Sealant and Waterproofing Submittal	Architects,	12/8/2004	12/15/2004	12/8/2004	-7	Approved
002-07161-0	The Water Company, Inc.	Elevator Pit Waterproofing Submittal	Architects,	12/14/2004	12/21/2004	12/14/2004	-7	Approved
001-07840-0	Coast Fireproofing Inc.	Fireproofing Submittal	Architects,	12/13/2004	12/20/2004	12/13/2004	-7	Approved
001-08110-0	O'Connor	Door, Frame and Hardware Schedules	Architects,	11/29/2004	12/6/2004	12/10/2004	4	Revise and Resubmit
001-08110-1	O'Connor	Door, Frame and Hardware Schedules	Architects,	1/4/2005	1/11/2005		0	
001-08411-0	Glass Co., Inc.	Traco Mirage Series Folding Glass Walls	Architects,	10/5/2004	10/12/2004	10/6/2004	-6	Appr as Noted, Resubmit
002-08411-0	Glass Co., Inc.	Kawneer Submittal	Architects,	11/17/2004	11/24/2004		48	For Review & Approval
003-08411-0	Glass Co., Inc.	Windows Submittal	Architects,	12/6/2004	12/8/2004	12/10/2004	2	Revise and Resubmit
004-08411-0	Glass Co., Inc.	Bay Glass Shop Drawings revised - Dated 12/15/04	Architects,	12/17/2004	12/24/2004	12/14/2004	-3	Approved as Noted
005-08411-0	Glass Co., Inc.	Custom Window Specifications, Performance Test Report and Sealant Color	Architects,	12/17/2004	12/24/2004	12/21/2004	-3	Approved as Noted
001-08520-0	Design and Construction	Custom Window Company - Aluminum Window Systems	Architects,	9/30/2004	10/7/2004		96	For Review & Approval
002-08520-0	Glass Co., Inc.	Bone White Sample for Custom Windows	Architects,	10/11/2004	10/18/2004	10/13/2004	-5	
003-08520-0	Glass Co., Inc.	PPG Coatings Protected samples		10/12/2004	10/19/2004	10/13/2004	-6	Approved
001-08710-0	Glass Co., Inc.	Bay Glass Hardware Submittal	Architects,	12/27/2004	1/3/2005	12/30/2004	-4	Approved
001-09250-0	Galvin, Inc.	Drywall Submittal	Architects,	12/27/2004	1/3/2005	12/30/2004	-4	Approved
001-10531-0	Flag	Awning Submittal	Architects,	11/11/2004	11/18/2004	11/29/2004	11	For Review & Approval
001-14240-0	Elevator Corporation	Hydraulic Elevator	Architects,	10/13/2004	10/20/2004	10/22/2004	2	Appr as Noted, Resubmit
002-14240-0	Elevator Corporation	Cab Panels Finishes	Architects,	10/27/2004	11/3/2004	10/28/2004	-6	Reviewed

FIGURE 5.3 The shop drawing log.

SECTION 9: FINISHES

SUBMITTAL PACKAGE NO.	ITEM NO.	DESCRIPTION	REFERENCE NO.	CONTRACTOR / SUB-CONTRACTOR NO.	SUBMITTAL NO.	SUBMITTAL DATE RECEIVED	NO. RECEIVED	TO	DATE SENT	NO. COPIES	DATE RETURNED TO PAID	NO EXCEPTIONS	APPROVED AS NOTED	REVISE & RESUBMIT	SUBMIT ITEM	REJECTED	DATE RETURNED TO CONST. COMP.
09250.001	001	DRYWALL SUBMITTAL: CERTAINTEED - INSULATION PRODUCTS	076		1	12/27/04	1	NONE	----	--	----	✓					12/30/04
	002	DRYWALL SUBMITTAL: USG - GYPSUM PANELS - FIRECODE CORES	076		1	12/27/04	1	NONE	----	--	----	✓					12/30/04
	003	DRYWALL SUBMITTAL: DIETRICH - SHAFTWALL PRODUCT DATA	076		1	12/27/04	1	NONE	----	--	----	✓					12/30/04
	004	DRYWALL SUBMITTAL: DIETRICH - INTERIOR METAL STUD & TRACK	076		1	12/27/04	1	NONE	----	--	----	✓					12/30/04
	005	DRYWALL SUBMITTAL: USG - SHAFT WALL LINER	076		1	12/27/04	1	NONE	----	--	----	✓					12/30/04
	006	DRYWALL SUBMITTAL: USG - JOINT TAPE	076		1	12/27/04	1	NONE	----	--	----	✓					12/30/04
	007	DRYWALL SUBMITTAL: USG - JOINT COMPOUND	076		1	12/27/04	1	NONE	----	--	----	✓					12/30/04
	008	DRYWALL SUBMITTAL: USG - CORNER BEAD	076		1	12/27/04	1	NONE	----	--	----	✓					12/30/04
	009	DRYWALL SUBMITTAL: USG - ANGLE	076		1	12/27/04	1	NONE	----	--	----	✓					12/30/04
	010	DRYWALL SUBMITTAL: USG - GYPSUM PANELS - WATER RESISTANCE	076		1	12/27/04	1	NONE	----	--	----	✓					12/30/04
	011	DRYWALL SUBMITTAL: DIETRICH - FURRING CHANNEL	076		1	12/27/04	1	NONE	----	--	----	✓					12/30/04
	012	DRYWALL SUBMITTAL: USG - 093 CONTROL JOINT	076		1	12/27/04	1	NONE	----	--	----	✓					12/30/04
	013	DRYWALL SUBMITTAL: HILTI - FASTENER PRODUCT	076		1	12/27/04	1	NONE	----	--	----	✓					12/30/04
	014	DRYWALL SUBMITTAL: NATIONAL GYPSUM - PERMABASE CEMENT BOARD	076		1	12/27/04	1	NONE	----	--	----	✓					12/30/04

SECTION 10: SPECIALTIES

SUBMITTAL PACKAGE NO.	ITEM NO.	DESCRIPTION	REFERENCE NO.	CONTRACTOR NO.	SUBMITTAL NO.	SUBMITTAL DATE RECEIVED	NO. RECEIVED	TO	DATE SENT	NO. COPIES	DATE RETURNED TO PAID	NO EXCEPTIONS	APPROVED AS NOTED	REVISE & RESUBMIT	SUBMIT ITEM	REJECTED	DATE RETURNED TO CONST. COMP.
10531.001	001	AWNING SUBMITTAL: IBIZA RETRACTABLE ARM AWNING PROD. DATA	015		1	11/11/04	1	NONE	----	--	----	✓					11/29/04
10531.001	002	SUNBRELLA FIRESIST PRODUCT INFO & SAMPLES	015		1	11/11/04	1	NONE	----	--	----	✓					02/04/05

SECTION 11: EQUIPMENT

SECTION 12: FURNISHINGS

SECTION 13: SPECIAL CONSTRUCTION

SECTION 14: CONVEYING SYSTEMS

SUBMITTAL PACKAGE NO.	ITEM NO.	DESCRIPTION	REFERENCE NO.	CONTRACTOR NO.	SUBMITTAL NO.	SUBMITTAL DATE RECEIVED	NO. RECEIVED	TO	DATE SENT	NO. COPIES	DATE RETURNED TO PAID	NO EXCEPTIONS	APPROVED AS NOTED	REVISE & RESUBMIT	SUBMIT ITEM	REJECTED	DATE RETURNED TO CONST. COMP.
14240.001	001	GENERAL LAYOUT HGB7116 SHEETS COORDINATION DR 1-8	007		1	10/13/04	4		10/13/04	1	10/21/04	✓					10/22/04
	002	GENERAL LAYOUT HGB7116 SHEETS COORDINATION DR 1-4	007		1	10/13/04	4		10/13/04	1	10/21/04	✓					10/22/04
	003	POWER DATA SHEETS	007		1	10/13/04	4	NONE	----	--	----	✓					10/22/04
	004	LIST OF QUESTIONS	007		1	10/13/04	4	NONE	----	--	----	✓					10/22/04
14240.002	001	SWL RIDDLIZED SAMPLE	009		1	10/26/04	1	NONE	----	--	----	✓					10/26/04

FIGURE 5.4 A shop drawing log that groups submittals according to the CSI's numbering system.

Some contracts require the general contractor to submit a shop drawing log listing each required shop drawing and its planned submission date, while other contracts require the general contractor to incorporate a shop drawing submission schedule into the Critical Path Method (CPM) Schedule for the project.

Insuring the prompt submittals of shop drawings. Many project managers face opposition from some subcontractors when requesting that they submit their drawings promptly. Too often, subcontractors will delay critical submissions because they continue to negotiate with their suppliers in order to seek the best possible price. Little do they realize that significant delays in key product submissions may create delays that will ultimately prove more costly to them. And the project manager should remind them of this fact if they do not submit their shop drawings in a timely fashion.

At the first job meeting, major subcontractors and/or material suppliers should be presented with a time frame for submission of a preliminary shop drawing schedule. This schedule should include the major pieces of equipment for which shop drawings are required and the date when each drawing will be submitted. This schedule should also indicate projected product delivery dates, taking into account the review period allowed by the architect/engineer. Refer to the contract specifications that frequently include a time frame for shop drawing review in the section dealing with that subject.

The project manager's review of shop drawings. It is tempting to merely pass shop drawings onto the architect or engineer at the end of a hectic day without reviewing them first. More and more architects and engineers are demanding that project managers actually *review* each drawing to insure compliance with the plans and specifications, and some contract provisions allow the architect/engineer to charge contractors for additional reviews beyond the first or second submission.

If it appears that the submitted shop drawing does not comply with the plans/specifications, contact the subcontractor or vendor submitting the shop drawing and discuss the matter. Is the variance minor or significant? Have the product specifications actually changed since the issuance of the contract specifications? Where deviations occur but still appear to conform to the specifications, highlight these variances before sending them on to the architect/engineer. The project manager can establish a high degree of creditability with the architect/engineer early on by showing them that drawings *are* being reviewed before further processing, and any variances noted.

When the shop drawings are submitted to the architect/engineer and an expedited review has been requested, state as much on the transmittal. As long as this privilege is not abused, an architect may process them more rapidly than required by contract. At each subcontractor's meeting, the shop drawing log should be reviewed with each attendee, and if a submission is late, the subcontractor should be placed on written notice in the meeting minutes that this late submission must be expedited to avoid any potential back charges.

At the project meeting attended by the architect/engineer, a similar review of the shop drawing log should be made with the architect/engineer to determine the status of outstanding drawings. Any appropriate notes should be included in that meeting's minutes. If this procedure is practiced religiously, one major weapon in the battle to assign responsibility in case of late product or equipment deliveries and related delays will have been developed and the flow of materials and equipment to the job will be enhanced.

Informational copies

Care must be taken to discern which subcontractors or vendors need to receive "informational" copies of shop drawings when approved. For instance, when an approved copy of the boiler shop drawing is returned to the mechanical subcontractor, the electrical subcontractor should receive an informational copy. All too often a piece of equipment is ordered with electrical characteristics at variance with the voltage requirements indicated in the contract drawings. If such an error is caught in the shop drawing stage, there may be little or no additional cost required to make the equipment compatible with the building's electrical system. If not caught at this early stage and the equipment is delivered with the wrong voltage, it is not difficult to envision the problems that will occur.

Remember that even if the architect/engineer mistakenly approves an equipment shop drawing with the wrong electrical characteristics, it does not relieve the general contractor of the responsibility to provide the correct (*contract*) equipment.

The RFI log

It is a rare project that doesn't generate Requests for Information (RFIs) either from the general contractor to the design consultants or from vendors and subcontractors to the general contractor. Depending upon the quality of the contract documents, the number and nature of RFIs can vary from a few to hundreds, thus a system must be instituted early in the project to record and monitor the flow of these requests. Since most of the queries will be generated shortly after vendors and subcontractors have been awarded contracts, the RFI log will be one of the first documents to be prepared.

Figure 5.5 shows a standard format used to monitor the passage of RFIs. It contains the important elements of this type of log, such as:

GENERAL CONTRACTORS INC.

RFI LISTING

STATUS

A - AWAITING RESPONSE
R - RESPONSE RECEIVED

Project: **United Hospital Addition & Renovation**

Project No.: **96001**

RFI No. Spec Sec	Originating Contractor / Vendor	RFI Subject	Date to A/E	Date Due from A/E	Date A/E Returned	Status	Days Late	Reference Remarks
001 3300.1.4	Manning Concrete Inc.	Can fly ash be added to the concrete mixes	03/14/96	03/17/96	03/18/96	R	1	
002 05100	Brown Steel Company	Structural Steel beam B-3 and B-4 discrepancy on plans.	03/15/96	03/20/96	03/24/96	R	4	PR-005 and CO No 7
003 04100	Bricker Masonry Construction Inc.	Brick Selection	03/19/96	03/24/96	03/26/96	R	2	
004 ALL	General Contractors Inc.	Provide a Color Schedule for all finishes	06/30/96	07/14/96		A	17	
005 15300	Upper New York Plumbing Co.	Sanitary Piping and Ductwork conflict Room 172	07/03/96	07/08/96		A	23	
006 16050	Sun Electric Company	Light fixture Type for Room 235 - Emergency Waiting Room	07/08/96	07/13/96		A	18	
007 06400	Golf's Woodworking Inc.	Plastic Laminate Color Selections	07/09/96	07/10/96		A	21	
008 15600	ABC Mechanical Inc.	Existing ductwork is not as shown on plans in 145 - Patient Lounge	07/10/96	07/11/96	07/15/96	R	4	T & M work proceeding.

FIGURE 5.5 A standard format used to monitor the passage of RFIs.

- The RFI, identified by number and specification section
- The originator—contractor/vendor/subcontractor
- The subject of the RFI
- The date submitted to the architect or engineer
- The date when a response was received
- The date when the originator was notified of the A/E's response
- The days required for review/response, or simply the "dates late"
- Remarks

When one or a series of RFIs are outstanding, the project manager can prepare a list of outstanding RFIs and send it on to the architect/engineer, alerting them to the fact that they are late in their response, that it's going on record, and that project time and costs may be in jeopardy (see Figure 5.6).

The proposed change-order log. In preparation for owner- or contractor-generated change orders, a proposed change-order (PCO) log will be produced to track the submission and disposition of any requests for changes that will probably occur during the life of the project. This form (Figure 5.7) should contain the following basic information:

- A numbering system to track and identify change orders
- A brief description of the requested change
- The current status of PCO: Pending, Approved (list the change order in which the PCO is included), Void
- A COR number to used when the PCO is accepted by the owner and a formal change order is issued incorporating the PCO
- A column indicating the lump-sum price of the proposed work, or in the case of T&M work, an estimated cost
- A column indicating acceptance or rejection of the PCO
- A reference column that can be used to insert the document that generated the PCO
- A remarks or notes column

A companion to the proposed change-order log is the change-order log (Figure 5.8), whose purpose is to track the submission, review, and approval of change orders once the owner has accepted the PCO generating the change in contract.

NOTIFICATION OF RFI'S NOT RETURNED

From: Doug Farr General Contractors Inc. 12345 Vernon Place Rd. Albany, NY 16723 (Phone) 717-678-4563 (Fax) 717-678-4564	**Notification Date:** 07/31/96 **Page No.:** 1 **Project No.:** 96001

To: Walter Brown
W B B Architects Inc.
789 111th St.
Sandusky, OH 43765
(Phone) 513-678-2345 (Fax) 513-678-2300

Project: United Hospital Addition & Renovation

Our records indicate that the following Requests for Information (RFI's) have not been returned. Please indicate in the space provided the anticipated return date. Then sign, date and return this notification to our office at your earliest convenience. If our records are in error, please indicate the date the RFI was returned.

RFI No.	Specification Section and Title RFI Subject	Date Sent	Response Due Date	Days Late	Anticipated Return Date
004	ALL FINISHES	06/30/96	07/14/96	17	
	Provide a Color Schedule for all finishes				
005	15300 PLUMBING	07/03/96	07/08/96	23	
	Sanitary Piping and Ductwork conflict Room 172				
006	16050 LIGHT FIXTURES	07/08/96	07/13/96	18	
	Light fixture Type for Room 235 - Emergency Waiting Room				
007	06400 MILLWORK	07/09/96	07/10/96	21	
	Plastic Laminate Color Selections				

The above listed RFI's will be returned to your office by the dates indicated.

W B B Architects Inc.

Signed By: _____

Dated: _____

FIGURE 5.6   Notification of late response to RFIs.

Job scheduling

A job progress schedule is initially roughed out as the job is being estimated to primarily determine the duration of construction for General Conditions time-related costs. Many bid documents require the general contractor to furnish a bar chart or CPM milestone schedule with their proposal. Now that the project has become a reality, a construction schedule needs to be prepared to serve many needs.

The specifications may require that an initial, detailed job progress schedule be submitted to the architect within a specified period of time after the contract signing and, as the revised General Conditions requirement indicates, this schedule is for the architect's *information*,

PCO Tracking Log
Update: _____

PCO Number	Description	Estimated Value	Date Submitted	DCI Review and Recom	CTDA Review and Recom	Review Date	BIC	Follow-up Action	Review Date	Accepted	Rejected	Deferred
500												
501												
502												
503												
504												
506												
507												
509												
510												
511												
512												
514												
517												
520												
522												
525												
526												
530												
532												
535												
537												
539												
541												
542												
544												
545												
549												
591												
592												
606												

FIGURE 5.7 A proposed change-order log.

UPDATED ISSUE LOG/STATUS
February 22, 2005

COR #	Issue	ISSUE DATE	WBI JM #	STATUS	Date Sub.	Comments	Budget Projection	Submitted for App.	Approved Amount	Close Date	WBI C.O.
	CLUBHOUSE										
1	Field Instruction #1	1/3/05	29	submitted	2/1/05	proceeding w/most items	$0	$298,568	$0		
2	FI #4; Lounge	1/26/05	49	submitted	2/1/05	need some pricing follow up	$0	$55,269	$0		
3	Premium for Custom Window Paint	N/A	9	submitted	2/7/05	Proceeded	$0	$8,216	$0		
4	Added Plates for Mini Piles (request by Ada)	N/A	19	submitted	2/7/05	Proceeded per Adam M. request	$0	$12,574	$0		
5	Add Window Screens	N/A	50	submitted	2/8/05		$0	$1,295	$0		
V	Change 13 door closers to exterior plated		51	VOID		not proceeding with change per					
6	Added Scope in Shop Dwg's 06400.01		74	approved	2/10/05	Proceeded per SBER email 2/9/05	$0	$0	$7,065		
	FI #5;Extend Roof over Emer. Gen. Room	2/1/05	61	pricing		need further details	$15,000	$0	$0		
7	FI #6; Sprinkler Add	1/31/05	59	approved	2/9/05	directed to proceed by RL on 2/15/05	$0	$0	$86,824		
8	Dehumification Unit Hand Access	N/A	101	submitted	2/17/05	Proceeded, necessary for coordination	$0	$605	$0		
	FI #7; Electrical Rev. to Media Room	2/11/05	76	pricing		reviewing	$10,000	$0	$0		
	FI #8; Relocate Patch Closet	2/15/05	94	pricing		preliminary drawings 2/15	$15,000	$0	$0		
	FI # ??; Fit out Dr.'s Office	tbd				haven't seen yet	?	$0	$0		
	SK-31; Slab Edge Details	2/8/05	79	pricing		Proceeding on T&M		$0	$0		
	TOTALS						$40,000	$376,527	$93,889		

COR #	Issue	ISSUE DATE	WBI JM #	STATUS	Date Sub.	Comments	Budget Projection	Submitted for App.	Approved Amount	Close Date	WBI C.O.
1	Field Instruction #1	1/3/05	29	submitted	2/1/05		$0	$45,768	$0		
2	Shoring for Masonry Wall Support	N/A	6	submitted	2/16/05	Work Done	$0	$8,767	$0		
3	Saw cut Additional Wall		7	Open			$2,500	$0	$0		
4	FI #2; Underside of Seating	1/20/05	47	Open			$30,000	$0	$0		
5	FI#3; Smoke to Heat Detectors	1/20/05	48	Open		should be zero cost to Sully/Mac	$0	$0	$0		
V	Change 4 door closers to exterior grade			Void		not proceeding with change per DAIQ	$0	$0	$0		
						haven't seen scope yet					
	TOTALS						$32,500	$8,767	$0		

FIGURE 5.8 The change-order log.

not *approval*. The schedule submission may be required as a precursor to submission of the first application for payment.

Although the primary purpose of creating a schedule is to provide all participants in the project with an orderly, time-related sequence of events to follow in order to bring about the timely completion of the project, it can become the general contractor's friend or foe when delays are encountered and the project completion date is in question.

The initial formal schedule submission will be referred to as the Baseline Schedule, and all future updates and revisions will draw on that initial presentation. Thus, the preparation of the Baseline Schedule is critical and must be assembled with great care, involving the informed input of the field, subcontractors, and vendors. This baseline will establish the time frame for each participant to commence work, and the time allotted for that work. When either time or durations change and subcontractors do not accept these changes, disputes and claims can follow.

The general contractor can hold a subcontractor to a schedule commitment, and so can a subcontractor hold a general contractor to that same commitment. Revised schedules must be presented for review and acceptance by all parties to avoid any potential claims for delays.

The critical path method (CPM)

Many software programs are available for the preparation and display of the critical path method (CPM) construction schedule. In fact, the subject is treated extensively in university project management courses. The Achilles heel of the CPM schedule is establishing the orderly list of predecessors and successors and, most importantly, the duration of each activity within the critical path.

As we all know, the CPM schedule provides

- Concise information regarding planned sequences of construction
- A means to predict with reasonable accuracy the time required for overall project completion and the time required to reach milestone events
- Proposed calendar dates when activities will start and finish
- The identification of critical activities
- A matrix that can be manipulated to change the project's completion time, if required
- A basis for scheduling subcontractors, material, and equipment deliveries
- A basis for balancing scheduling, manpower, equipment, and costs (if the schedule is resource loaded)

- A rapid evaluation of time requirements of alternative construction methods
- A vehicle for progress recording and reporting
- A basis for evaluating the impact of delays and changes.

CPM pitfalls to avoid. The critical element of the CPM schedule is the duration of each event or activity that contributes to the *critical path*. A common practice in preparing the CPM is to meet with subcontractors and vendors to obtain activity duration times and processor events that are required before commencing certain operations. One expert in CPM preparation states that initially at least 15 to 20 percent of the project's total activities should be on the critical path.

The question of "lag" before another activity can proceed needs to be addressed as well. For example, if cast-in-place concrete foundation walls are included as an activity, and the contract requires three days cure time before the forms can be stripped, do you include three (3) additional days in the form-pour activity, or create a separate activity designated as "curing time"?

Activity duration times

The success of the CPM process hinges on the ability to determine activity duration time since this is the system's basic building block. Subcontractors and vendors, committing to duration times, are prone to be risk averse and therefore may add duration time much above that actually required. According to one expert in the field, you should assume that all durations are at least twice as long as they need to be. This may be too much of a stretch, but a subcontractor not wanting to be the weak link in the chain of events may add a substantial cushion to increase the duration of their work. Duration times provided by vendors and subcontractors must be carefully reviewed to uncover any unrealistic estimates of time for various work tasks, whether they be much too low or contain too much "fluff."

Because the CPM is a living, breathing thing that changes frequently, predecessors and successors need to be kept up-to-date and advised of the status of the operations that will affect their performance. Frequent review of the status of each operation as it progresses is necessary to alert the next participant that they should be ready to step in and start their work, as planned, or sooner or later than the CMP schedule allows.

Communication is extremely important and must extend well beyond the weekly progress meetings where the CPM schedule is reviewed. Prior notification to "get ready" may not wait until that next weekly review.

The importance of float and who owns it

It's guaranteed that every event in the construction schedule will not take place as planned, chiefly because Murphy's law is alive and well in this business. Weather delays, manpower shortages, equipment and material delivery problems, and plain old mistakes or inaccurate data all take their toll on job progress. To compensate for the unknown or unanticipated, a contingency is added to each schedule in anticipation of occasion delays.

In the language of CPM, this contingency time is called "float," and the question of who owns this "float" can become an important issue, especially when the construction contract includes a liquidated damages clause with significant costs for late delivery of the project. Does the contractor "own" the float to be used to compensate for delays or missteps during the life of the project, or does the owner "own" the float to be used and should respond to designer or contractor issues raised during the course of construction? The determination or definition of which party owns the float time is important when either liquidated damages, penalties, or bonus clauses are included in the construction contract. A builder completing the project on time and not using any "float" time may be adjudged as having actually completed the project *ahead* of schedule, therefore qualifying for an early completion bonus, if such a provision was included in their contract.

Conversely, if by incorporating float, the completion of the project is extended by the number of float days, the "contract" completion time may arguably be defended, thereby avoiding liquidated damages. If the float issue is not included in the contract, this should be addressed and resolved at one of the first owner meetings.

The project meeting minutes

One of the most important documents, some would say *the* most important document, produced during the term of the project are minutes of meetings. The minutes documenting the events that transpire at project meetings, whether they be with vendors, subcontractors, inspectors, building officials, owners, or their design consultants are considered "official business records," and as such can be introduced in court as evidence—so bearing that in mind, the production of proper, concise, and complete meeting minutes can have a profound effect on the company's future if any disputes, claims, or law suits occur down the road.

The format for an effective project meeting is rather simple (Figure 5.9); the information to be placed in the minutes is not, however, which includes

- The project meeting number and the date of the meeting.
- The initials or the name of the person preparing the meeting.

017 – 1/25/05

MINUTES / AGENDA

PROJECT MEETING

CONSTRUCTION MEETING NO: **017**   PREPARED BY:

MEETING DATE: Office Phone

Attendees/Distribution:

Attendees: **CC:**

UPCOMING PROJECT MEETING SCHEDULE.
- Tuesday February 1, 2005   1:00 PM
- Tuesday February 8, 2005   1:00 PM

INSPECTIONS / PERMITS:

I.	*ITEMS:*	Responsible Party	Target Complete
1.01	**9/28/04 – Project Contract:** See Previous Minutes for history **11/30/04** – RWL indicated that the contract would be returned by 11/3. PWH to send billing form to RWL. **12/6/04** – Contract finalized on Monday December 6, 2004. Signatures to follow **1/4/05** – SDC to provide milestones for the project **1/11/05** – The milestones have been updated due to the evolution of the project and have distributed		

FIGURE 5.9   Format for an effective project meeting.

- The schedule of the next meeting posted either at the beginning or end of the minutes.

- A list of attendees obtained from the sign-in sheet passed around at the beginning of the meeting.

- A number assigned to each topic representing the meeting number where it was initially presented. This will allow the project manager to track the number of weeks a specific topic has remained "open" in the minutes. An alternate method is to indicate the date of insertion of the topic in the minutes, per Figure 5.9.

- Under the Items or Description column, a succinct statement regarding the nature of the item should be discussed.

- Most importantly, an action column, where the responsible party (Figure 5.9) is designated. Some refer to this column as BIC (Ball in Court) the party having the responsibility to respond to the item mentioned.

- A column indicating the date when the successful resolution of the item is achieved.

- A two- or possibly four-week Look Ahead Schedule attached to the meeting minutes as a separate exhibit, which should be updated at subsequent meetings.

- A closing statement to insure that all attending parties are in agreement with the contents of the minutes. A statement such as:

  All attendees are to report any errors, omissions, inconsistencies or misinterpretation of events contained in these minutes within three (3) work days after receipt. If no response is forthcoming, the non-response will be taken as acceptance of the meeting minutes as published.

Other forms to consider when getting organized

At some point during construction, a subcontractor may request payment for materials or equipment stored offsite. Many owner-contractor agreements allow these types of payment only when the general contractor submits a written request to the owner indicating the specific item(s) for which payment is requested and advising the owner that if approval is received, the following documents will be submitted along with the invoice for offsite stored materials and equipment:

- A list of the specific equipment or materials for which payment is requested

- Location of the stored materials (some owners require storage in a bonded warehouse)

- The value of the material, accompanied by an invoice(s) from the vendor
- A bill of sale (Figure 5.10), which will transfer the title to the owner once payment is received for the stored material
- Insurance certificates covering insurance while in storage and in transit

Tri-State Metals, Inc

Schedule "A" of Bill of Sale

January 5, 2006

Falls Construction Company
5402 Falls Road
Falls Church, Massachusetts 02109

Re: Sterling Hospital Project

The following material has been manufactured or purchased by Tri-State Metal ,Inc. and is specifically intended for use on the Sterling Hospital project and is stored at the Alliance Industrial Park, Newton, Massachusetts 023370

The total value of the material is $18,500

Description	% Number	Value
Wire Mesh guardrails-82 sections	25%	$18,500

Upon receipt of payment for the above, title will automatically be transferred to Sterling Hospital Corporation, 4585 West Highland Parkway, Sictuate, Massachusetts.

Attach to Bill of sale of Personal Property

582 Preston Highway, Newton, MA.   Tel: 617-577-6020   Fax: 617-577-6022

FIGURE 5.10   A bill of sale.

Tri State Metals, Inc

582 Preston Highway,Newton, MA 02111
T: 617-577-6020   F:617-577-6022

Bill of Sale

Name of Supplier: Tri State Metals, Inc.

Project: Sterling Hospital Project

Address: 5402 Falls Road, Falls Church, Massachusetts 02109

EFFECTIVE DATE OF THIS BILL OF SALE: January 5,2006

Know all these persons by these presents that the supplier in consideration of one or more dollars and other good and valuable consideration pending payment in full satisfaction by (General Contractor) does hereby grant, bargain, sell and transfer unto (General Contractor) and its successors and assigns, forever the materials and equipment listed in the attached with our January invoice, in the amount of $18,500.

And furthermore, the supplier, for itself and its successors and assigns, does covenant with (General Contractor) and its successors and assigns, that it is the owner of the Property, that is as good right, title and interest to transfer the Property to (General Contractor), that the property is free and clear of all security, interests, liens and other encumbrances, and that it will warrant and defend (General Contractor's) right, title and interest in and to the Property against all claims whatsoever.

In witness whereof, the supplier has caused its duly authorized officer to execute this Bill of Sale to be effective as of the date set forth above.

_____	By:_____
Witness:	John Jay, President

_____	_____
Date:	Date:

FIGURE 5.10   *(Continued.)*

Lien waivers—for progress and final payment

Owner-generated contracts usually include a form for a Final Waiver of Lien, required to be executed by all subcontractors and the general contractor prior to release of the project's final payment. This form warrants that all labor, materials, and equipment incorporated in the project have been paid for, and no outstanding sums are owed. Although the owner-contractor agreement may not have included a lien waiver form for progress payments, it will be required. Figure 5.11 shows a standard interim or progress lien waiver for use by the general contractor after their first payment to a subcontractor.

INTERIM WAIVER OF LIEN AND RELEASE

Subcontractor/Supplier: _____

Project: _____

Owner: _____

Scope of Work: _____

Statement Date: _____

Listed below is the current accounting of the above contract work:

    Original Contract Amount   _____

    Total Approved Change Orders (through CO #____)   _____

    Adjusted Contract Amount   _____

    Completed to Date   _____

    Less Retainage   _____

    Total Payable to Date   _____

    Less Previous Payments   _____

    Amount of this Interim Payment   _____

    Pending Change Orders   _____

    Disputed Claims   _____

In consideration of the receipt of all past payments received from incorporated in connection with the Project, if any, and receipt of the Interim Payment set forth above, the undersigned acknowledges and certifies the Subcontractor/Supplier has received payment for all sums currently due on account of labor, materials, equipment, and any other goods or services of every type and kind furnished by the Subcontractor/Supplier to or in connection with its work on the Project in accordance with the Agreement between the Subcontractor/Supplier and in connection with such Interim Payment, Subcontractor/Supplier waives any and all liens and right of lien on any of the Owner's real property arising from the Subcontractor/Supplier's work on the Project through the Statement Date listed above, except for the above listed unpaid retainage, unpaid agreed or pending change orders, and disputed claims.

FIGURE 5.11 A standard interim or progress lien waiver.

ATTACHMENT "D" CONTINUED   INTERIM WAIVER OF LIEN AND RELEASE

The Subcontractor/Supplier further warrants that, in order to induce to release this Interim Payment, it has paid in full each and every sub-subcontractor, laborer, labor supplier, material supplier, or supplier of any other goods and services of every type and kind with whom the Subcontractor/Supplier has dealt in connection with the Project, including, but not limited to all claims for labor, sub-subcontractors, material, insurance, taxes, and equipment employed in the execution of the work above through the Statement Date listed above.

The Subcontractor/Supplier hereby releases and agrees to defend, indemnify and hold harmless, at its sole cost and expense, and the Owner against any claims, demands, suits, disputes, damages, costs, expenses (including attorney's fees), liens, and/or claims of lien made by any sub-subcontractor, laborer, labor supplier, material supplier, or supplier of any other goods and services of every type and kind arising out of or in any way related to the Subcontractor/Supplier's work on the Project through the Statement Date listed above.

The Subcontractor/Supplier further guarantees that all portions of the work furnished and installed by them are in accordance with the Agreement and that the terms of the Agreement with respect to these guarantees will hold for the period specified in the Agreement.

The individual signing this Interim Waiver of Lien and Release represents and warrants that he/she is a duly authorized representative of the Subcontractor/Supplier, empowered and authorized to execute and deliver this document on behalf of the Subcontractor/Supplier, that this document binds the Subcontractor/Supplier and that document is signed under seal.

Subcontractor/Supplier   _____

Signature   _____

Name   _____

Title   _____

Date   _____

State of   _____

Sworn before me this _____ day of_____, 20____

Notary Public   _____

My commission expires _____

FIGURE 5.11   (*Continued.*)

Organizing in the field

Part of the project manager's responsibility is to assist the job superintendent in organizing the field office at the beginning of the project. A checklist similar to the one displayed in Figure 5.12 may be helpful in that regard. Along with assuming that ample supplies of pads, pencils, pens, and various payroll and field reporting forms are available, the field office must be organized to receive and store, for rapid retrieval,

FIELD OFFICE TRAILER CHECKLIST

Job Name: Date_____

Job #:   Requested by:_____

Location:

Item	ITEM DESCRIPTION	QUANTITY		Procured By
		Typical	This Job	
1	Trailer	12'x60'		
2	Security System			
3	Windows - bars			
4	Doors - lock and latch	Yes		
5	Steps	2		
6	Electric Service - 300 amp	300 amp		
7	Telephone Wiring			
8	Lines	2		
9	Jacks	3		
10				
11	Posters			
12	OSHA Requirements			
13	State Safety Requirements			
14	EEO Requirements			
15	First Aid Kit	1		
16	Fire Extinguishers	2		
17	Flashlight	2		
18	Red Traffic Cones	4		
19	FAX Machine	1		
20	Copy Machine	1		
21	Computer	1		
22	Printer	1		
23	Camera	1		
24	Radio w/ dual channels	4		
25	Radio w/1 channel			
26	Pagers - Supt., P.M., P.E.			
27	Cell Phones - Supt., P.M., P.E.			
28	Water Cooler	1		
29	Coffee Supplies/Service	1		
30	Waste Baskets	3		
31	Janitorial - Broom, cleaners, rags etc			
32	Postage Machine, if required	1		
33	Postage Scale	1		
34	Bookcase	1		
35	Desks	2		
36	Desk Chairs	2		
37	Stack Chairs	12		
38	Folding Table 30 X 72	1		
39	Legal 4-drawer File Cabinet w/lock	2		
40	Letter 2-Drawer File Cabinet w/lock	1		
41	Wall-mounted plan rack	1		
42				
43	Sanitary Unit	1		
44	Coin-operated phone	1		
45				

FIGURE 5.12 Checklist for organizing field office.

FIELD OFFICE SERVICES CHECKLIST

Job Name _____   Date_____

Job #: _____   Requested by:_____

Location: _____

Item	ITEM DESCRIPTION	QUANTITY		Procured By
		Typical	This Job	
1	Security Alarm Service			
2	Electric Service	1		
3	Water / Sewer Service			
4				
5	Telephone Service			
6	__ lines - (__ phone, _ FAX)			
7	Telephones	3		
8	Answering Machine/Service	1		
9	Rollover Service			
10	Sprint Long Distance Service			
11	Coin-operated telephone service	1		
12				
13	Sanitary Disposal Service	1		
14	Trash Removal Service	1		
15	Janitorial Service			
16	Guard Service			
17				
18				
19				
20				
21				
22				
23				
24				
25				
26				
27				
28				
29				
30				
31				
32				
33				
34				
35				
36				
37				
38				
39				
40				

FIGURE 5.12   (*Continued.*)

all the paperwork, reports, and drawings that will be forthcoming from the office, subcontractors, and design consultants.

Shop drawings for structural steel, precast or cast-in-place concrete structures stored in the field office will be voluminous. Ductwork, heating and plumbing, piping and sprinkler shop drawings will be required for construction purposes and also need to be maintained to record "as-built" conditions. Equipment and material catalogue sheets, and samples of products to be incorporated into the structure all require controlled storage space.

Unless there is organization in the field office, chances are that critical drawings and documents will become lost—and always when they are needed most. A condensed version of the office job files prepared for the field office can include correspondence from the architect and engineer, memos from the office, job meeting minutes, and letters to the owner and architect to be filed in their respective files—*promptly*.

Unless filing is current, papers pile up on desks, get misplaced, or are mistakenly discarded. Filing daily is key to prompt and accurate retrieval. Equipment catalogues for the mechanical and electrical trades can be placed in the folders designated for each trade.

Shop Drawing Organization

Shop drawings should be placed on a plan rack if full size, or in a file drawer if only letter size, but only when they are approved. Some superintendents prefer to receive an advance copy of a shop drawing for review either before or during its transmission to the architect or engineer. In many cases, this preview by the super may reveal problems that might have been missed by the project manager; therefore, this procedure should not be discouraged.

However it is important that all *unapproved shop drawings* be stored in an area that is not accessible to anyone but the superintendent, and that they are clearly marked "NOT APPROVED—DO NOT USE." There is a danger that unapproved shop drawings left on the plan table in the field office may be referred to by a subcontractor looking for information and not realizing that the information they are looking at is not current. This can spell trouble for everyone.

So Mr. Project Superintendent, put away those unapproved shop drawings and all those superceded plans before someone inadvertently refers to them for current, approved field information.

The Future of Project Organization

Linking and *integration* are two key words in shaping project organization for the twenty-first century. The shortage of qualified, experienced managers in the construction industry requires each manager to become more productive and reduce the duplication that exists in many areas of project management today.

Earlier an engineer's software program may not have had the ability to "talk" to the architect's CAD system, and most certainly not to the general contractor's project management software. This is now changing, and the interoperability of various software programs combined with wireless technology will unite all participants in the construction process instantaneously with a single keystroke or mouse-click.

Successful Project Completion Demands a Successful Start

Quite often, a project that has had a successful run fails miserably in an owner's eyes because of a project manager's inability to bring it to a swift and final conclusion. Punch lists that remain incomplete for weeks, failures to correct rejected work, and an inability to assemble all warranties, guarantees, and Operating and Maintenance manuals (O&Ms) promptly are just some of the reasons owners may become disappointed with their general contractors.

Owners get very upset with ongoing problems such as uneven air or water balancing, cracked sidewalks, improperly functioning door hardware, and other items that "the contractor just doesn't seem able to fix." A contractor's reputation is one of their most important assets, and it is the responsibility of the project manager to protect and enhance that reputation by successfully closing out the project.

What Owners Consider Important

In 2004, the Construction Management Association of America (CMAA) in conjunction with FMI, a Colorado-based construction consulting, management, and research company, conducted their fifth annual survey of owners. Although primarily directed toward the role played by construction managers, responses received from owners can certainly apply to general contractors working with cost-plus, lump-sum, or GMP-type contracts.

The following comments expressed by owners should alert the project manager to some of the concerns that can make or break a project, and which deserve attention as the job unfolds.

- A perceived decline in design documents leading to cost overruns was reported by 71 percent of the respondents. *Project managers who conduct a thorough review of the construction documents early on to assist both owner and design consultants in remedying any design deficiencies will certainly earn some points with not only the owner but also with the design consultants. A project manager embarking on an effective value engineering program to counter some of these potential cost overruns will surely gain extra credibility.*

- Problems with communication and lack of collaboration were also listed by owners as one of several reasons for cost overruns. *Project managers need to enhance their efforts to form an effective project team, open lines of communication between all team members, encourage the open sharing of information, and create an environment that avoids adversarial relationships.*

- Ethics and ethical behavior play a vital role in creating and maintaining the integrity of the project team, so said many owners. This CMAA survey indicated that 84 percent of the owners, architects, and contractors responding to questions of ethical practices said that they had had experience with unethical acts many times. *This is actually a three-way street, traveled not only by the contractor, but by the owner and their design consultants, each of whom is responsible for upholding ethical behavior when dealing with the other party. Unethical behavior by any one of the team members should not be met with a like response but countered with a renewed commitment to practice high standards of behavior. Ethics and trust go together.*

Starting Off on the Right Foot

Successful completion of a project begins with its commencement. When project managers work with a new group of design consultants and owners, each participant will form a distinct impression of the others at the opening salvo. The project manager, at that first meeting, must display a sound understanding of the unfolding project—the obligations of the contract, and knowledge of the plans and specifications (maybe not all the details, but enough to map out a narrative of its key components). A perception of organization, professionalism, and a well thought-out progression of construction activities will add to the owner's initial impression of the project manager.

Remember one of the key owner concerns expressed in the CMAA survey: cost overruns due to less than adequate plans and specifications. Also recall Article 3 of AIA A201—General Conditions: "The contractor shall carefully study and compare the various drawings . . . any design errors or omissions noted by the contractor . . . shall be reported promptly to the architect."

This review should take place before the initial meeting with the owner's group so that, if necessary, any requests for information can be passed on to the design consultants early on.

Controlling the Project Start

Although it would appear obvious, one of the first responsibilities of the project manager is to become familiar with the project, not only the plans and specifications but the construction site and the contract for construction with all of its modifications, exhibits, and addendums.

Review of the Contract with the Owner

Familiarity with the contract for construction is the first step as a new project unfolds. Read the contract from beginning to end, noting all modifications that may affect the performance of the general contractor and future relationships with the subcontractors and vendors. Most subcontract agreements refer to the contract with the owner using language similar to the following:

> The subcontractor agrees to perform all work described in accordance with the contract between the contractor and the owner, and assume toward the contractor in reference to the work all obligations that the contractor assumes toward the owner.

This tie-in between the contractor-owner contract and the contractor-subcontract agreement adds further importance to the understanding of all terms and conditions that affect this "pass-through" provision. There are several provisions in this contract with the owner that merit attention:

- Date when requisitions are to be submitted and the format and content of their submission

- Restrictions on allowable overhead and profit on change orders, both to the general contractor, prime and second-tier subcontractors

- Unit prices contained in the agreement that will impact the subcontractor/vendor negotiations

- Allowance and alternates, and methods of dealing with them

- Restrictions on use of the contingency, if one is included in the contract

- Provisions for liquidated damages or bonus arrangements

- Requirements for the general contractor to submit a list of personnel to be assigned to the project, and obtain the owner's acceptance of the same

- Requirements to submit names of proposed subcontractors for the owner's review and comment prior to the award of subcontract agreements

- Appointment of an owner's representative, and the authority and responsibilities vested in them
- Requirement for noise abatement and restriction during work hours
- Any other restrictive language that should be incorporated into subcontract agreements or notifications to the field

Review of the Project Specifications

Not only will project start-up and close-out provisions in Division 1 be of importance, but a thorough reading of the specifications should be made at least once to uncover any unusual requirements. Specific sections of the specifications ought to be read very carefully if the project manager has any responsibility for negotiating subcontract awards.

If the project manager will "buy-out" the job, this review of the specifications will include noting all items to be included in each subcontractor's scope of work, especially those that deviate from the norm. Too frequently, we glance through the specifications and assume that they are similar to the last project, even though they may have been prepared by a different group of design consultants. Sometimes these assumptions can prove costly.

The author of this book was managing an office project being built for an established developer who controlled several million square feet of commercial space. The owner's representative stated several times that "All we will require of you is what's in the plans and specs—no more, no less." This seemed like a fair approach. As the work progressed, there were some changes in the finish hardware requested by the owner that amounted to approximately $40,000, so a change order was prepared for that amount. At about that time, we had placed about 300,000 ft^2 of concrete slab-on-grade and concrete suspended slabs. The specifications required a survey and an as-built drawing to verify compliance with the specifications. The survey revealed level tolerances in the plus or minus range of approximately 3/16 in. as measured below a 10-foot-long straightedge. The subcontractor and the author thought the owner would be pleased with the results. And they were, to a degree, pointing out however that the specification required a 1/8 in. tolerance—unheard of for general commercial office space, but nonetheless designated "in the specs." At first, we were told that all 300,000 ft^2 would have to be replaced, but the owner relented, agreeing that this was too punitive. Just the same, they were willing to trade this off for the added $40,000 hardware costs. This author read *every* specification section after that episode!

**Specific items to look for when reviewing
the specifications**

A careful review of each section of the specifications can produce several checklists:

- One checklist should be for the project superintendent which can be used to serve as a reminder of items and activities requiring attention during the life of the project (for example, field inspections and mockups required at various time during construction). A copy of this list should be distributed to the field.

- There should be one checklist for each subcontractor concerned which should be distributed at the appropriate project meeting, highlighting the key provisions of their section. This is also the appropriate time to distribute close-out elements of the spec for which they are responsible: items to track as they develop their as-builts, coordination shop drawings, and other start-up requirements. The super can be given a copy of each of these checklists that he or she can review with the appropriate subcontractor from time to time to insure they are tracking them correctly.

Record Drawings

Depending upon the sophistication of the designers and the complexity of the project, record ("as-built") drawings may be prepared as paper or electronic copies. Whatever the format, it's important that they are prepared carefully and contain *accurate* information, including:

- Records of all changes, either those made because of field conditions or those caused by changes in scope (for instance, change orders)

- Records of all changes due to the acceptance of any alternates listed in the contract

- Dimensions—vertical and horizontal—that either confirm or correct the design dimensions of entire areas or components within those areas

- Elevations relating to site and site utilities work, line and grade for all underground utilities, manhole rim and invert elevations for storm, sanitary sewers, pipe inverts, duct banks, and their concrete encasements

- Floor-to-floor elevations and floor-to-ceiling elevations

- The locations of concealed items, MEP risers, branch piping, and wiring

- Structural changes

- The locations of plumbing valves (generally via a valve chart), sometimes noted by colored tacks in accessible ceiling panels

- Fire dampers and adjustable HVAC dampers
- Heat tracings, particularly when encapsulated under insulation

Inspections and Test Reports (Other than Those Required by Local Officials)

- Earth compaction inspections and tests
- Concrete compression tests—the number of cylinders to be taken; notes on the proper storage of the same
- Infiltration and exfiltration tests for underground storm sewers
- Mill reports from the structural steel supplier
- Weld, bolt up steel connections (if tension control—TC bolts are not required), shear stud testing
- Mortar cube testing
- HVAC and plumbing testing—includes water/air tests and pump performance duct leakage tests
- Acoustical batts or insulation batts, concealed in partitions/exterior walls
- Inspections of flashings—around exterior wall penetrations and fenestration, roof accessories, and penetrations
- Inspections of various substrates before being encased or enclosed
- Fire protection testing—including underground piping, fire pumps, and the pressure testing of filled lines

Responsibility for these inspections and tests are fixed in the specifications, and a checklist prepared after a review of the specifications will aid in alerting the appropriate subcontractor and design consultant that a test needs to be scheduled.

If you recall from the discussion of the General Conditions document, failure to conduct a test may result in the contractor having to expose the concealed item and then recover it at their expense. Local building departments or building officials may also require copies of compaction reports, concrete test breaks and steel bolt or weld inspection reports before issuing a certificate of occupancy.

Operations and Maintenance Manuals

Each subcontractor required to submit Operations and Maintenance manuals (O&Ms) should be familiar with the format in which they are to be submitted (for instance, in three-ring binders with tabbed sections), and the number required for submission to the architect and engineer.

The subcontractors might use this information as a guide during their procurement activities to insure that their vendors provide all required materials. It is important that the company submitting the manuals verify that the information is all inclusive and the project manager should review these submittals before sending them on to ensure that they, in fact, include all pertinent equipment.

Do the specifications require a video presentation of the operation of a specific piece of equipment? Is the owner entitled to a certain number of instruction sessions with the manufacturer or subcontractor? Check the specifications with the subcontractor to determine the extent of these instructional materials.

Commissioning and TAB

Few things can turn a previously satisfied owner into an angry bull more quickly than experiencing erratic heating and cooling after they move into their new building. Today's sophisticated HVAC equipment, coupled with very tight tolerances and complex direct digital controls designed to meet indoor air quality standards, make the commissioning of a building's systems a real challenge. Sound and vibration tolerances become more acute and require special testing equipment and highly trained technicians to achieve the necessary results. Not only are the American Society of Heating, Refrigeration, and Air Conditioning Engineers (ASHRAE) standards to be met, but the National Environmental Balancing Bureau (NEBB) founded in 1971 has further defined these commissioning standards.

The commissioning of equipment is often a lengthy process that begins prior to occupancy and continues as the building's tenants move in. In the case of HVAC equipment where occupancy during a cold period precludes start-up and testing of the air-conditioning system, this commissioning process may last for several months.

TAB—A Procedure that Requires Special Attention

The mechanical and electrical trades often have very special close-out procedures and this may require a separate meeting, held after the regular subcontractor meeting to review their specific requirements. None will be more critical than TAB.

TAB (Test-Adjust-Balance) is a critical operation that requires the project manager's close attention. According to ASHRAE, the following procedures define TAB.

- *Testing:* Determining the qualitative performance of the equipment.

- *Adjust:* Regulating the specified fluid flow rate and air patterns at the terminal equipment through operations such as adjusting dampers, and the fan and pump speed via sheaves and belts.

- *Balance:* Checking proportion flows within the distribution system (branch piping, sub-mains, and terminals) according to specified design quantities.

This process, commissioning, can make-or-break a project. An extended process where heating and cooling systems remain dysfunctional will bring lots of complaints if the building is occupied and the HVAC contractor is struggling to provide properly conditioned air. Some owners may become very belligerent, while others can be quite forgiving if their HVAC systems are awry. The author of this book was project manager in charge of a corporate headquarters' project for a major book store chain. A few days after occupancy, he received a call from the president's secretary, who was very upset and complaining that the Top Guy's office temperature was in the mid-90s (this was in mid-July). She wanted someone to check it out immediately, so I assembled the HVAC and control team as soon as the call ended, and when they reached the CEO's office an hour or two later, found him working away in his undershirt. A quick glance around the office and its furnishings revealed the problem. Someone had hung a large picture directly over the wall thermostat, preventing it from responding to heat loads generated through adjacent windows. With the picture removed, temperatures went down, and Mr. CEO put on his shirt and tie as his secretary apologized. But not all clients are so reasonable in their demands to fix a nonresponsive system—and fix it ASAP!

Depending upon systems and equipment, TAB can include

- Fan equations affecting speed, fan curves
- Duct system pressure losses
- Unit air measurements
- Heat transfer
- Indoor air quality
- Energy recovery
- Pumps, curves, and pump and hydronic equations affecting flow
- Flow measurements
- Control systems, including direct digital controls (DDCs)
- Sound and vibration transmission

When TAB requirements are reviewed at the beginning of the job and commitments to prompt and quality commissioning procedures are requested and acknowledged by the subcontractors involved in the process, a positive step will have been taken that may pay solid dividends at close-out.

The Punch List

It is everyone's intention at the start of a project that punch-list work during the closing phases of the project will be kept to a minimum. These stated goals often go the way of all good intentions, and unless this goal is actively pursued, not much will be accomplished. Completing the punch list promptly and properly can become a frustrating experience for a project manager who is unable to close out a project because of some lingering incomplete punch-list items that are the responsibility of one or more subcontractors.

Sometimes empty promises seem to be the order of the day: "I'll be there tomorrow" or "We're just waiting for that replacement part and we should be there Friday" or "Our guy is out sick," and so on. The effort to produce a zero-item punch list at the end of the job begins with that first subcontractor's job meeting, and is reinforced as other subcontractors are brought on board.

It must be made clear to all parties that there is a common interest in punching out a project. Retainage for the entire project may be withheld by an owner because of one to two subcontractors being unable to promptly complete their punch-list work. As a result, all subcontractors suffer the consequences. Certain procedures can be established upfront to make the punch-list portion of the project close-out finish more smoothly.

Consider implementing the following policies:

- A stated zero-tolerance policy toward the punch list with a goal of producing a punch list–free job.

- A formal inspection of each trade's work to be conducted by the project superintendent prior to a subcontractor's demobilization, at which time any punch-list items not already corrected by their supervisor will be prepared for immediate action (This is prior to an official inspection by the design consultants to prepare the "contract" punch list).

- When items remain incomplete beyond two weeks of an official notice (or one week if you choose), unless substantiating documentation can be provided for the delay, each incomplete item shall be valued at 300 percent of cost, and if not completed within an additional 72 hours, this amount will be deducted from the subcontractor's current requisition. The project manager may then elect to have another subcontractor complete the work and deduct such sums from the responsible sub's next payment. In the case where a permit may be required from MEP work, these costs will also be included in the cost of work.

Is it a punch list or a warranty item?

An owner or architect may, at times, consider an item punch list when it should more correctly be determined warranty work. Remember that a punch list represents contract work that was not completed or improperly installed, or that was rejected because of nonconformance with the contract documents. A warranty item, on the other hand, applies to a part, equipment, or material that has been furnished and installed and which complies with the contract documents, but that fails and must be either replaced or repaired.

An owner that withholds funds for outstanding punch-list work is certainly within their contractual rights to do so, but unless the contract is written otherwise, it is not contractually correct to withhold money awaiting warranty repairs or replacement. So, when that next punch list is prepared, review each item to determine whether any "warranty" items are included.

Attic stock, special tools, and spare parts

Requirements for attic stock are very clear, but problems may arise when a subcontractor has used all or a portion of their attic stock materials in the final phases of construction to either replace damaged material or make up for purchasing shortages. If attic stock is not readily available at the close of a project, the owner should be advised that it is on order and scheduled for delivery at a future specified date. The project manager must monitor delivery and when received by the owner, obtain a written record of receipt.

Items such as spare finish hardware components, key blanks, and such must all be catalogued when received from the vendor and stocked in a secure place so that it can be located and turned over to the owner when required.

Special tools necessary for special work during construction, if required to be returned to the owner, must be in serviceable condition, preferably in the original packing box and with all operating, maintenance, and warranty papers included.

Material safety data sheets

Material safety data sheets (MSDSs) pertaining to all hazardous materials delivered to the site or incorporated in building materials are often required to be collected in a binder and turned over to the owner as part of the close-out documentation. OSHA requires MSDSs for hazardous materials (hazmats) to be sent to the construction site prior to the delivery of the item that they refer to. As a result, these sheets are often lost in the shuffle. Obtaining duplicates is time-consuming and if close-out requirements include submitting a binder of MSDSs, a separate folder should be set aside at the jobsite to collect these sheets when they are received and reviewed.

Preparing for that First Project Meeting with the Subcontractors

All of the pertinent information extracted from the contract with the owner, and the review of the specifications as discussed in the earlier portion of this chapter, can now be distilled for presentation at that first subcontractor's meeting.

The time spent during this review will be repaid many times over, and the thoroughness and professionalism displayed at that first construction meeting will set the tone for the balance of the project.

The project manager at that initial subcontractor meeting, should proceed with the belief that everyone may not have completely read their section of the specifications, nor the Special and General Conditions in Division 1. This first meeting, where the subcontractor's project manager and their onsite supervisor are in attendance, is the time to briefly review the specification sections dealing with each subcontractor and the start-up and close-out procedures that apply to all.

Does it appear that the key subcontractors "know" their job? Are there some subcontractors that appear weak and in need of special attention?

Copies of selected portions of the specs can be made for distribution to all attendees. Sections dealing with coordination drawings (Figure 6.1) and submittal procedures (Figure 6.2) can then be distributed. After briefly reviewing these procedures and other general and special requirements, and distributing copies of the actual specification pages, entries in the meeting minutes detailing everything should later hold everyone accountable. As new subcontractors are brought on board in subsequent meetings, this procedure should be repeated.

Depending upon the nature of the contract, four procedures (although spelled out in the specifications) may require special attention: change orders, time and material work, premium or overtime work, and winter conditions. It is that rare project that does not generate a change order or two, or that doesn't require some time and material work (T&M) or premium-time work, or which has no need to take into account colder climates and winter conditions.

Requirements by the general contractor, when any of these conditions are encountered, are quite often much more precise than those in the architect's book of specifications. The project manager is well advised to provide a rather detailed procedure, which should be followed by a subcontractor preparing a proposal to request a change order in order to properly document their T&M work when directed by the general contractor, or justify overtime work and substantiate costs for work during winter. Figure 6-3 sets forth a set of procedures to be followed whenever any of these items of work are requested, and provides the subcontractor with explicit instructions and documentation that must be submitted with their work.

COORDINATION DOCUMENTS

A.   General: Prepare coordination drawings for areas where close coordination is required for installation of products and materials fabricated off-site by separate entities, and where limited space necessitates maximum utilization of space for efficient installation of different components.

    1.   Coordination Drawings include, but are not necessarily limited to:

        a.   Structure.

        b.   Partition/room layout.

        c.   Ceiling layout and heights.

        d.   Light fixtures.

        e.   Access panels.

        f.   Sheet metal, heating coils, boxes, grilles, diffusers, and similar items.

        g.   All heating piping and valves.

        h.   Smoke and fire dampers.

        i.   Soil, waste and vent piping.

        j.   Major water.

        k.   Roof drain piping.

        l.   Major electrical conduit runs, panelboards, feeder conduit and racks of branch conduit.

        m.   Above ceiling miscellaneous metal.

        n.   Sprinkler piping and heads.

        o.   All equipment, including items in the Contract as well as OFCI and OFI items.

        p.   Equipment located above finished ceiling requiring access for maintenance and service. In locations where acoustical lay-in ceilings occur, indicate areas in which the required access area may be greater than the suspended grid system.

        q.   Existing conditions, including but not limited to mechanical, plumbing, fire protection and electrical items.

        r.   Seismic Restraints.

FIGURE 6.1   Coordination drawings.

Special Requirements for a GMP Contract.   When a cost-plus-not-to-exceed-GMP contract is being administered, the owner usually has the option of auditing the contractor's books to validate the entire project's final costs. The time to make provisions for a final cost accounting begins as the project unfolds. Coordinating with the Accounting department to segregate all costs is a first step. Costs to be reimbursed must be kept separate from costs that will not be reimbursed. Changes in scope that increase or decrease the contract sum must be clear and concise and represented by fully executed change orders or other forms of correspondence. Logs or other means of documentation must be set in place before costs begin to accumulate.

As costs are presented each month in the form of applications for payment, any disagreements related to acceptable or nonacceptable costs ought to be resolved quickly and not allowed to remain open until the

B.   Timing: Prior to fabricating materials or beginning work, supervise and direct the creation of one complete set of coordination drawings showing complete coordination and integration of work, including, but not limited to, structural, architectural, mechanical, plumbing, fire protection, elevators, and electrical disciplines.

C.   Intent: Coordination drawings are for the Construction Manager's use during construction and are not to be construed as replacing shop drawings or record drawings. Architect's review of submitted coordination drawings shall not relieve the Construction Manager from his overall responsibility for the coordination of the Work of the Contract.

D.   Base sheets: Architect will provide CAD files for use by the Contractor for the development of building coordination drawing "base sheets". Contractor is responsible to prepare and provide one accurately scaled set of building coordination drawing "base sheets" on reproducible transparencies showing all architectural and structural work. Base sheets shall be at appropriate scale; congested areas and sections through vertical shafts shall be at larger scale.

   1.   Highlight all fire rated and smoke partitions.

   2.   Indicate horizontal and vertical dimensions to avoid interference with structural framing, ceilings, partitions, and other services.

   3.   Indicate elevations relative to finish floor for bottom of ductwork and piping and conduit (6 inches and greater in diameter).

   4.   Indicate the main paths for the installation, or removal of, equipment from mechanical and electrical rooms.

E.   Construction Manager shall circulate coordination drawings to the following subcontractors and any other installers whose work might conflict with other work. Each of these subcontractors shall accurately and neatly show actual size and location of respective equipment and work. Each subcontractor shall note apparent conflicts, suggest alternate solutions, and return drawings to Construction Manager

   1.   Elevator subcontractor.

   2.   Plumbing subcontractor.

   3.   Fire protection subcontractor.

   4.   Heating ventilating and air conditioning subcontractor(s).

   5.   Electrical discipline subcontractors.

   6.   Control system subcontractors.

F.   Review and modify and approve coordination drawings in cooperation with individual installers and subcontractors to assure conflicts are resolved before work in field is begun and to ensure location of work exposed to view is as indicated or as approved by Architect.

   1.   The Contractor shall stamp, and sign coordination drawing originals, Make coordination drawings available for Architect to review on-site.

FIGURE 6.1   *(Continued.)*

Section 01330

SUBMITTAL PROCEDURES

PART 1 - GENERAL

1.1   SUBMITTAL COORDINATION

    A.   Make submittals in a proper and timely fashion, allowing for administrative procedures, Architect's review, corrections to submissions and resubmittal, if necessary, and fabrication of products without delaying the project. Minimum processing times required by the Architect are as follows:

        1.   Review for Architect's Office only: Allow a minimum of 10 working days for review and processing.

        2.   Review by Architect and its consultant: Allow 10 working days for review and processing of submittals by Architect plus an additional 5 working days for review by each consultant.

        3.   Reprocessing of submittals: For submittals requiring resubmittal, re-processing time required shall be the same as first submittal.

        4.   No extension of Contract Time will be authorized due to failure to transmit submittals sufficiently in advance of scheduled performance of Work.

    B.   Make submittals of similar items, systems, or those specified in a single specification section together.

    C.   Make submittals for products which other products are contingent upon, first.

    D.   The Contractor is fully responsible for delay in the delivery of materials or progress of work caused by late review of shop drawings due to failure of the Contractor to submit, revise, or resubmit shop drawings in adequate time to allow the Architect checking and processing of each submission or resubmission.

1.2   SCHEDULE OF SUBMISSIONS

    A.   Schedule procedure: Immediately after being awarded the Contract, meet with the Architect to discuss the schedule of submissions and then prepare and submit within 14 calendar days for approval a schedule of submissions for the Work. The schedule of submissions shall be related to the entire Project, and shall contain the following:

        1.   Shop Drawing Schedule (for shop and setting drawings to be provided by the Contractor).

        2.   Sample Schedule (for samples to be provided by the Contractor).

        3.   With respect to portions of the Work to be performed by Subcontractors, such schedule of submissions for the work of each Subcontractor shall be submitted for approval within 30 calendar days after execution of a subcontract with such Subcontractor.

    B.   List all submissions required of each trade:

        1.   Include the Specification Section number, name of subcontractor or vendor, submittal type, item, description, type, quantity and size (where applicable) of each submission.

        2.   For each submission, provide the following dates, as estimated:

FIGURE 6.2   Submittal procedures.

a.   Scheduled date of submission.

b.   Required date of approval. (permit time for appropriate review and resubmissions as may be required).

c.   Estimated date of beginning fabrication or manufacture of product (where applicable).

d.   Required date of submission of product to testing laboratory.

e.   Required date of testing laboratory approval.

f.   Required date for delivery of product to site.

g.   Required date for beginning of installation of product.

h.   Required date for completion of installation (and in-place testing).

C.   For each submittal, schedule to allow adequate time for review by the Architect and its consultants. The Architect will not be responsible for Work performed in shop or field prior to approval. Long-lead items requiring expedited action must be clearly indicated.

1.   The schedule shall be reviewed and resubmitted as necessary to conform to approved modifications to the construction Project Schedule, and shall be updated as may be required by the Architect.

D.   Posting of submittal schedule: Print and distribute the submittal schedule to Architect, Owner, subcontractors and other parties affected. Post copies in field.

E.   Update schedule throughout progress of the Project, coordinated with scheduling changes in the Work, and redistribute monthly in conjunction with submittal of Application for Payment.

1.3   SUBMITTAL PROCEDURES AND GRADING

A.   Prepare and submit to the Architect a Construction Schedule, a Schedule of Values, and a Schedule of shop drawings, product data, and samples.

B.   Provide space for Contractor, Architect and engineering consultant review stamps, on the front page of each item's submittal copy. Apply Contractor's stamp, signed or initialed certifying that review, verification of products required, field dimensions, adjacent construction Work, and coordination of information, is in accordance with the requirements of the Work and the Contract Documents. The Architect's stamp shall contain the following data:

_____ NO EXCEPTIONS TAKEN
_____ MAKE CORRECTIONS NOTED
_____ REVISE AND RESUBMIT
_____ SUBMIT SPECIFIED ITEM
_____ REJECTED

1.   The Architect will insert the date of action taken and an identification of the person taking the action.

2.   Submittal grading:

a.   NO EXCEPTIONS TAKEN - No corrections, no marks.

b.   MAKE CORRECTIONS NOTED - Resubmission not required. Minor amount of corrections; all items can be fabricated without further

FIGURE 6.2   *(Continued.)*

Re: **Protocol for Change Orders, Premium Costs, Winter Conditions –**

Change Orders

1. Each proposed change order is to contain a brief explanation of the nature of the change and who has initiated it (owner, A/E, contractor).Attach all supporting documentation. i.e. letter from owner, SK from A/E, request from subcontractor, etc .
2. If scope of work is increased or decreased, state prior condition and proposed condition (i.e. railings added – col. 9-10 – "X" lineal feet-none existed in this area on contract drawings, or in the case of a deduct – 20 lf railings deleted between Col. 10-13)
3. All costs submitted by Contractor for self-performed work and costs submitted by subcontractors to be broken down into Labor (hours x rate) ,materials (number if applicable, lineal or square feet if applicable). Overhead and profit to be added to "costs" and percentage of OH&P to conform to contract requirements
4. Equipment – indicate whether rental by Contractor or from independent rental company. List number of hours/days x applicable rate. Provide receipt of delivery, return.
5. If work is T&M follow procedures indicated below
6. If, requested by owner, allow owner's representative to be present when change order negotiations with subcontractor(s) are taking place.

Time and Material Work authorized by Owner

1. Contractor's supervisor to obtain Daily Tickets for all T&M work self-performed to include, worker's trade category (carpenter, laborer, etc.) number of hours worked, task performed. Ticket to be signed by Contractor's supervisor
2. For subcontracted work, Daily tickets from Subcontractor listing tradesman by category (apprentice, journeyman, etc), number of hours each person work and task performed . This ticket to be signed by Contractor's supervisor. Receiving tickets for all materials and equipment to be attached.

Premium costs:

1. For Contractor's self performed work, follow procedures outlined above for T&M work, but include reason for premium time work, .i.e. weather delays, request by owner to maintain previously agreed upon schedule, failure of subcontractor to provide adequate manpower, late delivery of critical material, lack of response from owner or A/E. All subject to Owner's approval.
2. For subcontractor, follow procedures for T&M work outlined above. Contractor to indicate on tickets reason for overtime work, i.e. weather delays, late delivery of materials, etc. All subject to owner's approval.
3. Contractor to accumulate and present all such tickets to the owner on a weekly basis. Identify all known costs or hourly rate to be applied to each trade.

Winter Conditions (if applicable):

1. Indicate operation taking place requiring winter conditions
2. Provide log with temperature readings at 7:00 A.M., Noon, 2:00 P.M.
3. Provide daily tickets for labor as outlined above
4. Provide list of materials used, type of fuel consumed
5. Provide list of any equipment used
6. All such tickets to be signed by Contractor's supervisor
7. Contractor to accumulate and present all such tickets to the owner on a weekly basis with a running total for Winter Conditions- costs to date for each operation or task (cast-in-place concrete, steel ,etc.) requiring winter conditions.

FIGURE 6.3 Procedures for work order requests.

waning moments of the project. If monthly requests for payments are all processed with approved costs, the final accounting should be rather easy and rapid, allowing the project manager to proudly announce that there were final savings amounting to $*XXX,XXX* shortly after payment for the last requisition was received.

That Dangerous End-of-Project Syndrome

As the project nears completion, supervisors and managers alike start thinking about that new project. Your Vice-President may have already given you the plans and specifications for your next job and it looks really interesting. The superintendent may also have been given a set of plans and specs for the job he will be moving to, and key subcontractor personnel may be drifting offsite permanently or at least to do part-time work in connection with their next assignment. These are people that have an intimate knowledge of the current project.

Any new replacements may not be able to get a firm grasp of some of the older, ongoing problems. There is the normal malaise that occurs after a long period of intense effort to close out a project. Unless everyone's attention and efforts are directed to the unfinished job at hand, those last weeks of wrapping things up will undoubtedly stretch out much longer.

The project manager needs to recognize these symptoms and direct all energies back to complete the current project; thus, there's a need to "rally the troops." Any change in subcontractor supervision must be scrutinized carefully, and if an alteration would result in the development of a new learning curve for the supervisor's replacement, a call to the subcontractor's office expressing concern that the project demands more should be made.

Human nature being what it is, those last days or weeks trying to button-up loose ends, will test the project manager's mettle, but this is the time when that last final push needs to be made. Remember that the goal of every project manager is fourfold:

- The project must be brought in on time.

- The project's costs must be contained and the initial profit goal achieved.

- No outstanding claims or disputes should remain once the project is finished.

- The relationship with the design consultants and the owner should be a professional and rewarding one.

A successful completion occurs when a well thought-out plan of execution is formulated before the project commences.

Chapter

7

Estimating

In this day and age when estimating computer software with such cutting-edge technologies as onscreen take-offs are now in almost universal use, it is easy to neglect the basics of the estimating process. As a remedy, this chapter was written to review many of the basics that might have been overlooked.

First of all, a solid data base of historical cost information is necessary to achieve or retain the keen competitive edge required to survive in today's business climate, so let's talk about that next.

The 2004 Edition of CSI's *MasterFormat*

In mid-July 2004, the Construction Specifications Institute issued their latest update to the CSI specification numbering system, saying this revision reflects the complexity of information generated by the industrial, commercial, and institutional projects being built today. The first 14 divisions for activities from General Requirements to Conveying Equipment remain unchanged except for the addition of two more digits to their existing codes. Divisions 15 through 19 are now "reserved" for future use. Several new groups of classifications have been added: Divisions 20–29 and the designated Facility Services subgroup. Division 20 is "reserved," Division 21 is Fire Suppression, Division 22–Plumbing, and Division 23 is Heating, Ventilating, Air-Conditioning. The old Division 16–Electrical has been changed to Division 26 with separate divisions for Communications, Integrated Automation, and Security. A Site and Infrastructure subgroup and a Process Equipment subgroup round out the expanded version of this new CSI code.

Along with the increased number of categories, a new six-digit section numbering system has been given two added numbers, allowing future expansions to designate specific types of concrete. As more architects and engineers switch over to this new system, general contractors will also need to revise their database accordingly. Even the format for the Schedule of Values and the AIA Application and Certificate for Payment form G704 with its accompanying Continuation Sheet (AIA Document G703) will now reflect this new numbering system.

Acquiring a Database

A data base can be built by purchasing one in either a published manual or electronic form, and modifying it to meet one's individual needs, or by developing field-generated cost data and combining it with cost data acquired from vendors and subcontractors. Every proposal received from a subcontractor or vendor may contain data base information when dissected and analyzed.

The purchased database

A number of companies have been publishing unit price cost guides covering topics as varied as sitework to MEP for use by commercial and residential builders, union or open shop in CD format, or Internet provider subscriptions.

These purchased data-base costs come with a How-To-Use directory explaining adjustments to labor rates, material costs, and even productivity factors so that costs can be modified to suit local labor and market conditions. Even with all of these adjustments, purchased database costs need further consideration before preparing an estimate solely from those sources, such as:

- The difference in costs associated with local weather conditions and seasonal variations
- The effectiveness of the management team in both office and field
- Collective bargaining agreements, or the lack thereof
- Local market conditions affecting the availability of production labor
- Inflation

Acquiring the in-house database

For those general contractors operating mainly as "brokers" with no nucleus of permanent laborers, carpenters, or masons, tracking the costs of self-performed work may be limited to a score of General Conditions items; all other costs will be gathered from subcontracted work. As for general contractors who do employ crews of laborers, carpenters, and possibly masons and operating engineers, the accuracy of future estimates depends upon accumulating and quantifying unit costs for all self-performed work. The foundation for that database begins with the weekly field labor report.

The weekly field labor report

Many general contractors still self-perform work such as rough carpentry, and some types of excavation and concrete work, and because of those builders they need to constantly update their field-labor costs. Even the General Conditions portion of the estimate needs some field reporting to accumulate costs for such operations as cleaning, winter conditions, and temporary enclosure work.

The first step in the process of data-base building for any self-performed work is insuring that the weekly field labor report will supply the basic information required to produce a unit cost. The formula is an easy one:

$$UC = L/O$$

where UC is unit cost, L is labor cost, and O is output (productivity).

The weekly field report serves many functions, such as:

- It reports the actual cost of work performed by the contractor's individual field forces.

- It reports the actual costs which can be compared with the estimated costs for each operation.

- It reports the actual quantities of work in place versus the estimated quantities.

- It reports the percentage of weekly work performed to reflect job progress.

- It provides a basis for projecting completed costs for each task.

A page from a typical field labor report is shown in Figure 7.1.

This report lists all of the operations this general contractor is performing in the field, and each task is assigned a cost code and a unit of measure—for instance, LF = lineal feet, LS = lump sum, and EA = each. The estimated quantity and total cost of each task is listed along with the unit cost for the unit of work.

When the weekly costs are reported as the payroll is processed, they are displayed opposite their corresponding estimated value. Therefore, on a weekly basis, the project manager can monitor a specific operation and view how it is performing compared to the dollar value estimated. Also, actual quantities reported on the Daily Labor Report or a separate form can be compared with the estimated quantities for each work task.

CONSTRUCTION CONSULTANTS
WEEKLY LABOR REPORT

JOB #   354-79   NAME: PERIOD ENDING:

	ESTIMATED			ACTUAL TO DATE			THIS WEEK			% CMP	EST COST TO COMP	EST COST FIN PROJ	GAIN /LOSS
	QTY	UCOST	TOTAL	QTY	UCOST	TOTAL	QTY	UCOST	TOTAL				
12100 ROOF BLOCKING-1 MEMBER - LF	0	0.00 / 0.00	0 / 0	0	0.00 / 0.00	0 / 0	0	0.00 / 0.00	0 / 0	0	0	0	0
12110 MISC. BLOCKING - LF	0	0.00 / 0.00	1,200 / 0	0	0.00 / 0.00	0 / 0	0	0.00 / 0.00	0 / 0	0	1,200	1,200	0
12210 ROOF CURBS - LF	0	0.00 / 0.00	0 / 0	0	0.00 / 0.00	0 / 0	0	0.00 / 0.00	0 / 0	0	0	0	0
12300 WINDOW BLKG - 1 MEMBER - LF	0	0.00 / 0.00	0 / 0	0	0.00 / 0.00	0 / 0	0	0.00 / 0.00	0 / 0	0	0	0	0
12312 BLOCKING @ CEILING - LF	1,450	1.00 / 0.00	1,450 / 0	130	0.54 / 0.00	81 / 0	0	0.00 / 0.00	0 / 0	10	702	783	667
12313 BLKG @ TOILET ACCESSORIES-LF	2,300	1.43 / 0.00	3,340 / 0	0	0.00 / 0.00	0 / 0	0	0.00 / 0.00	0 / 0	0	3,340	3,340	0
12314 RAMP FRAMING - LS	250	0.82 / 0.00	205 / 0	0	0.00 / 0.00	0 / 0	0	0.00 / 0.00	0 / 0	0	205	205	0
12606 BASE CABINET BLOCKING - LF	4,700	0.60 / 0.00	2,820 / 0	0	0.00 / 0.00	0 / 0	0	0.00 / 0.00	0 / 0	0	2,820	2,820	0
13100 WOOD DOORS - EA	200	23.00 / 0.00	4,600 / 0	0	0.00 / 0.00	0 / 0	0	0.00 / 0.00	0 / 0	0	4,600	4,600	0
13103 BIFOLD DOORS-2 LEAF - EA	161	8.07 / 0.00	1,300 / 0	0	0.00 / 0.00	0 / 0	0	0.00 / 0.00	0 / 0	0	1,300	1,300	0
13104 BIFOLD DOORS-4 LEAF - EA	309	11.97 / 0.00	3,700 / 0	0	0.00 / 0.00	0 / 0	0	0.00 / 0.00	0 / 0	0	3,700	3,700	0
13200 BASE CABINETS - LF	202	54.95 / 0.00	11,100 / 0	0	0.00 / 0.00	0 / 0	0	0.00 / 0.00	0 / 0	0	11,100	11,100	0
13220 COUNTERTOPS - LF	0	0.00 / 0.00	0 / 0	0	0.00 / 0.00	0 / 0	0	0.00 / 0.00	0 / 0	0	0	0	0
13320 WOOD HANDRAIL - LF	1,100	3.00 / 0.00	3,300 / 0	0	0.00 / 0.00	0 / 0	0	0.00 / 0.00	0 / 0	0	3,300	3,300	0
13507 WINDOW STOOLS - EA	0	0.00 / 0.00	3,480 / 0	0	0.00 / 0.00	0 / 0	0	0.00 / 0.00	0 / 0	0	3,480	3,480	0

FIGURE 7.1   Page from a typical field labor report.

The field labor report can be one of the most effective means of establishing a database and monitoring costs for tasks performed by those general contractors who continue to self-perform traditional work such as excavation, concrete, rough and finish carpentry, and drywall with their own forces.

Combining cost codes and daily reports to produce the database

Recalling the basic cost formula—Unit Cost = Labor Cost divided by Output—the identification of the work task by using a specific CSI cost code and the reporting of daily/weekly hours of this work task requires only one additional document to complete the exercise and create the unit cost/quantity report.

Some contractors prepare a separate document to report weekly work quantities completed in the field, while other contractors incorporate this information on their daily time sheets. One important factor to remember: The unit of measure for each work task should correspond to the generally accepted practice in the industry, or at least the accepted practice in your company.

For example, when forming foundation walls, does the work task and corresponding cost code and reportable quantities include forms on *both* sides of the wall (total contact area) or only one side, which must then be multiplied by two in order to obtain the total "contact" area?

Reinforcing steel is generally reported on a per pound or per ton in-place basis, while concrete placement is reported by the cubic yard. Therefore, these units, per pound (or per ton) and per cubic yard, will be used.

Once quantities of work, identified by their respective identifying codes and labor costs are reported each week, it becomes a rather simple matter to either manually or electronically calculate the unit cost.

Let's say that J. Johnson, carpenter, worked 40 hours on concrete foundation wall forming last week and reported it on code 03278 (forming 8-ft-high foundation walls).

The area formed was 283 ft^2 of *contact area* (both sides of the form).

If John's weekly salary, at a rate of $55 per hour for 40 hours work, including fringe benefits, was $2200, then the unit square foot cost of forms set in place would be $7.77 [2200 (L) divided by 283 (O)].

Unit costs, however, may vary from week to week throughout the life of the operation. When a task is just starting, there is a familiarization period in which construction details are worked out, a pattern of work established, worker crews formed, and a rhythm created for each repetitive operation.

In this beginning mode, reported unit costs may be almost double their estimated value, but as the operation continues, costs should go down. When the operation has ended, the unit price based upon the completed cost and final quantities will represent *one* unit cost for that particular operation. If this unit cost is used in the preparation of a subsequent estimate, it will represent a good cost model. However, not until several costs for similar operations from other projects are accumulated will a more representative unit cost develop. As wages and fringe benefits increase, these unit costs will reflect those changes.

Analyzing Unit Costs

Unit costs for the same operation will vary from job to job for a number of reasons. Weather conditions can have an appreciable effect on costs. Concrete forming operations, for instance, should be less costly in May than in December when working in the northern part of the country. Forming very high walls (16 to 20 ft or more) will be more costly than forming 4 to 8 ft high walls because more scaffolding will be required and material handling demands will be more pronounced.

Repetitive operations such as straight walls will produce lower final costs than start-and-stop or intricately configured operations. More productive, qualified workers, or course, will also affect the cost.

Displaying unit costs

When unit prices are compared from one job to the next, not only the costs but the similarity and dissimilarity of each project must be taken into account. One way to view unit costs for the same operation or task from various projects is to create a spread sheet either manually or electronically.

Figure 7-2 shows such a spreadsheet in which a series of tasks or operations are listed along with their respective costs culled from four different theoretical projects. Cost code 03254—Form and Strip Footings on a square-foot-basis—is being analyzed in this spreadsheet. Costs vary from $1.04/square foot for the Elcon project to $1.39/square foot for the B&G Tool Company building. By considering the average of all costs, it is reasonable to assume that $1.21/square foot can be used to estimate future projects. However, the cost from the B&G Tool project is 33-percent higher than the Elcon project. There may be a reason for this disparity and a more accurate unit cost may be arrived at by eliminating that unit price and averaging the Apex, Elcon, and Onex jobs, resulting in a unit cost of $1.15.

SPREAD SHEET ANALYSIS OF UNIT COSTS*						
DESCRIPTION	COST CODE	UNIT	JOB-537 APEX MAF.	JOB-486 B&G TOOL	JOB-502 ELCON	JOB-498 ONEX CORP
FORM AND STRIP FOOTINGS	03254	SF	$ 1.24	$ 1.39	$ 1.04	$ 1.17
FORM AND STRIP WALLS	03262	SF	2.21	2.75	2.01	2.34
FORM AND STRIP DOCK PITS	03273	EA	375.00	N/A	N/A	425.00
PLACE AND FINISH FOOTINGS	03304	CY	20.00	24.00	19.00	27.00
PLACE - FINISH FOUNDATION WALLS	03312	CY	65.00	72.00	84.00	69.00
SET ANCHOR BOLTS	03352	EA	2.35	1.95	2.70	2.10

FIGURE 7.2   Tasks and their costs.

Monitoring the reporting of costs

The project manager should be familiar with which operations are currently taking place on the job and can periodically review the time sheets to determine whether the correct code numbers are being assigned to the work tasks. If he or she finds that rough carpentry cost code numbers are being mistakenly applied to concrete formwork, the field should be notified of this error immediately.

Are the square-foot quantities being reported correctly? Does everyone understand how quantities *are* to be reported? If a 4-foot-high foundation wall is being formed, for example, and its total length is 8 feet, does the superintendent report 64 square feet (32 feet on each side) or a total of 32 square feet of contact area? It may not matter so much which reporting system is used as long as the same unit costs are reported in a consistent manner.

Similar questions may arise when drywall work is being performed. Do the reported areas include the square feet of gypsum board installed on each side of the partition, or the total square footage of the assembly? Is the metal stud framing to be reported on the basis of square feet of wall area or linear feet of the full height wall, whether it be 8, 10, or 12 feet high?

It is the project manager's duty to verify uniformity in reporting after reviewing the forms generated in the field.

When disparities in costs appear

When field-generated unit costs substantially exceed the estimated costs, and the superintendent is of the opinion that those operations were performed efficiently, the project manager may determine that the disparity between estimated and actual costs lies with the estimate and not the actual costs. Conversely, if actual costs are considerably less than the estimate, the estimating department will also welcome the news.

Periodic monitoring should take place as the work progresses. If costs exceed the estimate by a significant amount, maybe something can be done if discovered early on. At least the warning signs have been raised and closer scrutiny could be paid to this losing operation to see if there are ways to mitigate losses.

Bundling to create a unit cost

Some subcontractors "bundle" various components of a particular operation in order to obtain a unit cost. Several concrete subcontractors have done away with the cumbersome task of reporting each individual task involved in placing a cubic yard of concrete.

By tracking all costs such as labor, forming materials, tons of reinforcing steel, and cubic yardage of concrete placed, a final job cost is tallied, which when divided by the total number of cubic yards of concrete placed provides a unit cost per cubic yard of concrete. When reinforcing steel is required, the total cost of labor and materials will also be added to the concrete number.

As other jobs are completed and additional costs per cubic yard are developed, a "measured mile"—the average of several projects' unit costs—is established, which might be something like $500 per cubic yard.

When a new job is estimated, a detailed take-off is produced to determine the total number of yards of concrete and tons of reinforcing steel required. The human factor then enters the equation. Meeting with their field supervisors and estimators the project manager will assess this new project: Is it similar to the last one where we estimated a cost of $500 per cubic yard? If so, then this job with 2000 cubic yards is worth $1,000,000. If this new project is more like a recent job that was more difficult, where final costs were $550 per cubic yard, then the price will be quoted as $1,100,000.

This "measured mile" approach provides a baseline of costs that can come in handy if their project flow is interrupted or work is delayed by events beyond that subcontractor's control. If the baseline cost of concrete is, say, $500 per cubic yard, but due to delays the cost is now $793 per cubic yard, the subcontractor has the makings of a claim for additional money—an extra $293 per cubic yard in this case.

This same technique could theoretically be applied to other repetitive subcontractor operations such as metal framing and drywall work, painting, and flooring, to name but a few. This is also the technique that forms the basis for conceptual estimates—the square-foot costs of a completed structure, or a component of that structure.

Conceptual Estimating

The ability to "conceptualize" a project based upon preliminary design concepts and create some "ballpark" costs is the first step toward developing the ability to negotiate work with an owner. This is an invaluable asset for any project manager.

Let's say you're having lunch with a client who switches the discussion from the NFL playoffs to talk about a pet project that the president of their company is interested in pursuing. The client sketches what appears to be a four-story office building on the back of a napkin (what else!) and says to you "What do you think? What should we budget for construction of a project like this?" How do you respond? Do you have enough knowledge about unit costs to pick up on this comment? Ask some further questions and say, "Well, I'd need more information, but it looks like you're talking about $300 to $400 per square foot"—that's an accuracy range of 33 percent, but if you provide a very brief idea of what this will buy, chances are it could get your foot in the door! Using your mental database of building costs, a casual remark like this could provide you with an opportunity to attract a new business relationship and enhance your career.

Some consultants in the field call this a feasibility estimate, one that provides a prospective owner with enough information to begin to develop a pro-forma that may result in asking your company to provide a conceptual estimate after meeting with their staff to flesh out some of the basic design issues.

Feasibility estimates can have an accuracy range of +/–20 percent while a well thought-out conceptual estimate can develop costs within a +/–10 percent range. Conceptual estimating was defined by Jeff Beard when he headed the Design-Build Institute of America (DBIA) in Washington, D.C. as "the skill of forecasting accurate costs without significant graphic information, or sometimes no graphic information, about a project." This is about as good a definition of conceptual estimating as you can get.

So how does one get to the point where they can respond to that hypothetical question by replying, "I would say that square-foot prices for a typical office building consisting of your four-level 40,000 square foot building is between $300 and $375 per square foot, exclusive of any sitework costs. For this price, you can get a steel structure with a masonry veneer exterior wall system, two elevators, and an efficient variable air volume heating and cooling system. Of course, these numbers can vary considerably depending upon building skin and interior finishes."

As a first effort in displaying your conceptual estimating skills, you probably have now caught the attention of your client. Conceptual estimating is not just making an educated guess as to the cost of a project sketched out by a new or prior client, it requires an understanding of structural systems, building envelope options, mechanical, electrical, and plumbing systems indigenous with certain types of construction and the basic cost structure of various interior finish levels.

We'll discuss how to gain this knowledge later, but right now let's talk about the process called conceptual estimating.

The various stages of the process

Stage 1.   That luncheon conversation or a phone call from an owner looking for "ballpark" costs for a project can be called Stage 1, and a meeting that afternoon to discuss it can be looked upon as the beginning process to develop a conceptual estimate. In this stage, often referred to as a *parametric estimate*, three basic building components must be considered:

- The type of structure: commercial, institutional, warehouse, and so on
- The square footage of the proposed structure
- The number of floors in the proposed structure

Stage 2—The Feasibility Study.   If that initial contact generated enough interest for the client to carry things a step further—with no commitments—they may want a little more detailed information. More of the owner's requirements are needed such as alternate structural systems, material selections, the configuration and size of the structure, and special requirements like unusually high electrical demands. Is that range of $300 to $375 realistic? Given an hour to mull over the more detailed information, the project manager ought to be able to respond and possibly offer some clarity regarding the low-end and high-end pricing.

Stage 3.   The client may progress from the phone call stage to the next meeting stage—Stage 3—indicating that they'd like to provide even more detailed information to see if that spread between $300 and $375 per square foot can be narrowed. The next meeting with the owner's staff will be a fact-gathering session, and upon returning to their office will require the project manager to spend some time with their estimating department. Unit prices for various MEP systems will come into play as well as exterior wall options, and some degree of interior partitioning and finish levels will be established. A contingency should be considered at this point, possibly 5 to 10 percent of the conceptual estimate.

Stage 4.   Stage 4 can take one of several forms. If there is enough interest on the part of the owner in pursuing this project, the question of a project delivery system can be discussed. The pros and cons of lump sum, GMP, and CM can be presented. Has the owner been discussing this project with an architect who might be persuaded to produce some very preliminary sketches, an elevation, or a typical floor plan that might permit even closer scrutiny of the estimated cost? Some design development drawings will be required at an advanced level of conceptual estimating, and it would appear at this point in the relationship that a discussion regarding developing some form of commitment between owner and contractor is appropriate.

This is the roadmap for a negotiated contract.

The Postconstruction Project Review

Many general contractors make it standard practice to conduct postconstruction meetings in order to review what went right and what went wrong. Both estimating and purchasing people can learn a great deal from these meetings, where the project manager and the project superintendent review strategies, subcontractor performance (or lack thereof), and specific problems and resolutions that evolved as the job went forward.

Among other items to be reviewed: How did the final project costs compare with the initial estimate, taking into account any scope increases or decreases? It is just as important to analyze costs that were much lower as to examine those which were higher.

If any of the final costs were substantially *lower* than estimated, the method by which these costs were assembled might reveal why some previous bids were lost by a small margin when the company should have been competitive. The unit costs obtained in this manner will be broad-based, but can fulfill several objectives when put to use in conceptual estimating.

Successful conceptual estimating depends upon having this kind of broad-brush cost data readily at hand. When a similar project is being estimated in-house, costs obtained from a recently completed comparable job can be used to check against component costs for the new job, when updated for increased labor and material costs. This follows the previous example of the concrete subcontractor practice of using unit costs from a prior similar project but making adjustments to upgrade costs for a new bid.

Mining Completed Projects to Enhance the Database

Completed projects can supply a wealth of cost information—individual building component costs, unit costs, and purchased materials and equipment costs can all be reviewed with an eye to adding to the company's database.

One form, quickly prepared as a template can be used to store basic project information that will be of assistance to the sales and marketing people: a Project Parameter Cost Model worksheet (Figure 7.3). It is much easier to extract and compile this project information at the completion of the job rather than digging through the archival boxes in the storage room.

Project Parameter Cost Model Worksheet

Project: _____

Location: _____

Start Date: _____   Completion date:_____

Brief Project Description:_____

_____

_____

Number of floors: _____ Gross square footage: _____

Excavation and sitework: Size of site:_____	$_____
Site Utilities:	$_____
Site improvements/landscaping:	$_____
Paving:	$_____
Total Site:	$_____
Foundations: Type:_____	$_____
Concrete slab on grade: SF:_____ S.O.M.D _____	$_____
Structure: _____	$_____
Exterior wall type:_____ SF_____	$_____
Roof: type:_____ SF_____	$_____
Drywall: Exterior: type:_____ SF_____	$_____
Interior: type: _____ SF_____	$_____
Carpentry: Rough:_____	$_____
Finish:_____	$_____
Fenestration: Type:_____	$_____
Finishes: ACT: type_____ SF_____	$_____
Painting:_____ SF_____	$_____
Flooring:_____ SF_____	$_____
_____ SF_____	$_____
Elevators: Type and number:_____	$_____
Plumbing: No of fixtures: _____	
_____	
_____	$_____
HVAC: Type:_____	
_____	
_____	$_____
Sprinkler: _____	$_____
Electrical: _____	
_____	
_____	$_____
Other costs:_____	
_____	
_____	$_____
Subtotal:	$_____
General Conditions: (percent)	$_____
Overhead and profit (percent)	$_____
Total contract amount:	$_____

Comments: see attached sheet

FIGURE 7.3   A project parameter cost model worksheet.

The Project Parameter Cost Model Form

This Project Parameter Cost Model form should contain actual costs of the completed project that can be compared to estimated costs. When those final costs are broken down into building components and converted to unit costs, they will serve as the basis for evaluating the performance of the recently completed project. This information added to the Estimating department's unit cost data base will also enhance the project manager's conceptual estimating knowledge.

The Project Parameter Cost Model sheet for a hypothetical project called the Woodbridge Plaza Office Building is shown in Figure 7.4 and

Project Parameter Cost Model Worksheet

Project: Woodbridge Plaza Office building
Location: Litchfield Turnpike, Woodbridge, CT.
Start Date: June 15,2005 ____ Completion date: June 15,2006
Brief Project Description: Two story office building, brick veneer exterior,operable fenestration,
Aluminum glass storefront entrance, flat EPDM roof

Number of floors: 2 stories Gross square footage: 40,000 square feet	
Excavation and sitework: Size of site: 4 acres	$ 95,000
Site Utilities:	$ 20,000
Site improvements/landscaping:	$ 18,000
Paving:	$ 45,000
Total Site:	$178,000
Foundations: Type: 4 foot high x 1" concrete	$ 21,511
Concrete slab on grade: SF: 14,400 S.O.M.D : 14,400	$ 100,000
Structure: Struct Steel,joist, 20 gauge metal deck and roof deck	$388,000
Exterior wall type: Modular brick veneer SF: 9408	$164,600
Roof: type: EPDM-mechanical fastened SF: 14,400	$ 79,000
Drywall: Exterior: type: 6"x20 ga. studs,gypsh SF: 9408	$ 98,000
Interior: type: 3 ½" x 25 ga studs,5/8"FCSR	$ Included above
Carpentry: Rough:	$ 11,000
Finish:	$ 14,000
Fenestration: Type: 4'x4' operable,bronze glass-Alum tubing	$ 125,000
Finishes: ACT: type – 2 x 2 grid and tiles SF: 24,000	$ 60,000
Painting: SF	$ 15,000
Flooring: ceramic/resilient SF	$ 6100
Carpet –28 ounce SY: 2,000	$ 30,000
Elevators: Type and number: One hydraulic	$ 37,500
Plumbing: No of fixtures: 20 roof drains, 3 interior roof drains,	
Four electrical water coolers	$ 75,000
HVAC: Type: Gas fired boiler,hot water baseboard heat,	
RTU for cooling, VAV boxes	$390,000
Sprinkler: Throughout	$ 70,000
Electrical: 800 amp service, 2 x 2 troffers, tel/data wiring,	
Decorative lobby chandelier, 10 x 20ft site light poles	$332,000
Other costs: permits,fees	$ 14,000
Subtotal: exclusive of site work	$2,030,751
General Conditions: (7%)	$ 142,152
Overhead and profit (8% on $2,172,903)	$ 173,832
Total contract amount:	$2,523,542
Comments: Contract amount includes site work w/ GCs and OH&P added	

FIGURE 7.4 A project parameter cost model form for the hypothetical Woodbridge Plaza office building.

illustrates the various types of cost information that can be gleaned from its preparation.

The preparation of a project cost model data sheet

The start and completion dates are inserted at the top of the page of the cost model worksheet and will be used in future comparisons to compensate for inflation and seasonal weather factors. A brief description of the building, the number of floors, and total square footage is included. More descriptive information can be incorporated if necessary and if the project manager wants to expand on the type of components incorporated into the building.

Dealing with Site Costs.   Excavation and related sitework costs are kept separate since they will vary considerably from job to job, and as such, can be used only in the most general way. By keeping these site-related costs separate, the true *building costs* can be analyzed.

Approximate values such as cost per "disturbed acre" can be culled from the site, information which can provide a general overall sitework unit cost. And, of course, bituminous concrete paving costs can be "unit priced," taking into account the nature of the sub-base and the thickness of the paving. Walks and curbs costs can be added to the site improvement category.

The building and its components.   The costs in the model will be of two types: the cost per square foot of the actual component (for example, the cost per square foot of roofing, fenestration, and flooring); and structural and finish materials, as well as unit prices and the square-foot-of building cost for components such as electrical, heating, ventilation and air-conditioning (HVAC), the plumbing system, and the fire protection system.

The square-foot measurements or calculations for the Project Parameter Cost Model sheet do not have to be exact to be meaningful. Citing the exact floor area as 11,578 square feet, or stating that it's actually 11,600 square feet will be sufficient for our purposes.

If any unusual foundation construction was employed, such as pile foundations, mud slabs, or extensive subsurface drainage, these conditions should be isolated and noted. When a description of the building's structural system is prepared, bay spacings or unusual loading requirements should be noted. In the mechanical and electrical portions of the cost model, there should be space devoted to a brief description of these systems.

The plumbing fixture count can be used to establish unit costs. Alternatively, plumbing cost as a function of the area of the building can also be used. If the cost per fixture is to be the basis for comparison, the other significant plumbing costs such as interior storm drains for roof leaders can be isolated and expressed as a cost per leader.

Heating, ventilating, and air-conditioning costs, as well as fire protection or sprinkler costs are most effectively handled on the basis of cost per square foot of building area. A brief description of the HVAC system is helpful in comparing costs for one general system type to a similar or different system on another project.

Several factors may influence the electrical costs. One is the size of the incoming service. Another is the number of metering devices inside the building (if installed for individual apartments/condominium use or in multitenanted office buildings), building equipment loads, or specialized equipment such as emergency generators or uninterrupted power sources (UPSs).

The cost model contains a line item "Other Costs," which is just a catchall for the costs of miscellaneous or special components that make up the total project cost, but are so unique in nature that they need to be segregated for cost comparison purposes.

A separate category on the form is established to record the general conditions or common requirements in both the percentage of the contract sum and the actual cost. By segregating these costs, which are generally substantial, and analyzing them from project to project, normal conditions costs, can be expressed as a percentage of total project cost when assembling a conceptual estimate.

General conditions can range from a low of 6 to 8 percent of construction costs to a high of 11 to 13 percent, with the range being dependent upon the size and complexity of the project and what resources the general contractor must devote to its particular aspects. Analyzing overhead and profit is rather simple since it will involve comparing the difference between final costs and the contract sum—the delta being the profit earned.

Sources of Conceptual Estimating Data

Conceptual estimating can be based upon costs obtained from any one of the following sources, or a combination thereof:

- Costs developed in-house from previous projects
- Cost index data published by companies specializing in compiling and selling construction cost data on a national or regionally adjusted basis
- Component estimating with in-house or purchased cost data
- Square foot estimating
- Order-of-magnitude estimates

Cost index date

As inflation accelerates or recedes and as labor rates change throughout the country, building component costs and related materials change accordingly, revealing trends that are important to project managers. The rate of inflation in the building industry varies from the standard Consumer Price Index (CPI) published by the federal government. During the period 2003 to 2005, the Fed was reporting inflation running at a rate considerably less than inflation in the construction industry. Spikes in the cost of structural steel, as much as 25 percent, as well as substantial increases in cement, gypsum drywall, and lumber and plywood prices caused quarterly construction cost increases far in excess of the government's CPI.

A good guide to price indexes is available in *Engineering News Record* magazine and on the McGraw-Hill web site.

Order-of-Magnitude Estimating

The order-of-magnitude method of establishing preliminary budgets or estimates is commonly used when contractors discuss specific types of projects with prospective clients.

Hotel and motel owners prefer to discuss project costs in terms of cost *per room*, a cost that would include all other ancillary space such as meeting and conference rooms, swimming pools, and the lobby and registration areas. On a "per room" basis of, say, $100,000 for a hotel, a total *project cost* of $20 million for a 200-room complex represents an "order of magnitude" estimate.

Owners of apartment buildings, parking garages, and nursing care facilities, on the other hand, will refer to total project "cost" on the basis of *cost per apartment, cost per parking space,* or *cost per bed,* respectively.

A project manager with experience in building any or all of these structures will want to first convert total project costs into *cost per apartment, cost per parking space,* or *cost per bed* before meeting with a prospective client.

When discussing order-of-magnitude costs with a client, reviewing cost *per parking space* (for a new parking garage project, for example) is necessary to define the method of construction included in that order-of-magnitude number. Does the cost per space relate to a precast concrete, cast-in-place concrete, or structural steel framework? Does this unit cost include precast or masonry spandrel panels, and are the slabs post-tensioned or precast structural concrete members? A contractor experienced in parking garage construction can provide an owner with the pros and cons and cost differentials between the various types of structures, based upon costs acquired form recently completed projects.

Order-of-magnitude issues to consider

When dealing with order-of-magnitude estimates, you should recognize the following adjustments and limitations:

- Price escalation must be factored into the "cost"

- When the "cost per room" or "cost per bed" or "cost per parking space" from one project is compared with a per-unit cost for the proposed new project, similar construction elements of the prior project is implied and should be stated. This includes structure, building envelope, and interior finishes.

- As in other types of conceptual estimating, since site costs vary so much, they ought to be clearly defined, treated as an allowance, or excluded.

- When submitting an order-of-magnitude proposal, include a brief narrative which will include scope definition.

Project managers planning to meet with a client whose business relates to any of the preceding type of endeavors may wish to review some similar projects completed by the company and become more conversant in the order-of-magnitude approach when discussing costs.

Special Requirements Associated with Office Building Estimating

Several special conditions apply to any contracts between owner (or developer or landlord) and the general contractor when commercial office construction is involved.

Whenever a general contractor is discussing the cost of a commercial office building with a developer, the basic cost structure may involve what is known as "core and shell." A rather detailed document, referred to as a Tenant Work Letter, will also develop as further contract discussions unfold.

Core and Shell

Speculative office developers will request the cost to erect the building "shell" and basic "core" features—a reception lobby, elevators, stairwells from floor to floor, and bathroom facilities on each floor (at a minimum). HVAC generating equipment will be required with riser ductwork to each floor. Electrical systems will feed this core building equipment—HVAC, elevators, and so on—and provide distribution to each floor terminating at an electrical closet with a series of panel boards for continuation of the circuitry. The interior of exterior walls will be sheetrocked and a suspended ceiling grid system installed. With the exception of completely finished employee bathrooms on each floor, as well as finished stairwells and possibly elevator lobbies, the balance of partition work (the installation of doors/frames/hardware, light fixtures, distribution ductwork, and floor and wall finishes) should be completed as tenants sign leases and prepare to configure the space to fit their needs. The tenant will usually be given an "allowance" by the developer to be spent to complete their leased office space. This is known as a Tenant Work Letter.

Since costs of these "improvements" can amount to as much as 20 to 25 percent of the base building costs, it's important to obtain a clear understanding of how the owner plans to handle the costs of items classified as "tenant improvements."

The Tenant Work Letter

The Tenant Work Letter shown in Figure 7.5 applies specifically to tenant space in an urban building designated as a historic structure. Most of the items included in this work letter are applicable to those for new office construction as well, but would exclude such listings as "enhancing existing masonry, wood, etc."

The tenant in the work letter is "allowed" a certain number (or a certain square footage) of items based upon the total number of square feet leased. For example, if 10,000 square feet of office space is being rented at $25 per square foot and the landlord includes a "fit-up" allowance of $7.50 per square foot, the tenant will be credited $75,000 against the total cost of their "fit-up" improvements.

Some developers require the tenant to sign an agreement with the contracted builder for the core and shell structure to perform all the tenant's improvements. By doing so, the completion of HVAC, plumbing, electrical, and fire protection systems will be more easily monitored, and since the developer is familiar with that contractor's supervision capabilities and schedule compliance they will be more comfortable complying with the owner's building and quality standards. However, larger tenants, having had experience with other contractors on previously leased space, may object to using the developer's builder, and if their space requirement is large enough and of considerable importance to that developer, they may prevail and be permitted to bring in their own contractor.

The Developer's Responsibility

Contracts between the developer and general contractor frequently contain a fairly complete list of unit prices, ranging from painting to door and hardware costs to electrical outlets and even the number of water coolers.

STANDARD OFFICE TENANT IMPROVEMENTS
OPEN CEILING
MINIMUM 10,000 S.F. TENANT

INTRODUCTION
The scope of work described below will provide tenant improvements that enhance the existing masonry, wood and steel structure, which are the most distinctive interior elements of the existing buildings. An open ceiling and a relatively open floor plan will take advantage of the large windows and tall floor-to-floor heights, resulting in dramatic, airy space. Due to considerations for retaining the historic fabric of the existing buildings, the open ceiling approach outlined below will be required in designated locations.

ASSUMPTIONS
- General contracting by
- Minimum 10,000 square foot tenant.
- Landlord will pay for initial space planning to prepare a test fit. All design costs after the test fit are the Tenant's responsibility. If Tenant desires design costs included as part of turnkey program, then the cost of these designs will be included within the Tenant Improvement Allowance.

FLOOR FINISH
- 28-ounce direct glue down carpet (usually priced around $15/sy); one color from a selection of Landlord's standard colors. At tenant's option vinyl composition tile or vinyl sheet goods selected from Landlord's standard colors may be substituted at selected locations. Four-inch vinyl cove base will be provided from Landlord's building-standard selection.

CEILING
- Ceilings will be scraped clean of any lead-based paint as part of the Shell building; any other ceiling finishes (i.e. painting, drywall or ACT) will be at tenant's cost.

DOORS
- Suite entrance door- 3'-0"x7'-0" solid core wood, 1 ¾" thick, stained or painted from Landlord's standard colors.

- Interior- 3'-0"x6'-8" hollow core wood, 1 ¾" thick, stained or painted from Landlord's standard colors, equipped with cylindrical lever lockset. One door will be provided for each 20 linear feet of partitioning.

PARTITIONS
- Corridor partitions will be constructed from slab to slab. Partitions between suites will be constructed from slab to slab and will be sound insulated. Exterior walls and (unfinished) corridors are not included when measuring partitioning, but 50% of partitions between suites are included.

FIGURE 7.5   A tenant work letter.

- Maximum partition allowance is one linear foot for each 15 square feet of rentable area.
- Interior partitions to be constructed of 2 ½ " metal studs with ½" gypsum wallboard with taped and spackled joints. The height of typical partitions is to be 9' - not extended to roofdeck.
- Up to 15% of the allowed partition length will be constructed with sound attenuation batts.

WALL FINISH

- Interior face of exterior wall- tenant may elect to paint or clean masonry walls at tenant's cost; base building has included wall repair only
- All other interior walls will be painted with a prime coat and two finish coats of flat latex paint selected from Landlord's standard colors.

CASEWORK

- For tenant 10,000-15,000 square feet: 2 linear feet of 24" deep base cabinet, counter-top and 12" deep wall cabinets per 1,000 rentable square feet.
- For tenant 15,000-25,000 square feet: 1.5 linear feet of 24" deep base cabinet, counter-top and 12" deep wall cabinets per 1,000 rentable square feet.
- For tenant 25,000-50,000 square feet: 1 linear foot of 24" deep base cabinet, counter-top and 12" deep wall cabinets per 1,000 rentable square feet.
- For tenant over 50,000 square feet: .75 linear foot of 24" deep base cabinet, counter-top and 12" deep wall cabinets per 1,000 rentable square feet.

SHELVING

- 4 linear feet of shelf and rod.

WINDOW TREATMENT

- Adjustable 1" narrow slot horizontal mini-blind will be installed at each window, Landlord Standard

HEATING AND COOLING (HVAC)

- Tenant provides for all VAV boxes required and all ductwork, louvers and grilles which are downstream from the VAV boxes. Typical office configuration is for 1 VAV Box per 1,000sf. Return air devices will be provided only where partitions are extended to the underside of the deck above and are not included in the allowance.

PLUMBING

- One stainless steel sink with piping. Landlord may limit location based upon reasonable access to plumbing.

SPRINKLER SYSTEM & FIRE ALARM

- Tenant provides for any additional sprinkler and fire alarm work which is over and above that required by code for a general office space plan designed in accordance with the Tenant's maximum partition allowance. Automatic pendant heads directed upward at open ceiling areas.

FIGURE 7.5 (*Continued.*)

The contractor's base contract may include the installation of electrical service to a specific area or electric closet on each floor via a separate panel, excluding branch wiring, lighting fixtures, telephone, and data communication systems. The developer may require the contractor to provide unit prices to furnish and install a specific type of ceiling fixture or fixtures, and a wide range of wiring devices, and data and telecommunication outlets.

If the core building HVAC system consists of perimeter baseboard heat, and includes a variable air volume (VAV) system for tempered and conditioned air distribution, it is standard practice to exclude the installation of these VAV boxes and associated ductwork, leaving those costs separate from base building costs and including them instead in the unit-cost schedule for tenant fit-up work. The base HVAC contract may include the heating and cooling source, but for tenant fit-up purposes, unit prices to furnish and install a VAV box with X feet of associated ductwork will be included in that unit cost schedule.

The sprinkler and fire protection systems, as required, can be estimated in several different ways: either as a completed system with sprinkler heads which will be located to meet the tenant's space, or one with only risers and mains with branch piping, drops, and heads which are then estimated as part of the tenant improvement package.

Depending upon the local fire code regulations, an entire building or just the floor above and the floor below may be required to be fitted with sprinkler heads before occupancy.

Subcontractor Responsibilities as They Relate to Tenant Fit-Ups

Not all tenant work will take place as the base building work progresses and while the nucleus of key subcontractors are onsite. The unit price schedule prepared by the project manager for the developer or landlord must take into account that some work will be required of the general contractor after the base contract work has been completed and some or all subcontractors are demobilized.

If leases are signed and tenant work commences when most of the required subcontractors are still onsite working on the base building, their unit costs for tenant improvements will be based upon adding work to already productive crews. But when leases are signed after these subcontractors have completed their base building work and have left the site, the cost to have them remobilize for additional work will be higher. These higher prices will increase once more (or possibly several times) if work is required, say, three months after substantial completion and six months after substantial completion.

As an example, let's take an electrical outlet and price it according to an installation time frame:

Price for installation while base building work is ongoing	$85
Price for installation up to three months after base building work has been completed	$125
Price for installation six months after base building work	$200

When entering into a contract with an owner where tenant fit-up work will be required, take into account this potential for delayed improvements installation. The owner/contractor contract should be specific in its application of unit costs based upon the number or square footage required, and equally as important, the time frame in which the work is to take place.

This escalation of unit costs will be reflected across the entire spectrum of tenant fit-up work, and likewise each purchase order and subcontract agreement issued for this work must take this potential for phased installations into account.

8

Buying Out the Job

The start of this chapter warrants a short story. The author of this book, early in his career, was a project manager with a small general contractor and did a fair amount of hard-bid work. The owner of the company would often come back to the estimating department and make last-minute adjustments to the bid, which generally meant arbitrarily reducing each low subcontractor's bid by at least 5 percent. The notion was that this would give us a better chance to get the job, and since most subcontractors expected to have their price "negotiated," we'd end up okay. Well, after a while, all the subcontractors submitting bids to our company actually *increased* their prices by 5 percent because word had gotten out in the local construction community of what we were doing; thus, we actually became *less* competitive as a result.

There are no shortcuts to intelligent purchasing or buy-outs. Buying out the job is more than negotiating a specific price for a specific amount of work. A relationship of trust and respect between both parties and the desire to work together harmoniously, efficiently, and ethically will ultimately yield the best price from a subcontractor. Whether the project manager or a central purchasing agent has the responsibility to issue purchase orders or award subcontract agreements, it is a process that requires a great deal of work to achieve the most cost-effective end product: the buy-out.

Awarding Subcontracts

The process of negotiating subcontracts is more complicated than merely selecting the apparent low bidder and writing a contract—the *apparent* low bidder may not be the most competitive, if and when a proper analysis is made of all the bids received.

Purchasing or "buy-out" responsibility varies from company to company. According to a recent survey conducted by the Construction Finance Management Association (CFMA), more than 70 percent of all respondent companies indicated that their project managers negotiated subcontracts and purchase orders. This allows the project manager to understand the intricacies of the project early because they must research the plans and specifications, and vendors and subcontractors will often alert them to any plan and/or specification design inadequacies, inconsistencies, or omissions. Specialty contractors and material and equipment vendors can be invaluable in the early stages of a project if they point out inadequacies in those plans and any specs that need coordinating before finalizing the deal.

The project manager must be thoroughly familiar with all aspects of the project requirements in order to produce the most cost-effective subcontract or purchase order. If it is company policy that the Purchasing department rather than the project manager "buys out" the job, they should consult with project management to review the scope, terms, and conditions of the work before any awards are made, and take advantage of the project manager's experience.

A key point that should *always* be remembered: Don't assume that the plans and specifications are perfect. Merely purchasing a subcontract based upon "plans and specs" will, most likely, not represent all that's required. During the buy-out process, a *full* scope of work should be developed and included in the contract, which could mean that additional "related" or missing work items are required for a complete job.

The subcontract interview form

Construction projects are becoming more complex, something which is reflected in both the specifications and today's more detailed drawings. Although the specifications often contain lots of "boiler plate," as each subcontract scope is defined, it is important to read every page of its "related" specification section. It is not unheard of that a specification from a previous project will be cut and pasted in its entirety into a newer one, even though it may not apply to the new project.

For instance, how about a specification on a K–12 school project that requires the wall-mounted TV brackets "*to be adjusted by the patient.*" Obviously, this section was lifted from a hospital spec. Interlacing provisions must be sought out; all too often, one specification section will include a clause stating "See other sections for Related Work." This related section(s) should therefore be investigated.

To extract all of the important requirements from the plans and specifications and be able to recall them during subcontract negotiations is difficult for many people, but there's a rather easy way to review and note important scope issues before such meetings. Figure 8.1 is a subcontractor interview form prepared in advance of a meeting with the subcontractor. By checking off the key issues for review, a thorough examination of the plans and specifications can be made before discussing the points at the meeting. Customized checklist templates such as this can be prepared beforehand for all trades.

SUBCONTRACTOR NEGOTIATION FORM

PROJECT

TRADE	SPECIFIED SECTIONS	DATE
Miscellaneous Metals		

SUBCONTRACTOR	REPRESENTED BY	(AREA CODE) TELEPHONE NO.

BASE BID AMOUNT	ADDENDUM NO.	

ALTERNATES	UNIT PRICES	UNIT PRICES
(1)	(1)	(6)
(2)	(2)	(7)
(3)	(3)	(8)
(4)	(4)	(9)
(5)	(5)	(10)
SALES TAX	INSURANCE	

SCOPE OF WORK Including but not limited to the following:

This form must be completely filled out and signed by the subcontractor and general contractor's representative to provide a record of this negotiation meeting.

ITEM	YES	NO	EXPLANATION AND/OR COMMENTS
1. All miscellaneous metal items are to be furnished and installed, unless otherwise noted.			
2. Galvanized as required/ Shop primed as required			
3. Bituminous coating as required			
4. All accessories such as bolts, screws, clips, anchors, etc to fabricate and install work			
5. Furnish inserts/anchoring devices set into concrete/ masonry for attachment to miscellaneous metal items			
6. Furnish inserts and/or anchoring devices for other trades as specified			
7. Shop assemble to greatest extent possible			
8. Field measurements as required			
9. Coordinate delivery, setting drawings, diagrams, templates, instructions, and directions for inserts, anchors, bolts or anchorages set by others			
			FINAL AGREED AMOUNT

GENERAL CONTRACTOR'S SIGNATURE   SUBCONTRACTOR'S REPRESENTATIVE SIGNATURE

ORIGINAL

FIGURE 8.1   A sample subcontractor interview form. (with permission from McGraw-Hill, New York)

CONTINUATION SHEET

SUBCONTRACTOR NEGOTIATION FORM

ITEM	YES	NO	EXPLANATION AND/OR COMMENTS
10. Subcontractor has accepted GC's safety program and all OSHA safety regulations			
11. Subcontractor to provide all temporary power unless general contractor agrees to provide			
12. If lay-down area is required, provide drawing for general contractor's approval			
13. Specified manufacturers or "Or Equal" only (burden of proof on Subcontractor)			
14. Insurance certificates to be furnished to meet limits as set forth in the contract specifications			
15. Subcontractor has received, read, and accepts terms/conditions of subcontractor agreement			
16. A payment and performance bond will be forthcoming, if requested			
17. For additional work _____% overhead and ____% profit will apply			
18. Subcontractor has received all dwgs, including A,M,E,P and includes all related work			
19. Subcontractor has received GC's base line schedule and accepts portion for their trade			
20. All items requiring factory finish to be in strict accordance with specification requirements			
21. Submit shop drawing schedule and delivery of each item upon receipt of approval			
22. Acknowledge attached detail list(s) to be appended to this form			

FIGURE 8.1 *(Continued.)*

The form in Figure 8.1 pertains to the miscellaneous steel trade, one that encompasses work on many different drawings from sitework to the MEP trades. A checklist like this is helpful in ensuring that all items required of the particular subcontractor are discussed during the interview process, including the scope of work to be completed and its corresponding cost. Prior to interviewing this miscellaneous steel subcontractor, a page-by-page review of each contract drawing is appropriate in order for this preprinted detail sheet to be completed.

Miscellaneous Metal - Detail Lists

Steel Stairs

- ☐ Steel-framed stairs including metal framing, hangers, columns, struts, clips, brackets, bearing plates, treads, risers, platforms
- ☐ Temporary supports
- ☐ Metal pan units as specified or shown
- ☐ Metal safety nosing as specified or shown
- ☐ Steel floor plate treads as specified or shown
- ☐ Diamond plate as specified or shown
- ☐ Grating construction
- ☐ Railings of size - amount, gauge, and material
- ☐ Kick plates
- ☐ Fabrication as specified or shown
- ☐ Open riser stairs

Handrails and Railings

- ☐ Material of type, grades, finishes, weights, construction, tolerances as specified
- ☐ Steel pipe
- ☐ Galvanized pipe
- ☐ Gray iron castings
- ☐ Malleable iron castings
- ☐ Stainless steel pipe
- ☐ Aluminum pipe
- ☐ Other types of railings
  - a._____
  - b._____
  - c _____

Expansion Joints

- ☐ Materials of type, grade, finish, weight, construction, and tolerances as required
- ☐ Floor expansion joints - angles with anchors
- ☐ Floor expansion joint covers
- ☐ Wall expansion joint covers
- ☐ Ceiling expansion joint covers
- ☐ All accessories of equivalent construction or grade as specified for a complete installation

Wall, Floor, and/or Ceiling Supports and Reinforcements for:

- ☐ Handrails
- ☐ Grab bars
- ☐ Toilet accessories
- ☐ Toilet partitions
- ☐ Television sets- wall or ceiling mounted
- ☐ Miscellaneous hospital equipment
- ☐ Wall-mounted equipment

FIGURE 8.1   (*Continued.*)

☐ Exhaust hoods
☐ Roof openings
☐ Knee wall partitions
☐ Countertops- toilet
☐ Countertops- kitchens
☐ Wall shelving

Miscellaneous Supports for Precast Concrete, Cast Stone, Limestone, Granite, etc.

☐ Bolts
☐ Clip angles
☐ Struts
☐ Bracing
☐ Relieving angles
☐ Inserts
☐ Wedges
☐ Tie backs
☐ Support framing

Miscellaneous Supports for Curtain Wall, Storefront, Window Wall, etc.

☐ Bolts
☐ Clip angles
☐ Struts
☐ Bracing
☐ Head supports
☐ Jamb supports
☐ Sill supports
☐ Inserts, wedges

Exterior Miscellaneous Metal Requirements

☐ Manhole, catch basin, trench drain, drain inlet frames
☐ Exterior ladder rungs for manholes, catch basins, drain inlets
☐ Exterior access hatches, frames, and rungs for mechanical, electrical, and other trades
☐ Bench supports
☐ Wall rail supports
☐ Abrasive nosing for concrete stairs
☐ Abrasive nosings for door sills
☐ Metal saddles
☐ Metal thresholds
☐ Roof hatches
☐ Roof scuttles
☐ Catwalks – framing, stairs, railings, gratings
☐ Dock bumpers

FIGURE 8.1   (*Continued.*)

Other such forms are equally valuable. Figure 8.2 shows one used for painting, another trade that may include work in "related" spec sections from carpentry to masonry to mechanical, electrical, and plumbing trades where the color-coding of piping and conduits is required.

SUBCONTRACTOR NEGOTIATION FORM

PROJECT

TRADE	SPECIFIED SECTIONS	DATE
Painting		

SUBCONTRACTOR	REPRESENTED BY	(AREA CODE) TELEPHONE NO.

BASE BID AMOUNT	ADDENDUM NO.	

ALTERNATES	UNIT PRICES	UNIT PRICES
(1)	(1)	(6)
(2)	(2)	(7)
(3)	(3)	(8)
(4)	(4)	(9)
(5)	(5)	(10)
SALES TAX	INSURANCE	

SCOPE OF WORK Including but not limited to the following:

This form must be completely filled out and signed by the subcontractor and the general contractor's representative to provide a record of this negotiation meeting.

ITEM	YES	NO	EXPLANATION AND/OR COMMENTS
1. Include all labor, materials, equipment, hoisting, and scaffolding to complete painting work in accordance with the all contract documents- A,S.MEP drawings and specifications.			
2. Structural steel & metal deck painting/touch up and miscellaneous metal (interior and exterior) painting/touch-up (List)			
3. Hollow metal work, frames, doors, borrowed lites			
4. Millwork finishing, cabinetry, built-ins, wood railings, running trim, miscellaneous architectural woodwork			
5. Drywall- subcontractor to inspect and accept substrate prior to commencing work			
6. Masonry, where required, to include block filler, special coatings			
7. Mechanical work – piping, pipe covering, ductwork, hangers, supports, equipment, color coding, as required by all applicable spec section(s)			
8. Electrical work – conduit, hangers, junction & pullboxes, panelboards, equipment, color coding, as required by all applicable spec sections			
9. Wood doors, frames and trim			
			FINAL AGREED AMOUNT

GENERAL CONTRACTOR'S REPRESENTATIVE SIGNATURE	SUBCONTRACTOR'S REPRESENTATIVE SIGNATURE

ORIGINAL

FIGURE 8.2 Form used for painting. (With permission from McGraw-Hill, New York.)

CONTINUATION SHEET

SUBCONTRACTOR NEGOTIATION FORM

ITEM	YES	NO	EXPLANATION AND/OR COMMENTS
10. Materials and colors as specified. What manufacturer will be used? Provide samples and mock-ups as required.			
11. Number of coats specified for each substrate. List.			
12. Preparation of surfaces to include putty and spackle as required to make them acceptable to the architect			
13. Finishes on wood and metal surfaces to be sanded between coats to assure smoothness and adhesion of prior coat			
14. Mill thickness on all coats to be as specified			
15. All work in strict accordance with manufacturer's instructions			
16. Special coatings - glazed wall coating, fire resistant, epoxy			
17. Wall coverings - manufacturer, thickness, weight, roll width, per specifications			
18. Application: Roll, brush, spray. Identify for each substrate type.			
19. Protection as required for floors, walls, ceilings, diffusers, convectors, installed equipment. All masking of surfaces as required.			
20. Clean up, remove masking, deposit daily in dumpster in location provided by GC			
21. Subcontractor has received complete set of drawings and specifications including general and special conditions and assumes responsibility for all requirements pertaining to their			
22. Subcontractor to repair all defective and rejected work as directed by the Architect			
23. Additional work percentage: ____% overhead ____% profit			
24. Subcontractor has received, read, and accepts general contractor's safety program.			

FIGURE 8.2   (*Continued.*)

Figure 8.3 is another form for subcontractor bid analysis, this one based upon the renovation of a roof. This portion of the review focuses upon compliance with the general requirements of the project and requires the subcontractor is respond with "Y" for yes or "N" for no regarding acceptance of the items discussed.

Once each interview is finished, the subcontractor should review the completed form and, if they agree with all of the statements and the scope of work included in their bid, sign the form as to acknowledge their agreement. This prevents a subcontractor from later stating they misunderstood the points during the interview and that they did *not*

SCOPE SHEET (SUBCONTRACTOR BID ANALYSIS)		2005
PROJECT NAME:	Renovation of Roof	
ESTIMATE #:		
ESTIMATOR:		

Y-S = yes included but deduct this subs cost	Bidder Number	n/a
Y-A = yes included but deduct another subs cost	Subcontractor :	
Y-W = yes included but deduct WBI estimated cost	Phone:	254-3140
N+S = no, not included so add this subs cost	Fax:	254-3069
N+A = no, not included so add another subs cost	Contact:	
N+W = no, not included so add Walsh's estimated cost		

The following scope must be included (Y) or excluded (N) from the scope of work - all listed scope is considered " including but not limited to "	
Performance and Payment Bond	N
All costs to employ Job Stewards (as required)	Y
Conformance with all OSHA Rules and Regulations	Y
Conformance with safety program	Y
Conformance to Owners Safety Policy and Rules	Y
All insurance requirements with named as the additional insured	Y
A Supplemental OCP (Owners Contractors Protective Liability) policy	Y
Testing and Inspections	N
All costs and provisions of the Nextel Phone System (connected to the Builders Network) for your Project Foremen, Project Manager and key personnel	Y
Submittals and shop drawings, mill test reports, certifications and design calculations (as required)	Y
Attendance at the weekly project meetings by a representative that has the authority to make binding decisions	Y
All costs for printing and reproduction	Y
Commissioning, O&M manuals	Y
Union Labor	Y
Taxes	Y
Premium portion of overtime labor required to maintain schedules due to the delays of others or for circumstances beyond this subcontractors control	N

Scope of Work Subtotal:		
Confirmed by Subcontractor letter dated:		

VE, Alternates or Material substitutions		
Sub-section 1 3: See the document entitled		
		add

FIGURE 8.3   Form for subcontractor bid analysis.

include or exclude specific items. If the subcontractor is unprepared to discuss some of these items at the initial interview, the project manager may ask the subcontractor to review their bid and the scope of work back at their office and later return for a final interview. When all bids for a particular trade have been received and analyzed, another form (called a bid summary sheet) is useful in comparing one subcontractor's quote with another to ensure the sameness of scope or to highlight differences in scope and corresponding cost.

Many GMP contracts require the general contractor or construction manager to tabulate bids, make a recommendation, and then forward it to the owner for review and approval. Look for the following owner modification to the Standard AIA A201 contract provision relating to subcontract awards that may take this form:

> The Contractor shall obtain bids for at least three prospective subcontractors whom the Contractor determines, subject to the Owner's approval, to be qualified to perform the work in question. Contractor shall determine which bids will be accepted, subject to Owner's approval. Owner will

promptly notify Contractor if Owner has a reasonable objection to any sub-contractor proposed by the Contractor.

The Bid Summary Sheet

A bid summary sheet containing the project name, the scope of work represented, the bid price, and budget estimate for the item is a document that can be transmitted to the owner.

Figure 8.4 is an example of a simple form for drywall work, including batt insulation to be installed within the partitions. Whether this insulation has been included in the drywall section or not, the general contractor has requested all drywall subcontractors include this in their bid, for obvious reasons; it is preferable to have the drywall subcontractor assume

---

BID SUMMARY SHEET

JOB: GRISWOLD MANUFACTURING

TRADE: DRYWALL - BATT INSULATION

BUDGET: $140,000

BIDDER	DRYWALL	INSULATION	ADD #1	SALES TAX	TOTAL
BEST DRYWALL	$122,000	$11,000	$5,000	Incl.	$138,000
CGM ACOUSTICS	135,000	10,000	Incl.	Incl.	145,000
GYPSUM ASSOCIATE	133,000	No-($9700)*	Incl.	Incl.	142,700
L & M DRYWALL	141,000	9,800	Incl.	Incl.	150,800
R & S INSULATION	–	* $9,700	–	Incl.	
ACME INSULATORS	–	10,400	–	Incl.	

FIGURE 8.4   A bid summary sheet.

responsibility for any insulation to be installed within their partition work no matter the specification section in which this insulation appears.

By using the form and having to review "included" or "excluded" items, the project manager will have yet another check on the completeness of each subcontractor's bid. When variables do exist, if no spreadsheet tabulation is done, comparison analysis will be difficult.

Figure 8.5 is another form of bid analysis. This one shows the progression of bids for selective demolition based upon preliminary drawings submitted to subcontractors for "pricing," all the way up to the final quotes after updated drawings were issued.

Some contractors use this bid analysis to rate the subcontractors, a process that may prove helpful if the bidding has been opened to all subcontractors and not a select list.

Unit Prices

Some project managers consider it helpful to solicit unit prices for various items of work during their subcontractor interviews, whether or not the contract with the owner requires such unit prices. This is a good idea and may prove helpful in dealing with change orders during the construction process. When owner-directed change orders are requested and appropriate unit prices have been obtained from the subcontractor, the cost of the proposed change can be assembled rather quickly, using these unit prices, thereby speeding up the change-order approval process. Since the cost of each unit of work will decrease as quantities are increased, it is standard procedure to request units prices from subcontractors on a sliding scale—as quantities increase, the unit cost decreases.

The hauling offsite of, say, 4 cubic yards of clean fill will be much more costly than a full truckload of 8 or 10 cyds or more, so the unit prices for haul-off and disposal could start with a less-than-full truckload quantity and continue with increased numbers of full truckloads. And as the quantities increase to a certain point, the price per cubic yard should decline.

The same is true for many other items of work. An electrician, asked to price the installation of one 20-amp receptacle will charge considerably more than if requested to install ten such devices. Again, a sliding scale of unit prices will be fair to subcontractor, general contractor, and owner.

Other items to consider during the solicitation of prices

It is the rare project that doesn't generate a change order or two, and costs are always scrutinized by diligent owners. When change-order work is to proceed on a time-and-material basis, one item that always shocks an owner is current labor rates—particularly when dealing with union labor where the basic wage rate is often augmented by fringe

DEMOLITION SUBCONTRACTOR BID RESPONSE SUMMARY

PROJECT NAME:

ESTIMATOR:

SPEC SECTIONS: 01739 SELECTIVE DEMOLITION

_/2005_

See May 14, 2005 Invitation to Bid and Scope of Work documents faxed to Demolition Subcontractors

Subcontractor	Invited to Bid	Will Bid, Requested Drawings	Attended Mandatory Walk-thru	Comments:	Received Bid	Bid Date	Comments:
FIRST PRICING ROUND							
1	14-May-05	yes	no	sent drawings for pricing	$439,000	6/6/2005	
2	14-May-05	yes	yes	sent drawings for pricing	no bid		withdrew from bidding on 6-13-05 due to project being outside his market niche
3	14-May-05	yes	yes	sent drawings for pricing	no bid		interested in bidding but is holding back quote until we "talk"
4	14-May-05	yes	yes	sent drawings for pricing	$1,050,000	6/23/2005	
5	14-May-05	maybe	yes	sent drawings for pricing	$2,100,000	6/23/2005	
6	14-May-05	maybe	yes	sent drawings for pricing	$1,795,000	6/15/2005	
SECOND PRICING ROUND							
1					$995,000	7/8/2005	CLAIMED ERROR ON FIRST BID
THIRD PRICING ROUND							
1				verbal quote	$1,350,000	9/10/2005	CLAIMED ERROR ON SECOND BID
FINAL PRICING ROUND - architectural demo only							
1			yes - 9/15/05	rec'd scope dated 9-19-05	$917,000	9/21/2005	In accordance with final WBI scope of work and site walk-thru
3			yes - 9/15/05	rec'd scope dated 9-19-05	no bid		project too complicated and too "fast" to man properly
4			yes - 9/15/05	rec'd scope dated 9-19-05	$1,187,000	9/21/2005	In accordance with final WBI scope of work and site walk-thru
FINAL PRICING ROUND - steel demolition only							
1					$210,000	9/21/2005	
4					$43,000	9/21/2005	unrealistic value
7					$156,000	9/19/2005	

FIGURE 8.5 Another form of bid analysis.

benefits (burden) that drives a $40/hour rate to $80 or more. By requesting labor-rate breakdowns from each subcontractor and upon, selection, including that labor rate in their subcontract agreement, an owner reviewing subcontractor bids will at least be made aware of the cost of labor if and when any time and material transpires. Figure 8.6 is a

OPERATOR/OILER PREMIUM RATE

As of 3-15-06 the rates will be:

Foreman	$35.87
Journeyman	$33.87

FIELD LABOR BREAKDOWN				LABOR BURDEN		
Assignment:	Manhours:	Rate:	Cost:	*Misc. Union Benefits Per hr.	$1.89	$ 1.89
General Supervisor:		$55.00	$ -	Pension Fund Per Hr.	$4.95	$ -
General Foreman		$36.87	$ -	Welfare Fund Per Hr.	$6.75	$ -
Foreman		$34.64	$ -	Annuity Fund Per Hr.(OT Doubles)	$6.00	$ 12.00
Journeyman		$32.64	$ -	**Fica, Med., Futa, Suta, (See Below)	19.03%	$ 13.79
Crane Operator		$44.53	$ -	Workmens Comp. Ins	42%	$ -
Oiler		$27.96	$ -	Automobile / Equipment Ins. Per hr.	$0.54	$ -
Prem Time Gen. Foreman		$36.87	$ -	Gen. Liability:	10%	$ -
Prem.Time Foreman		$34.64	$ -	Umbrella Per Hr.	$0.97	$ -
Prem.Time Journeyman		$32.64	$ -	Holiday Pay (Foreman/Operators Only)	$3.85	$ -
Crane Op. / Oiler Prem	1	$72.49	$ 72.49			
			$ 72.49	FICA-6.2%, Med-1.45%, FUTA-.08%, SUTA-10.58%		$ 27.68

* Edu. Fund, Bldg. Fund, Ind. Fund, Pens. Suppl., Labor/Mgmnt. Trust

EQUIPMENT RENTAL COSTS						
TYPE	COST	HOURS	DAYS	WEEKS	MONTHS	COST
100T HYD. CRANE	$180.00					$ -
70T HYD. CRANE	$140.00					$ -
140 CONV. CRANE	$125.00					$ -
888 CRANE	$275.00					$ -
						$ -
TORCHES	$3.50					$ -
GAS WELDER	$26.00					$ -
Boom Lift 80' or Less	$75.00					$ -
SCISSOR LIFTS	$25.00					$ -
Tractor w/ /Driver	$125.00					$ -
						$ -

MATERIALS / HARDWARE					SERVICES			
	PRICE	Qty	PER	TOTALS		PRICE	HR. / EA	TOTALS
Paint			gal.	$ -	Trucking	$75.00		$ -
Structural bolts			each	$ -	Survey Crew (2)	$175.00		$ -
Expansion bolts			each	$ -	Engineering	$125.00		$ -
Chemical Anchors			each	$ -	Shop Rate			$ -
Wedge Anchors			each	$ -	Galvanizing			$ -
Anch.bolts 3/4 x 12"			each	$ -	Bending			$ -
Nelson studs			each	$ -	Rolling			$ -
Turnbuckles			each	$ -				$ -
W.F. Beams			lb	$ -				$ -
T.S			lf	$ -				$ -
Pipe			lb	$ -				
Angle			lb	$ -				
Flat			lb	$ -				
Misc Material			lb	$ -				
				$ -				

OTHER					TOOLS / CONSUMABLES			
Item	PRICE	QTY	PER	TOTALS	ITEM	Price Hr / Ea	Qty / Hrs.	Total
Welded Moments			Ea.	$ -				$ -
Bolted Moments			Ea.	$ -				$ -
Framed Openings			Ea.	$ -				$ -
Beam Penetrations			Ea.	$ -				$ -
				$ -				$ -
				$ -				$ -
				$ -	*Standard tools ** Consumables	2.80	0	$ -
				$ -				$ -

*May include some or all of the following: •   Hand tools / rentals (pins, hammers, wrenches, impact guns, TC guns,

TOTAL
$104.47

grinders, chainfalls, comealongs, electric welders etc.)
** Oxygen, acetylene, welding wire, grinding discs, Saw blades, rigging gear, gloves, glasses, tips etc.

FIGURE 8.6   Typical wage rate breakdown (for a union operator/oiler).

typical wage rate breakdown, this one for a union operator/oiler. The sheet also contains equipment rental costs and material costs.

But before accepting these labor rates, they should be reviewed and analyzed. Figure 8.7 is a labor rate breakdown for an HVAC/sheet metal mechanic. It looks pretty straightforward, but wait a second, look at the breakdown for premium time and double time. This subcontractor included $2.92/hour for travel time for each hour worked during a normal 8-hour day. But would working an additional 2 hours at premium rates also merit that same $2.92/hour travel add-on? The mechanic has already been paid for travel to and from, and no further travel pay is

Class (5538)	Desc.	Journeyman			Foreman/Cordinator		
		Straight time	Premium time	Double time	Straight time	Premium time	Double time
Base rate	Base hourly rate	$34.56	$51.80	$69.12	$39.56	$59.34	$79.12
Taxes	FICA (7.65%)	2.64	3.96	5.29	3.03	4.54	6.05
	FUTA (.80%)	.28	.41	.55	.32	.47	.63
	SUTA & SUMI (13.02%)	4.50	6.74	9.00	5.15	7.73	10.30
Ins.	Work comp. (14.54%)	5.03	5.03	5.03	5.75	5.75	5.75
	Umb. (3.0%)	1.04	1.04	1.04	1.19	1.19	1.19
Union benefits	Natl. pension	5.91	5.91	5.91	5.91	5.91	5.91
	SASMI	1.63	1.63	1.63	1.63	1.63	1.63
	NTF/NEMIS	.18	.18	.18	.18	.18	.18
	Local ins.	8.32	8.32	8.32	8.32	8.32	8.32
	Local training	.86	.86	.86	.86	.86	.86
	Local ind.	.40	.40	.40	.40	.40	.40
	Annuity	5.00	5.00	5.00	5.00	5.00	5.00
	Equality	.50	.50	.50	.50	.50	.50
	Suppl. pension	.56	.56	.56	.56	.56	.56
	Vac./	--	--	--	2.50	2.50	2.50
	Holiday				4.40	4.40	4.40
	Travel				2.92	2.92	2.92
	Sub total	$71.41	$92.34	$113.39	$88.18	$112.20	$136.22
	Overhead	7.14	9.23	11.34	8.82	11.22	13.62
	Subtotal	$78.55	$101.57	$124.73	$97.00	$123.42	$149.84
	Profit	3.93	5.08	6.24	4.85	6.17	7.49
	Total	$82.48	$106.56	$130.97	$101.85	$129.59	$157.33

FIGURE 8.7 A labor rate breakdown for an HVAC/sheet metal mechanic.

warranted. It is better to catch and correct any irregularities in wage add-ons before sending them off to the owner.

Combining Work to Best Advantage

There may be some merit, other than a monetary one, in combining items of work in a particular subcontract agreement, such as having the mason apply the water-repellent coating to the exterior brick or block-work, a work task possibly defined in a Division 7–Moisture Protection of the specifications rather than Division 4–Masonry. All too often, one subcontractor will blame another for the failure of a system. By combining masonry and water repellents, the responsibility for water penetration application through the masonry wall squarely falls in the hands of one subcontractor.

Other tasks can be combined to advantage, and the drywall subcontractor presents a very visible one—installation of hollow metal door frames set into stud partition walls and the mason contractor installing frames in their CMU walls.

Having the drywall subcontractor install hollow metal frames in their drywall partitions, and having the mason contractor install hollow metal door or window frames in masonry walls combines two operations that, if awarded separately, can lead to arguments and unwanted problems. A door or window installer, starting their work, may claim that the frames, whether for doors or windows, have been installed out of plumb or out of square and are not level, thus requiring additional work at additional cost. This scenario is always followed by an argument over who performed the substandard work. A combined award to the mason and/or drywall subcontractor will avoid another potential problem.

Consider having the drywall subcontractor install all blocking required in their partitions for subsequent trades such as wall-mounted cabinet installers and toilet accessory installation crews. It may be possible to include this blocking in the drywall subcontractor's scope of work at little additional cost by convincing that subcontractor that they can now maintain full control over both blocking, framing, and drywall, thereby making their overall operations more efficient and productive.

Even if the combination of separate bids for work such as those just described are slightly higher than the individual prices from other subcontractors, think twice before making an award, and consider the advantages of having fewer subcontractors responsible for more operations involving related tasks.

Subcontract or Do It Ourselves?

Depending upon the capabilities and availability of the general contractor's work forces, the project manager may consider their options:

What work will be performed by the company's own forces or be subcontracted? For those general contractors who regularly self-perform certain types of work, any number of factors can come into play in this decision-making process:

- Does the amount of workload currently being performed by the general contractor's own forces, and *supervisors* (remember someone has to supervise the work!), permit committing those forces to the new project? Note the word *supervisor* was accented and noted. Even if skilled workers are available, unless a trained supervisor or foreman is available to oversee the work, you should consider subcontracting it.

- Is the new project such a distance from the home base of operations that travel time, with all its added costs, makes local subcontracting a more reliable and more cost-effective way of doing business?

- Is it possible to incorporate work normally performed with the company's own forces into a subcontract that combines other work so as to offer a "package" to the subcontractor and thus receive a more competitive price for both operations? For example, if the company purchases the finish hardware and generally uses their own carpenters to install it, consider incorporating hardware installation with the drywall subcontractor's tasks since they also employ carpenters and may welcome the opportunity to expand their scope of work.

- If the work is of such a nature that it presents the potential for future problems and/or liability, will it be less expensive in the long run to award the work to a more experienced company to which accountability can be transferred?

All of these considerations have to be weighed before subcontract negotiations commence.

The "we can do it cheaper" syndrome

Sometimes a project manager will say, "All of the bids exceed our budget; let's do this job with out own forces." This can be an ill-advised approach unless all of the facts are reviewed and discussed.

In normal market conditions, subcontractors will generally add 15- to 20-percent overhead and profit to their costs, and if this is the difference between the company's estimate and the low subcontractor bid, the decision to self-perform the work may be somewhat easier to make. But bear in mind that unless the company's own workforces perform these operations on a regular basis, the subcontractor's forces, who do this work every day, may be more efficient and even their "bare cost" price may be difficult to match.

Even after discounting the subcontractor's price by an amount *assumed* to represent overhead and profit, the project manager should carefully scrutinize their own estimate. If expensive rough hardware is required for the job (some expansion bolts cost $5 to $7 each) were these costs included in the estimate? Are manlifts, scaffolding, and ladders required, and if so are these costs included in the estimate? What about material waste factors—a reasonable amount of waste and its related cost must be factored into the cost. Ask the question, "Can our workers, who perform this kind of work intermittently, achieve the unit costs included in the estimate? And do we have current documentation of actual unit costs from our own forces performing this operation?"

The answers to these questions, and possibly more, should be considered carefully before embarking on the path to self-perform work when faced with the potential of higher subcontracted prices.

Key Questions to Ask Subcontractors during Negotiations

In addition to the need to ensure that subcontractors have incorporated the proper scope of work in their bid, along with any exclusions, it is important that other criteria be acknowledged or clarified. If the subcontractor interview form is used and has been prepared carefully and thoughtfully, these questions will have been answered during the interview. If not, you should run through them again, as outlined in the following:

- Have the latest contract drawings been reviewed, and do the subcontractors accept all revision dates as being part of their bid? (Note: Prepare an exhibit listing the "contract" plans and specifications, with revision dates and all issued addenda. This list should be presented to the subcontractor to ensure that they have reviewed the appropriate documents. This list will be appended to their subcontract agreement.)

- If the project is exempt from state sales tax, does the bid reflect this exemption, and conversely, if the project is not tax exempt are the appropriate taxes included in the bid? This seems like a rather simple question, but you may be surprised by the number of subcontractors that have included or excluded sales tax inappropriately.

- Are the subcontractors aware of the project schedule? It is important that they be apprised of the construction schedule—both overall start and completion so they can anticipate when their work will commence and when completion is anticipated in order to incorporate potential wage, material, and equipment price increases. They should be given a copy of the baseline schedule to review and comment on, or accept. Remember, by doing so you are committing to the time slot and duration that this schedule represents and if deviations occur as work progresses, document the reasons for these deviations and attempt to get the subcontractor's acceptance—at no added cost.

- Are any exceptions being taken to the scope of work as shown on the contract drawings or the specifications? A subcontractor may have based their price on their ability to substitute what they consider an "equal" product, only to discover as the project progresses that their "equal" product has been rejected by the architect/engineer. It must be established during the negotiation stage that the subcontractor is bound to furnish only those products specifically called for and if "equals" are submitted and rejected, the subcontractor will be obliged to furnish the specified product at no additional cost. The project manager must take the initiative in this matter. Some subcontractors will not divulge their intent until their product submission has been rejected, at which point they will state "Well, our price was based upon getting this substitute product approved. The architect approved it on another project we worked on." This happens all the time, so ask the question! Remember few things are "equal." A better term is probably "similar to."

Pitfalls to Avoid in Mechanical and Electrical Contract Negotiations

A project manager must be completely familiar with all aspects of the contract documents in order to make the most intelligent, informed decisions on contract awards. In the case of the mechanical and electrical trades, a number of items should be addressed beyond the basic work contained in their respective specification sections.

An architect may have selected one electrical engineering firm and another mechanical engineering firm to develop their respective drawings and specifications. Thus, there may be gaps or overlaps between those prepared by the engineers for Division 15 (Mechanical) and Division 16 (Electrical) and the General and Special Conditions sections (Division 0 and Division 1) prepared by the architect. The project manager should review these four specification sections to insure there are no gaps, overlaps, or omissions in the work prepared by three different design consultants. Some things to watch out for:

- Will the electrical contractor be required to provide temporary utilities for all trades during construction?

- Will the electrical contractor be required to provide temporary power to the general contractor's field office?

- Will the electrical contractor be required to obtain and pay for all permits required for both temporary and permanent power?

- Who will furnish starters for all mechanical equipment? Is this in the mechanical specification section or the electrical section?
- Does each trade perform its own trenching and backfill operations, or should the general contractor provide these services?
- Does each trade perform its own cutting or patching, or is this provided by the general contractor? What about fire-safeing all required penetrations—who is responsible?
- If concrete work is required—the encasement of underground conduits or housekeeping pads—who will provide this?
- Is clean-up and trash removal clearly assigned to these trades?

With respect to these and other similar questions, the project manager may have to read the fine print and boilerplate in several specification sections to unearth the answers.

Who Is the Contractor?

Even more confusing in some specifications is the term "contractor." In some spec books, there is a separate "Definitions" section defining whether the term "contractor" refers to the *general contractor*, or the *subcontractor* when used in a specific section.

For example, does the phrase "contractor shall perform all cutting and patching" included in Division 16—Electrical refer to the *electrical contractor* or the *general contractor*?

The answer to this simple question might not be so clear, and if in doubt, you should request interpretation from the architect prior to completing negotiations with the respective subcontractor. Each party, during negotiations, will obviously assume that the definition applies to the other party, and if that be the case, clarification upfront will save a great deal of controversy as the project gets under way.

Issues To Be Addressed

Temporary light and temporary power

The question of furnishing temporary light and power is one of those items that sometimes gets "lost" in the specifications—the electrical engineer assumes that the architect will include this in Division 0 or Division 1, while the architect assumes the electrical engineer will include it in Division 16—Electrical. This needs to be clarified, and even though there may not be any question about the electrical subcontractor furnishing temporary lights or power during construction, some questions may remain unanswered.

If the specifications do not stipulate the intensity of the temporary lighting, such as the number of 100-watt lamps to be installed per 100 ft^2 of floor area, for example, that is another issue to be resolved during negotiations with the subcontractor.

In the absence of any specific commitment to provide a level of lighting intensity, the project manager should direct the electrical subcontractor to provide, *and maintain, sufficient levels of lighting as required for proper working conditions for all trades.* This phrase should be included in the subcontract agreement.

Installation of underground utilities

When the project requires the installation of new utilities such as electrical primary and secondary service, incoming water mains, gas lines and so forth, related costs may require some clarification to determine who is responsible for:

- Trenching and backfilling
- Bedding materials for underground pipes and conduits
- Cutting and patching
- Tap-in and tie-in costs, including any required permits
- Temporary and permanent patches in existing paved roadways, including permits and bonds
- Utility company charges and fees
- Concrete encasement of underground utilities
- Concrete pads for related equipment—for instance, meters and transformers

Designing to local utility standards

Another problem can occur when an out-of-state architect or engineer is preparing mechanical and/or electrical plans and specifications. These designers may not be familiar with all utility company regulations in the state for which the design is being prepared, and their specifications may not conform to that utility company's requirements. The specifications also may not reflect local trade practices. For instance, in one state the plumbing subcontractor may have, by custom, assumed responsibility for trenching and backfilling for their underground installations, but in an adjacent state this work is customarily performed by the general contractor.

Specifications prepared by an out-of-state architect/engineer team should be carefully reviewed by the project manager with these concerns in mind. The following story clearly illustrates how such problems can arise.

The author of this book was hired by a general contracting firm as a consultant to resolve a dispute with their fire protection subcontractor. The dispute involved the subcontractor's obligation under the "perform- ance specification" section of the fire-protection specifications. The proj- ect was being built in the Commonwealth of Virginia, while the design engineer was based in Maryland. The subcontractor requested addi- tional money to add some system components to comply with Virginia's regulations, but the general contractor refused, stating that the speci- fication was a performance spec. However, a careful reading of that specification section showed that the subcontractor was to comply with all rules, regulations, and local and state codes of the *State of Maryland*. Needless to say, this put the general contractor on the defensive as we worked through the process of resolving this matter.

Warranties and Guarantees

Another subject requiring investigation during subcontractor negotia- tions with both electrical and mechanical subcontractors is that of war- ranties and guarantees. One-year warranties/guarantees are standard in the industry for most products, but when do these warranties commence?

If the general contractor is given permission to use the permanent HVAC equipment during the finishing stages of the project, rather than furnish temporary heat or cooling by use of portable equipment, the one-year warranties and guarantees will probably begin when this per- manent equipment is operational and functioning. And depending upon the length of time required to complete the structure and turn it over to the owner, the one-year warranty/guarantee period, which started when the equipment was accepted and made operational may have expired. In order to satisfy the owner, the general contractor will be obliged to purchase an extended policy, at a substantial cost.

If the projected period of use for the building's permanent HVAC system had been determined with some degree of certainty at the begin- ning of the job, then the project manager could have attempted to include an extended warranty in the negotiations with the subcontractor. The subcontractor, in turn, while negotiating with their equipment suppli- ers, could then possibly obtain the extended warranties at no cost, or at the least, at some small incremental additional cost.

Job Cleaning and the Contract

A clean project is a safe project, and a clean project is more likely to have higher-quality levels than a trash strewn job. But maintaining a clean job can be an expensive proposition, and enforcing cleaning operations can be a frustrating one.

This rather simple idea often causes rather strong disagreements between general contractors and subcontractors, leading to backcharges, claims, and counterclaims. But it doesn't have to be that way. The specifications should include provisions for cleaning, and responsibilities should be assigned to each trade to clean the debris they generate. If the contract specifications do not assign cleaning operations to a specific subcontractor, the general contractor's subcontract agreement will most certainly do so. Enforcing this provision at times may be a problem, however.

Various options are available to the general contractor to clean the building and site. Some contracts stipulate that the general contractor must provide a large central trash container or several small containers located around the building or site, and each subcontractor is to deposit their debris in them. Other subcontract agreements place the responsibility on the subcontractor to provide debris containers for their own trash. A common provision in most subcontract agreements is that cleaning will be performed on a daily basis. But none of these procedures work smoothly without a strong, continuous campaign waged by the project superintendent.

Arguments from subcontractors range from "We didn't generate that trash; it belongs to the electrician (or carpenter, or drywall sub)" to "Don't worry, we plan to have a crew out here on Wednesday (or Tuesday, or Friday) to clean out the entire area." Some cleaning must be performed daily—such as with gypsum drywall debris—while other cleaning can wait a day or two.

The project manager should address cleaning responsibilities and assignments at the first project meeting and at all subsequent project meetings. The job minutes should include the nature of the discussion, for instance: "I call your attention to the provision for daily cleaning, and this provision will be enforced (date of first project meeting). The drywall subcontractor has been advised to clean up their debris within the next 48 hours (or whatever the time frame stipulated in the subcontract agreement), or the general contractor will provide the cleaning and backcharge their account." This creates a paper trail so that back charges can be enforced.

Subcontractor arguments about cleaning and back charges have two aspects:

- The subcontractor was not notified that they would be backcharged if their cleaning was not completed by a specific date.
- The subcontractor denies responsibility for generating all, or some, of the debris the general contractor claims.

Objection No.1 can be overcome by a timely notice—in writing. Objection No.2 is more difficult to deal with—how to identify debris generated by a specific subcontractor. Methods frequently used to resolve this issue are photographs taken periodically of the piles of debris to help identify its source, walkthroughs with the subcontractor's foreman to point out the debris they have generated, or an apportionment to each subcontractor of the general contractor's total cleaning costs.

Whichever one is employed, promptness of notification and action is important. The building and site should be inspected daily as part of the superintendent's walkthrough, and the debris-generating culprits thus identified and notified. If the offenders fail to clean, the general contractor should do so promptly and issue a backcharge to show their instructions about the cleaning provision will be enforced. When principals of the subcontracting firms receive substantial backcharges for cleaning, they will most certainly get involved in resolving these matters.

Communicating the Terms and Conditions of the Subcontract Agreement

After subcontract awards have been made, the language and terms and conditions of the agreement must be made clear to all persons not present during that final negotiation session. For example, the project superintendent should be able to determine, with clarity, the exact scope of work included in, or excluded from, the subcontract agreement. The primary definition of contract scope and obligations will be the plans and specifications, but if an agreement that has been reached with the subcontractor deviates from those plans and specifications, or expands the scope of work, any such deviation, addition, or vagueness should be clarified in the written agreement.

Too often, a subcontractor's foreman will advise the project superintendent, "Oh, I know the specifications call for us to install the blocking, but my boss took exception to that and we excluded it from our work" or "Your boss and my boss agreed that it was okay to eliminate that."

Changes made to the work must be specific and include all such inclusions, exclusions, modifications, and changes, possibly in a separate contract exhibit. Some project managers think they can "slip" something by the subcontractor if, in fact, it was not included in the negotiations and subcontract agreement. But it usually doesn't work out that way and the disagreement always surfaces at the wrong time—when measures are urgently needed.

Importance of lien waiver requirements in the subcontract agreement

Whether stipulated in the contract with the owner or not, subcontractors should be required to submit lien waivers with each request for payment. The subcontractor's lien waiver stipulates and warrants that all proceeds from the previous payment have been used to pay for all labor, materials, and equipment installed in the project during that period. Lien waivers are not required with the submission of the subcontractor's first application for payment since no prior payment was received for their disbursement. However, lien waivers for all subsequent payments should be a contract requirement, including a final waiver containing a statement that all outstanding claims have been either resolved by the subcontractor or withdrawn when final payment is received.

This lien waiver requirement must be rigidly enforced—that is, *no lien waiver, no payment.*

The second- and third-tier subcontractor lien waiver problem. Most subcontract agreements contain a provision requiring the subcontractor to advise the general contractor of any second- or third-tier subcontractors they plan to employ on the project. Often, subcontractors ignore this provision and fail to notify the general contractor, which may create a problem.

Typically, mechanical subcontractors will employ lower-tier subcontractors—possibly sheet metal duct fabricators, pipe and equipment insulators, control wiring contractors, and air and water balancing contractors. Other subcontractors may also "sub" out a portion of their work to another contractor. The project manager must be alert to, and advised of, these lower-tier subcontractors since they will also be required to submit lien waivers. There is a danger that the prime subcontractor, once paid, may fail to pay their lower-tier sub, and if so, that second- or third-tier subs may lien the job.

In such a case, a demand by this lower-tier subcontractor may have to be honored by the general contractor who having already paid the prime subcontractor for this work, in effect will be paying twice for the same goods and services. A subcontractor filing a fraudulent lien waiver could face legal action, but that lower-tier subcontractor must be paid by either the prime sub of the general contractor.

Avoiding the problem. To lessen the possibility of such a problem, a detailed schedule of values should be requested from each subcontractor performing significant portions of work, and all potential lower-tier contractors identified. The field superintendent should also be on the alert to sort out and identify any subcontractors that come to work on the site and have them identify their company and who hired them.

The joint check. Monies received from the general contractor for a specific project are expected to be paid by them to either a subcontractor for labor, materials, and equipment incorporated into that project, or materials and equipment purchased from vendors. But cash-flow problems are not unknown in the construction business, and funds received from one project may often be disbursed for use on another project—a process known as the "co-mingling of funds." If a subcontractor's financial status is perceived or known to be weak or shaky, the general contractor can suggest that joint checks be issued to their subcontractor or vendors for major or expensive pieces of equipment or materials. This will ensure that the appropriate funds are paid to that vendor. The joint check is made out to two parties—the subcontractor and their vendor/supplier—and both will have to endorse the check before cashing it: the subcontractor first, the vendor second. This provides an assurance that the funds dispersed by the general contractor reach the parties for whom these funds were intended.

If the subcontractor is amenable to this procedure, a formal joint check agreement should be prepared for their signature. If the subcontractor balks or is reluctant to sign a joint check agreement, that should send up a red flag. Why are they resisting? Some subcontractors say that the issuance of a joint check by the general contractor casts an aspersion that the subcontractor's financial position is weak and they don't want that stigma. If that's the case, ask them to furnish receipted and paid invoices from their suppliers, along with their lien waivers.

Purchase Orders

Purchase orders are written for materials or equipment where only the material or equipment, but no labor to install, is required. Purchase orders, while not as detailed as subcontract agreements, should include certain basic elements, such as the following:

- The vendor's name, address, telephone, fax number, e-mail address, and the contact person for product or delivery information.

- The project name, project number, purchase order number, and cost code for the product or material with instructions to include all three designations on all receiving tickets and invoices.

- The project address, field office phone, e-mail, fax number, the superintendent's name, and their cell phone number.

- The quantity of materials and equipment ordered, a brief description of the material and/or equipment, a unit price (if applicable), and the total price. Trade or payment discounts should be included along with freight charges defined as either FOB job site or shipping point. Taxes, if applicable, should also be included. If tax exempt, the tax exemption number or a copy of the tax exemption certificate should be included/attached.

- A preprinted statement on the purchase order form should designate that signed receipts for all deliveries must be attached to the vendor's invoice and that all invoices must include the project name, project number, and the product cost code (These duplicate receiving tickets will help the accounting department verify that deliveries have actually been made since receipts from the field are often misplaced.)

- Another preprinted statement on the purchase order form should stipulate that the price(s) included in the purchase order are not subject to change without prior notification, in writing, and presented before the delivery of the purchased material or equipment.

- There should be a line for the signature of the person issuing the purchase order; some contractors require the recipient (vendor) to sign one copy acknowledging receipt of the terms and conditions.

- Any special delivery instructions should be appended to the purchase order, such as "All deliveries must be made between the hours of 7:30 A.M. and 3:30 P.M." or "All deliveries to be made via Elm Street entrance to project."

The job superintendent should be consulted before establishing a delivery date, and if deliveries are scheduled for several weeks or months ahead, a statement such as the following should be added: "Do not release for shipment unless authorized by Mr. Joe Purcell, Project Superintendent" (or whoever is authorized to make releases).

A copy of the purchase order should be distributed to the Accounts Payable department so it can be compared to the invoice when submitted for payment. Another copy can be filed in the project Purchase Order file or in the file pertaining to the product or material purchased.

A copy of each purchase order will be sent to the jobsite so that the superintendent is fully aware of all purchases and can either anticipate their delivery or schedule deliveries as required. Some materials such as extruded polystyrene (ESP) and batt insulation are so bulky that adequate areas must be cleared and set aside for their delivery. The super must be prepared to make these arrangements before shipment.

Ordering when exact quantities are not known

When orders for some items such as framing lumber or lumber for blocking are placed, the actual quantities required are not known but only estimated. Careless storage and handling procedures or unanticipated waste factors can increase the actual quantity consumed on the job site, or, conversely, proper storage, handling, and the skillful use of materials may result in substantially smaller quantities than originally anticipated.

Approximate quantities can be used to obtain pricing, but rather than stating a specific amount of the product in the purchase order, an approximate amount can be included within the range of the pricing discount, and a minimum-size shipment can be stipulated. For example: 1000 board feet of No.1 Hem/Fir studs—minimum shipment: 250 bft, $245.00/mbf

This gives the superintendent some latitude when releasing an order and avoids large shipments that may remain in storage for some time. A proviso may be added that if quantities released exceed those specified in the purchase order by, say, 25 percent, a lower price will be subject to negotiations. Partial releases provide the superintendent with better control over waste and theft. When smaller quantities of framing lumber or sheet rock are stored on the job site, it is easier to spot excessive waste or theft of materials. It may be slightly more expensive to have purchases delivered in a series of partial releases, but in the long run that might be more economical, considering waste and theft.

Price protection and the purchase order

Often, what goes up must come down, and vice versa. Commodities such as framing lumber, plywood, and drywall prices rise and fall over the short term, and are often influenced by the residential housing market, and recently by worldwide demand for certain basic building materials such as cement.

There are several ways in which purchase orders can be negotiated to obtain a full or partial measure of price protection. When purchasing transit or ready-mix concrete, it is standard to request that the supplier guarantee the price for the duration of the project. Other factors, some of which might not be so apparent, should also be considered when purchasing ready mix concrete.

Projects constructed in those parts of the country where winter means cold weather require "winter concrete." For those in southern climates, the term *winter concrete* refers to concrete shipped from the batch plan during certain times of the year when temperatures are likely to fall below, or remain close to, freezing. The ready-mix supplier will generally begin shipping winter concrete on a pre-established date, and publish this date to all customers. At that time, the supplier will heat the sand and aggregate before it goes into the mix and will add hot water to prevent the mixture from forming ice crystals before the heat of hydration process begins. Standard procedures call for suppliers of winter concrete to automatically add an up-charge for each cubic yard shipped after the specified date; not all plants establish the same date for the start of winter concrete shipments.

By estimating how much concrete will likely be placed during the winter months, the project manager may be able to negotiate a more favorable price for winter concrete during their initial dealings with the ready-mix suppliers. With respect to lumber and sheetrock, negotiating a firm price for extended deliveries can be a two-edged sword; if prices advance in coming months, savings will accrue, but in a falling market, the supplier will benefit. For large quantity orders, pricing arrangements with suppliers can usually be made with provisions to adjust prices in changing market conditions, or the project manager can lock in a firm price and gamble on making the correct decision; however, suppliers of basic construction materials are very aware of market trends and will price their products accordingly.

Is price the only consideration?   After reviewing the plans and specifications, and discussing the job's progress with the field superintendent, a shopping list of materials required (with approximate delivery dates) can be established.

Price is important, but other variables enter into the equation. For example, a better grade of framing lumber may result in less waste and discards, and therefore cost less overall than a lower grade, where waste factors may be greater.

Or less expensive material may not be readily available and using a more expensive grade will result in continued, uninterrupted operations. No need to be reminded of the high cost of labor and what a crew's downtime costs. So look beyond price to evaluate other options that may ultimately be more cost-effective.

The Domino Theory in purchasing.   The purchase of one item may force the expeditious purchase of another item. If architectural or custom-grade wood veneer doors must be bought, and the intention is to have these doors "prepped" (prefitted and premachined to accept hardware installation in the field), obviously other items must be purchased concurrently. Doors to be installed in hollow metal or aluminum door frames require frames to be purchased along with finish hardware for coordination with the frame manufacturer and the door supplier.

So one delivery—wood doors—will depend upon receipt of approved shop drawings from hardware and door-frame vendors before the production of wood doors can begin—a process that often requires 14 to 16 weeks lead time.

The purchasing of major HVAC roof-top equipment will be required to produce information for the structural steel subcontractor since they will need confirmation of the weights and sizes of the equipment in order to properly locate roof beams and supports for roof-deck penetrations, as well as provide the proper dunnage for the equipment. Other roof-top accessories, such as roof hatchs, will also require prompt purchase in order to complete the roof-top penetration scheme for the steel fabricator, miscellaneous metal, and the roofing contractor.

Embedments in concrete foundations or slabs may necessitate the early purchasing of miscellaneous steel, even though these steel items are not required for many months down the road. It is important to think through the chain of events that must be followed in order to provide a prompt, well-coordinated flow of materials to the jobsite.

Pitfalls to Avoid When Issuing Subcontracts and Purchase Orders

The following problems should best be avoided during the issuance of subcontracts and purchase orders:

- Intelligent buying is based upon knowing as much as possible about the item or items to be purchased. Rather than hurriedly attempting to buy out a subcontractor, purchase materials, or pieces of equipment, spend some time researching exactly what's required, and how it might impact other related items. Review the specifications carefully and examine all the drawings to uncover items to be purchased that might appear on drawings where they shouldn't be. Make lists of key requirements or use some form of subcontractor interview form to jog your memory when discussing these items with the subcontractor or vendor during negotiations.

- When a project is just getting started, there never seems to be enough time to select all of the subcontractors and vendors needed. Thus, set priorities and goals that can reasonably be met—first, second, third, and so forth. This priority list can be updated on a daily or weekly basis.

- Create an "A" list—activities that *must* be accomplished that day— and make a "B" list to be used if all of the A list items have been completed. The A list should not be overly ambitious, but once made, make it a rule to *always* complete those items—no matter what. This type of discipline will pay off.

- Immediately after a construction contract has been awarded, subcontractors who have submitted bids will begin to call, requesting a meeting to review their proposal. Calls from landscape contractors, flooring contractors, and painting contractors are often received at this early period—items that are fairly far down on the project manager's list of priorities. Avoid being rude or abrupt, but tell these less-than-urgent subcontractors or vendors to call back in two weeks, one month, or whatever time may be more convenient. Don't even discuss their proposals at this early date, you have more important things to do.

- Avoid *crisis management* whenever possible so that enough quality time can be spent on each purchase order and subcontract agreement. Again, this relates to time management and doing what is necessary to take care of immediate and near-term requirements. By reviewing job sequences and the project schedule with the project superintendent, one can develop a list of priority items during those first hectic weeks of construction activity.

- Recognize the long lead items and concentrate on them. Lumber prices can be obtained and analyzed, and a purchase order negotiated, all within a relatively short period of time, but hollow metal doors and frames require considerable coordination, along with purchasing several related items.

- Although the elevator installation will not commence for months and months, this equipment must be purchased early on since shop drawings will verify or change hoistway openings, miscellaneous items such as sill angles, roof openings, and the configuration of the elevator machine room. By reviewing construction sequences, other such priority items will undoubtedly be discovered.

- There are difficult decisions to make when potential awards exceed the budget and possibly generate losses. It is easier to award contracts and purchase orders when they fall within budget or result in savings. The tough decisions arise when the costs of certain subcontracted items have been drastically underestimated, for whatever reason, and all bids exceed the budget. The natural tendency is to delay making these awards in anticipation that somewhere, somehow, a subcontractor or vendor will appear magically with a price that matches the budget. This seldom, if ever, happens. When four or five qualified subcontractors submit pricing that exceeds the estimate, it is logical to assume that their bids are correct and the estimate is not. If these overbudget items of work are not dealt with on a case-by-case basis, and analyzed and finalized in a timely manner, additional costs will occur since job progress will likely to be affected by these delays.

These are difficult problems to deal with, and there are no easy answers to them, but positive steps must be taken to resolve them promptly in order to mitigate potential losses. Placing the problem at the bottom of the stack accomplishes nothing.

Do Your Subcontract Agreements Include These Key Provisions?

All subcontract agreements have legal consequences, and any alterations to existing subcontract formats should be reviewed with the company's legal counsel. But the following offer some suggested provisions:

- If not already a part of the subcontract agreement, insert a clause binding the subcontractor to the contractor by the same terms and conditions of the contract between the owner and general contractor. Although this provision gives the subcontractor the right to be privy to the terms and conditions of the general contractor and client contract, this clause is worthy of inclusion in the subcontract agreement. It thus enhances the "flow through" concept, as it relates to "paid when paid"—the general contract can withhold payment from the subcontractor if payment is not received from the owner. If the owner is still reviewing a subcontractor's change order and has not yet approved it, the general contractor does not have an obligation to pay the sub for work included in that change order. If there are penalties for failing to meet the construction schedule, and the subcontractor is at fault, any such penalties assessed by the owner can be passed on to the subcontractor. There are other reasons why the general contractor would like to apply flow-through owner contract provisions to the subcontractor. Think of a few.

- A performance clause that affords the general contractor more control over the flow and pace of the subcontractor's work is critical. This type of clause generally stipulates "The Subcontractor agrees to commence and complete the Subcontractor's work by the date or within the time frame set forth in Article X (the article that includes the project start and completion date) and to perform the Subcontractor's work at greater or lesser speeds and at such times and in such quantities as, in the Contractor's judgment, is required for the best possible progress of construction of the job or as shall be specifically requested by the Contractor, and the Subcontractor shall so conduct the Subcontractor's work as to facilitate and so as not to interfere with or delay the work of the Contractor." This clause effectively gives the contractor control over the pace of the subcontractor's work, which can then be directed to be "sped up" or "slowed down." However, one must keep in mind the construction schedule, which if accepted by the subcontractor may make enforcement of any "speed up" or "slow down" requests more difficult. The subcontractor may take the position that they have been directed to deviate from the initial (baseline) schedule to which they were bound contractually. One way to possibly avoid such a dispute or claim by the subcontractor is to issue a revised construction schedule with a revised time sequence for their work and have them accept this revision with no time or cost impact.

- A nonperformance clause is as essential as a "performance" clause in the subcontract agreement. "Should the Subcontractor fail to prosecute the Subcontractor's work or any part thereof with promptness and diligence, or fail to supply a sufficiency of properly skilled workers or materials of proper quality and/or quantity or fail in any other respect to comply with the contract documents, the Contractor shall be at liberty after seventy-two (72) hours (*or 48 hours if you so chose*) written notice to the Subcontractor to provide such labor and materials as may be necessary to complete the work, and to deduct the cost and expense thereof from any money due or thereafter to become due to the Subcontractor under this agreement."

- An article dealing with changes and change orders is also important. Although most subcontract agreements require that the general contractor provide written authorization for a subcontractor to proceed with extra work, some subcontractors will do so only if they have established the cost of such work *before starting it*. A general contractor can always fall back on the Construction Change Directive (CCD), which is basically a time-and-material instrument, or they can insert a suitable clause in their subcontract agreement. "Should the Contractor and the Subcontractor be unable to agree as to the amount to be paid or allowed because of any alteration, addition, or extra work, if so ordered to do so in writing by the Contractor, the work shall go on under the order or orders of the Contractor with the understanding that the *reasonable* value of such work shall be paid or allowed, and in case of the failure of the parties to agree, the determination of the amount to be paid or allowed, as the case may be, shall be referred to arbitration as hereinafter provided." *This assumes that an arbitration clause is part of the subcontract agreement.*

Speaking of change orders, there is often confusion about the authority to authorize change order work and authority to approve related costs. Consider the following clause:

> Any extra work must be authorized by the contractor's "office" not the contractor's "field." The "field may only verify that work has been performed, not be responsible as to payment."

Job cleaning, as previously discussed, is often a contentious issue at the site, but a clause included in the subcontract agreement may make it somewhat easier for the general contractor to keep the project clean. Subcontractors often claim that they require *a written* demand to clean up their work before a general contractor can claim them in default and backcharge their account for all such associated costs. Occasionally, the project manager or project superintendent doesn't have the time to prepare a written notice. Therefore, just insert the following paragraph in the subcontractor's agreement:

> The Subcontractor shall at all times keep the job site orderly and free from dirt and debris arising out of the Subcontractor's work. At any time, upon the Contractor's request, the Subcontractor shall immediately clean up and remove from the job site anything which it is obligated to remove hereunder, or the Contractor may, at its discretion, and *without* notice, perform

or cause to be performed such clean up and removal at the Subcontractor's expense.

Note: Even though no written notice is required, the Subcontractor may challenge the verbal one, conveniently stating that they were "unaware" that such direction was given. So to be safe, even though a verbal directive was issued as required by the subcontract agreement, follow up with a written one.

- Subcontracting by the subcontractor, without advance notice to the general contractor can present problems such as the lien waiver situation previously described. Also the general contractor must be aware of the identity of all subcontractors working on the site for safety and security reasons. To lessen the chance that these types of situations will occur, insert the following in the subcontract agreement:

  The Subcontractor shall not subcontract or delegate all or any portion of its work, nor shall it assign any amounts due or to become due or any claim or right arising in connection with the subcontract agreement without the prior consent of the Contractor." (Alert the project superintendent to this subcontract provision so they can be sensitive to the presence of any lower-tier subcontractors working onsite without the general contractor's knowledge and approval).

- Payment issues are always important and the "pay when paid" clause is standard issue in most subcontract agreements. Of late, there has been increased resistance to this method of payment. Some states have passed legislation that voids this "pay when paid' provision, stating that it is not in the public interest. In some tight labor markets, subcontractors are demanding to be paid when their work is installed and accepted, not when the general contractor receives payment from the owner. Various subcontractor organizations have mounted campaigns to eliminate this clause from subcontract agreements. If your company has a "pay when paid" clause in the subcontract agreement, review any recent court rulings to determine whether this provision is still enforceable.

- Delays frequently occur in construction projects and these delays may impact subcontractors who share no responsibility for the delays but may have incurred additional costs due to the actions of others. These "damages," generally referred to in legal terms as *consequential damages*, include such items as loss of productivity, working out of planned sequence, extended corporate overhead, an inability to procure other work because their supervisors are tied up on the delayed projects, and so forth. To fend off such potential claims, insert a "no damages for delays" clause in the subcontract agreement, such as:

Contractor shall not be liable to Subcontractor for any damages or extra compensation that may occur from delays in performing work or furnishing materials, or other causes attributable to other subcontractors, the owner, or any other persons."

Relationships in this business are very important, and good relationships with a nucleus of knowledgeable, hard working subcontractors working in an ethical environment will produce buy-outs that are beneficial to all parties.

The Change Orders

The change-order process can be one of the most difficult aspects of the entire construction undertaking, but it doesn't have to be that way. Of course, it's almost inevitable that at some point change orders will be issued during a construction project, but the manner in which the change orders are proposed, submitted, and resolved can either be contentious or it can be a fairly smooth operation.

The AIA A201 General Conditions document defines a change order as a written order to the contractor, signed by the architect and the owner, that's issued after the contract has been executed, authorizing a change in the scope of work or an adjustment in the contract sum, contract time, or both.

The contentious part of the process, however, involves three basic elements:

- An interpretation of what constitutes a change in the scope of the contract work
- The cost of the work
- How this change or changes affect the completion time of the project

Owners and their design consultants seem to have the following problems with contractors' change orders:

- Contractors don't clearly define the nature of the change order when submitted except when it is in response to an A/E-generated revision, complete with drawing revisions/specification changes or detailed instructions.

- The contractor does not provide sufficient cost information with detailed breakdowns, thus allowing the reviewing party to thoroughly examine and understand all costs associated with the change.

- The contractor (project manager) has not scrutinized all accompanying subcontractor/vendor proposals to insure that they have correctly identified the changes, that the requests are legitimate, the costs reasonable, and that the terms and conditions of the contract, such as allowable overhead and profit percentages, have been met.

Each of these elements that cloud the change-order process, if approached professionally, will result in a prompt and equitable closure. If approached haphazardly, a feeling of mistrust among the participants will develop and pervade the entire project. Therefore, this chapter deals with the procedures that allow for effective change-order processing, and points out the many pitfalls to avoid.

A Cardinal Rule

An important rule-of-thumb when preparing a change-order request is to first change your perspective. View this request as though you were the recipient and *you* were being asked to pay out a substantial amount of money for this change in work. Would you approve such a request?

Starting off on the right foot

Establishing the proper procedures for the submission of change orders needs to be presented at the beginning of the project. Even though the project manager has successfully worked with the owner and their design consultants, as well as various subcontractors on previous projects, it's wise to review the steps required to present and implement a change order. At the commencement of the project, the project manager can review the *Protocol for Change Orders* referenced in Chap. 6 with all the subcontractors, and let everyone know what is to be expected if they intend to respond to a change-order request.

Change-order work will generally involve the following:

- Changes in scope requested by an owner, design consultant, or a general contractor or subcontractor for which a lump-sum proposal will be required

- Time and material work requested by an owner for additional work

- Premium time costs for work that must extend beyond normal working hours

- Winter conditions, if not included in the contract as a lump sum, or included as an allowance item requiring documentation of costs prior to reconciliation

Reviewing the Important Contents of a Change-Order Request

The following portions of any change order should be reviewed:

- Each proposed change order should contain a brief explanation of the nature of the change, and the party initiating it (for instance, the owner, A/E, contractor, subcontractor). All supporting documentation such as letters from the owner, drawings/sketches from the A/E, requests from the subcontractor, and so on should be attached.

- If the scope of work is increased or decreased, the prior condition and revised conditions should be stated (for example, 20-lf railing added between Cols 9–10 per A-3.4 dtd 06/04/06: 5-lf railing indicated on A-3.4 dtd 04/23/06—add of 15 lf).

- All costs submitted for self-performed work should be broken down into labor hourly rates multiplied by the number of hours required. The detailed breakdown of each hourly wage rate should include the burden for trades involved. Materials should be listed in lineal or square feet, as appropriate, and the unit cost for each should be noted (attach copies of vendor invoices if applicable). Allowable overhead and profit should be added as a separate entry, not included in each line item.

- Equipment costs should indicate whether the machines are contractor-owned or rentals. The hourly rate and the number of hours idle or active should be listed. Signed receipts documenting the delivery and return of the equipment should also be included.

Note: Some owner contracts stipulate that if an hourly rate is used but equipment is rented for more than 5 hours, the daily rate will apply. Likewise, if the equipment is used for more than three days, a weekly (not daily) rate will be applied. Read the owner's contract to determine if such a provision is included.

One further note about equipment: There can be both active and idle equipment rates for contractor-owned equipment. Idle rates for contractor-owned equipment will not only reflect deducting costs for the operator and fuel, but should also include a deduction for maintenance since the equipment is not operating and therefore will require less maintenance.

- If work is time and material, follow all the previous procedures.

- If requested by the owner, the owner's representative may elect to be present when change-order negotiations with subcontractors/vendors take place.

(This last statement will provide the owner with some assurance that all such negotiations are open and above board—if they had any doubts to the contrary.)

Time and material work

The project superintendent should obtain daily tickets for all self-performed T&M work and sign a copy of each ticket indicating acceptance of the number of hours worked, the number of workers, and the equipment provided (both active and idle time and the materials consumed).

For subcontracted work, daily tickets from each subcontractor should list the number of hours each tradesman worked and the task performed. These tickets must be signed by the project superintendent at the end of each day. Tickets should include the equipment and materials used and a brief description of the work task.

It is imperative that these tickets be signed at the end of each day when the work was performed and not gathered together at the end of the week for signing. Any discrepancies noted daily may have been forgotten by week's end.

On both self-performed work and subcontracted work, a description of the work being performed is not only necessary for the owner's edification but for the contractor as well. Some proposed change-order requests linger at the architect or owner's office for months and when they are finally reviewed, raise a number of questions that tax memories: "What exactly was that carpenter doing for the 32 hours you're billing us for on October 23, 24, and 25?" "Where was that underground conduit extension that was required but not shown on ES-1.1?" A brief description of work included on every daily work ticket will prove invaluable in explaining work performed four months earlier that's just now being reviewed by the owner.

The verbal authorization to proceed

Although verbal authorization to proceed with change-order work is used frequently, written confirmation must follow promptly to avoid any misunderstandings of what was authorized and their associated costs. And there's no time like the present to do so. If confirmation is not forthcoming from the architect within a reasonable period of time, the project manager can prepare a short and simple letter to the owner/architect such as:

As of (*date work was authorized*), (*general contractor*) is proceeding with (*work task*) based on your authorization to do so. The cost of this work, as mutually agreed is $_____ (*or state that it will proceed on the basis of time and material*) and includes *X* percentage for general contractor's overhead and profit.

It does not appear that this work will have any impact on the construction schedule (*or if it will have an impact but the number of days can't currently be determined, state "The impact on the schedule cannot be determined at this time.* We would appreciate receiving your formal change order at your earliest convenience".

The project manager should remember that although there has now been a verbal authorization to proceed, followed by a written confirmation, it is still important to obtain that formal change order from the architect. Although the extra work may be completed promptly, it cannot be included on a requisition for payment until the formal change order has been issued and signed by all parties. For that reason alone, it is prudent to press for prompt processing of the formal change order.

What Constitutes "Cost"

Certain elements of change-order preparation require some rethinking. One of which is: What constitutes allowable, reimbursable costs?

Article 7 of the AIA 201 document is a good guide to follow and states that *costs* include

- The cost of labor, including fringe benefits
- The cost of materials, supplies, and equipment, as well as the cost of transportation
- Rental costs of machinery and equipment, excluding hand tools, whether rented from the contractor or from a rental agency
- The cost of premiums for bonds and insurance, permits, fees, sales, and use tax
- The additional cost of *supervision* and *field office personnel* directly attributable to the change

Other considerations can also impact cost, such as:

- What effect will this work have on the completion of the project and will any general conditions costs be affected?
- Will the work in question require premium-time labor to complete in order to keep pace with the schedule, or can it be accomplished during normal working hours with the manpower available?

- Will the materials or equipment to be incorporated into the work be unloaded and distributed throughout the building? If so, what costs will this entail?

- Will additional drawings or other reproducibles be required, and will they need to be distributed to subcontractors and/or field personnel so this work can be performed? If so, the costs for these reproducibles and any express delivery costs are legitimate costs and should be incorporated into the change order.

- Will additional winter protection be required to complete this work, and if so, what are the costs?

- What about the additional costs of the utilities required to complete the work? (For example, if the building has reached the point where permanent utilities are operable and the contractor is paying the cost of their operation, the daily cost of these utilities will be substantial. A change order at this point should thus include costs to cover a portion of those utilities, particularly if the contract time will be extended.)

- What about the cost of expensive, depreciable tools such as diamond blades for masonry and concrete cutting? If the work involved requires considerable cutting, a $500 diamond blade may have its useful life shortened considerably.

- When hoisting or lifting equipment is required, remember to add the cost of the equipment and the wages of the operator. Also, don't forget fuel costs in this age of soaring gas and diesel prices.

- Consider whether additional cleaning costs will be incurred and what the added costs will be to dispose of any debris generated by the change-order work.

The project manager may wish to develop a checklist of all potential costs to be considered when preparing a change order to insure that *all* costs are incorporated in future change orders.

Completion Time and the Change Order

Is the nature and timing of the change-order work apt to affect the contract completion date of the project or impact any subcontractor's work schedule? Will the work extend the completion date or will it actually decrease construction time? In the case of completion time, even if a liquidated damages clause is not part of the contract, the general conditions portion of the budget may be impacted, causing delays of subcontracted work not associated with the change-order work.

Consequential damages

Although *consequential damages* may have been excluded from the contract with the owner and from agreements between the general contractor and subcontractor, a brief discussion of the subject will illuminate some concepts important to change-order work.

There are several types of costs involved in delayed completion time: direct costs or losses, and indirect or consequential damages or losses. Direct costs are rather easy to understand: the increased cost of rental equipment, increased salaries of management personnel stationed on the jobsite, extended field office expenses, and so forth. But *consequential damages* are not so easy to define. They include an owner's lost profits due to late delivery of their project, increased interest on their construction loan, perhaps increased rental at their current location, taxes, and so on.

A general contractor or subcontractor may also incur added expenses when the project's completion is delayed through no fault of their own. A contractor or subcontractor may be prohibited from bidding on new work because their management staff is still tied up on the delayed project or their bonding capacity is affected, thus their corporate overhead will also be impacted and their reputation possibly tarnished.

Will contract time remain the same, be reduced, or be extended?

There is a line item in AIA Document G701 (change-order form) that refers to contract time: *The architect is to confirm that the contract time will be increased, decreased or remain the same with the execution of this change order.* A space is included for the contractor to insert the number of days for the anticipated increase or decrease due to the proposed change in work. Another line directly below that one has a space in which to insert a revised date of substantial completion, if that date will change. The project manager should include in their proposed change-order request to the architect any potential for change in the contract completion time. If unknown at the time, the change-order proposal submitted by the project manager can write in that space "To Be Determined."

Most architects are reluctant to sign a document with such an open-ended condition—understandably so. But at that time, the contractor doesn't really know if the change order *will* impact the schedule.If that "TBD" is not acceptable to the architect, insert a number of days that will be more than adequate.

Theoretically, any increase in contract completion time, if accepted, can result in a contractor's request for extended general conditions costs covering the period of time for that extension, such as:

- Additional field supervision

- Additional field office expenses, including field office rental and supplies

- Added temporary heat and/or power

- Added temporary sanitary control

- Additional cleaning costs and debris removal associated with the work

When a request for an extension of time is strongly debated by the architect, try another approach. Get the architect to agree to the job completion extension time with the proviso that, if the project is complete within the original contract time frame, no claim will be made for extended general conditions. If the completion extends beyond the contract date, the contractor will receive a reimbursement of the general conditions for the cumulative time extensions included in the change-order proposal(s).

Get it in writing.   One word of caution about requesting time extensions, members of the project team can change jobs during the life of a construction project. Although the project manager may have developed a harmonious working relationship with the original Owner-A/E team, new faces may appear, and previously friendly relationships can turn cool or outright cold. If the architect or owner's representative had previously stated that formal time extensions were not necessary, because when the time came they would agree to them if need be, the new "team" may completely disavow any previous verbal commitments. To avoid such situations, it's wise to prepare for the worst and request confirmation of all time extensions if, and when, they are justified. A project manager can always use the "what if I'm hit by a car" scenario so as not to imply that the owner-architect representative can't be trusted. Simply stated, if there is mutual agreement that a job extension is warranted, it should be so indicated in the change order. After all, suppose one of the team members *is* hit by a car tomorrow and there *is* no documentation to support the verbal commitment?

Small-tool costs

Some subcontractors submit proposals in which they include a "small tools" expense as a function of a tradesman's hourly rate. Some are rather generous with these numbers, perhaps adding $2.55/h \times 40$ hours for a total of $102/week. That buys an awful lot of hammers and screwdrivers, and both owners and architects are sensitive to charges like this. There is, of course, some validity to "small tools" charges if they're reasonable and proper.

Will some hand tools need to be purchased specifically for this additional work? Generally, hand tools are excluded from "reimbursable" costs, except where a case can be made for reimbursement of special tools needed for the change-order work in question.

Residual value of tools and equipment.   When a cost plus or GMP type contract is in effect, the purchase of some expensive power tools or other equipment may be required for the work. Depending upon the length of the project and the use to which these tools or equipment have been put to use, they may have considerable residual value.

Power tools have limited life but may have some residual life after the job is complete, so a cost representing depreciation of the power tool is a reasonable one to be included in the change order as a cost of work.

If a contractor decides to purchase such tools or equipment, and the purchase price is included in the change order, then a residual value should be established and included in those costs. For example, if a miter saw is required for some detailed millwork, the change-order entry for this equipment might be entered as follows:

Makita—Model 485 Miter Saw $485

Less residual value   100

Cost of equipment in this PCO $385

If this procedure is rejected by the owner, forego the purchase and rent the equipment, possibly at a higher rate to that owner.

When there's a chance the extra work might extend over a lengthy period of time, the general contractor may consider purchasing a tool on a lease-purchase agreement and then charging the owner sufficient rental costs which would ultimately pay for the tool.

What costs other than bricks and mortar should be considered?

Building permit and bond costs.   Building permit costs will generally grow as the cost of a project increases from the value included in the initial permit. Many building departments monitor the cost of a project as it proceeds and establish procedures that require builders to pay additional fees when costs increase significantly. It is rather easy to establish a percentage of the total cost of each change order in anticipation of having to pay increased permit costs when the project has been completed.

The same is true of bond costs. The cost of the initial bond will rise incrementally as change orders increase the original contract sum. At the end of the project, the bonding company will review the total project costs and assess additional premiums based upon those increased costs. A call to the bonding company will determine the percentage to be added to each change order to cover these added project costs.

Temporary utility costs.   When substantial completion of a building has been reached and the architect issues a Certificate of Substantial Completion, the building's temporary utilities—gas, water, electricity, and telephone systems—are transferred from the builder's account to the owner's, unless indicated otherwise in the contract. Change-order work authorized after this transfer of utilities will obviously have no impact on light, heat, or power costs.

But when substantial completion is near but not achieved, the project manager must assess the potential for increased heating/cooling or power costs if change-order work is requested that may extend the original completion date of the project. At that point in time, whether it be summer or winter, the permanent heating and cooling systems will be in full operation and, depending on the size and nature of the project, utility costs could be running at $1000 a day—or more.

Get together with the owner and their architect and establish a daily rate for utility costs by presenting invoices from each local utility company in order to substantiate those costs. And don't forget the cost of those fasteners—nails, screws, and expansion bolts! More than one project manager has been shocked to find a $2000 bill for stainless steel expansion bolts purchased for change-order work which was not included in the cost of the work.

What overhead and profit fees can be included in change-order work?

The application of allowable overhead and profit percentages to change orders is normally established in the bid documents and carried through to the contract. Typical restrictions, frequently included in the general or supplementary conditions of a contract, or in Division 1 of the specifications, will state

> The undersigned (*general contractor*) agrees that the total percentage for overhead and profit which can be added to the net cost of the work shall be as follows:
> For work performed by the general contractor's own forces _____%
> For subcontracted work _____%

Some requirements limit the percentage of overhead and profit on a sliding scale.

> For the contractor, for work performed by their own forces:
> Up to and including $100K   allow 15%
> $101K to and including $200K   allow 10%
> $201K and over   allow 5%

The same fee structure usually applies to subcontractors, but with additional restrictions on allowable fees that can be added by their lower-tier subcontractors, such as:

> For each subcontractor, for work performed by the subcontractor's subcontractor, 5 percent of the amount due the sub-subcontractor.

There may even be a provision that limits the *total allowable* overhead and profit percentage for the entire change order, such as: *Total fees to the project are limited to 20 percent.*

A project manager should review all subcontractor and vendor proposals to ensure that they have complied with the "contract" overhead and profit percentages in order to avert questions from the owner and architect if any OH&P percentages exceed those limits. The architect may also insert a clause in the specifications that dictate the way in which overhead and profit is to be computed on certain types of change orders.

When credits and charges both apply

A statement may be included dealing with the method of establishing net cost when changes in the work result in cost increases or credits. Scrutiny of subcontractor proposals should include their method of applying overhead and profit percentages. Some subcontractors will add overhead and profit to the add portion of their proposal before deducting their credits, which will surely be detected by the architect. If no such procedure is established in the contract, and even if it is, this information needs to be communicated to subcontractors to avoid future misunderstandings.

The construction change directive—The CCD

Article 7 of the 1997 edition of the AIA General Conditions document A201 contains a term that first appeared in the 1987 edition: the Construction Change Directive.

This directive is an order prepared by the architect and signed by the owner directing a change in the work affecting the contract sum or contract time, or both. The CCD allows the architect to authorize extra work when there is either no time to compile and submit costs or when there is a lack of agreement on the terms of the change order, which usually means a disagreement over cost.

The cost of the work to be incorporated into the CCD is to be determined by either a mutually accepted lump sum, by any unit prices in the contract or by "costs to be determined in a manner agreed upon by the parties," which is loosely interpreted to mean a negotiated sum.

The distinct advantage of this CCD provision is that it breaks the logjam when neither the architect/engineer nor the general contractor can agree on a cost of work but the change-order work must proceed. In such cases, the general contractor can start the extra work and provide the architect with documented costs that include

- The cost of labor including all fringe benefits
- The cost of materials and equipment, including transportation expenses
- Rental costs of machinery and equipment, *excluding* hand tools, whether rented from the contractor or others

- Costs of premiums for bonds, insurance, permits, and fees
- Additional costs of supervision and field-office personnel directly attributable to the change

This article specifically states that in the event of an increase in one item and a decrease in another, the contractor's overhead and profit shall be calculated according to the net increase. Another advantage of performing work under the CCD guidelines is that the architects will accept additional *field supervision* costs which they might otherwise question if components of a lump sum proposal are submitted.

Be alert to other contract provisions relating to change-order work

Standard contracts modified by the owner can include provisions relating to change-order work in sections that appear to be unrelated to the change-order work—yet another reason to become familiar with all the provisions of the construction contract.

A "should have known" provision can be tricky if worded like the following:

> The contractor has constructed several projects of this type and has knowledge of the construction and finished product. The contractor shall immediately notify the architect and owner of any details that do not meet good construction practices. By proceeding with the work, the contractor indicates that all details, construction procedures, and materials shown or specified in the contract documents are consistent with sound, standard, and acceptable practices.

Try to get a reimbursement for some extra work with this provision in the contract! Then there's the provision that doesn't allow the contractor to request extended general conditions when being granted a time extension:

> Any extension of time in which to complete the work that has been granted by the owner for items beyond the contractor's control shall be the contractor's sole remedy for any delay, hindrance in performance of work, loss of productivity, impact damages, or other similar claims.

This effectively bars any requests for relief of damages that may surface from subcontractors impacted by these changes. This is yet another reason for the pass- through provision in the subcontract agreement that will pass this owner-imposed restriction to the general contractor.

Public Works and the Change-Order Process

A number of government agency equitable adjustment contract clauses (change orders) exist, which fall into 10 basic categories:

- Change orders to the work issued by the owner with full agreement by all parties regarding time, cost, and scope

- Constructive changes—changes that are accidental or unintended (for example, correcting a defect in the contract drawings or specifications)

- Differing or changed site conditions

- Suspension of work by the owner

- Constructive suspension of work—acts or omissions by the owner which have the effect of unreasonably delaying the contractor's work

- Delays (explained more fully later in this chapter)

- Acceleration—owner demands contract completion, but recognizes that justified delays have occurred to warrant an extended completion date

- Constructive acceleration—an unintended shortening of the completion time

- Termination for convenience—where the owner can terminate a project prior to its completion or delete major portions of the work

- Termination for cause or default—generally due to poor or nonperformance by the contractor

When work is performed for municipal, state, or federal agencies, additional restrictions, qualifications and procedures can be expected. Some governmental authorities will honor a change order only if one of the following conditions exist:

- The time required to complete the project is to be increased because of conditions beyond the control of the contracting parties.

- Revisions in the plans and specifications, ordered by the local authority, will result in an increase or decrease in the cost of construction.

- Unpredictable underground or superstructure conditions are encountered which will increase or decrease the cost of construction.

Some government agencies may only accept change orders that result in project "betterments," thus it is incumbent upon the contractor to prove that the change order request meets that criteria. The project manager must carefully read the "boilerplate" in all public works contracts to familiarize themselves with the terms and conditions required for change-order work.

Roadblocks to Acceptance of Change Orders

Why is the change order process so fraught with problems and frustrations? Aren't change orders inevitable in the design and construction process?

After all, what could be simpler: a change in the work under contract is requested, the nature of the work is discussed, a price is negotiated, and the work is completed and paid for on the next requisition. However, each participant in the change-order process has their own perspective on these events.

The owner's perspective

Owners of construction projects select a team of design consultants to turn their building program into a set of plans and specifications. A contract is then awarded to a builder to build the project. The owner has bargained for a complete package, so why is the builder submitting change orders when the owner has not made any changes to the project? Sophisticated owners know the answer, even if they sometimes refuse to acknowledge it. The plans and specifications are not perfect and a contractor may not be aware of those imperfections until various components begin to come together as construction progresses. If known at the time the project was estimated, these drawing deficiencies would have, most likely, generated a series of RFIs to the design consultants, and also probably added to the cost of the work. So it's almost like a "pay me now or pay me later" scenario. Although a few contractors have exploited the change-order concept to their advantage, most contractors would be content to never issue a change order unless the owner has directed them to change the initial scope of work.

When owner resistance to change orders appears, the contractor needs to clearly explain the events that created the need for this additional work.

The contractor's perspective

The contract prepared by the owner will usually include a general conditions exhibit, and that exhibit will frequently be AIA A201-General Conditions of the Contract for Construction. This document, prepared by the owner's architect, is specific as to the contractor's obligations.

Article 3.2.1 requires the contractor to "carefully study and compare the various drawings. These obligations are for the purpose of facilitating construction and not for the purpose of discovering errors, omissions, or inconsistencies . . . however, any errors, inconsistencies, or omissions. . . shall be reported promptly to the architect."

Article 3.2.2 goes on to state that "the contractor is not required to ascertain that the Contract Documents are in accordance with applicable laws, statutes, ordinances, and building codes."

And following through with Article 3.2.3, the architect states that any added costs that arise from the contractor's review and queries to the architect subsequent to these reviews may be subject to a contractor's claim for additional compensation.

Thus, the architects, acting in their capacity as the owners' agents or representatives, have made a clear case for the creation of change orders if any of these criteria are met.

On the other hand, the contractors have represented themselves as professionals, and some of these errors, omissions, and inconsistencies that should have been picked up by a responsible contractor should probably not be presented to an owner as change-order work.

The architect and engineer's perspective

In the preparation of plans and specifications for today's complex project, by constraints of time and/or money, or both, it is nearly impossible for an architect and engineer to produce that *perfect* set of documents. The language in the standard contract documents prepared by the American Institute of Architects recognizes these limitations.

Some "errors and omissions" are minor in nature—for instance, a door shown on a floor plan as a 3070, but included in a Door Schedule as a 2868. Other design deficiencies may be more significant in scope and cost and the contractor should request clarification before determining whether a claim for an adjustment to the contract price is warranted. A whole series of minor inconsistencies throughout the drawings may have an impact on costs, but a single one may not. Thus, both contractor and architect need to review these types of matters in their proper context.

The contractor must show the architect that they will approach such matters with a degree of reasonableness, and the architect, in turn, must grant the contractor the same degree of reasonableness when proposed change orders are prepared due to design deficiencies.

That hard-working, conscientious architect, aware that there may be some missing details or missing dimensions in their plans and specifications, needs to be protected from that unscrupulous contractor who might be waiting to exploit these document shortcomings by generating lots of change orders. By the same token, that hard-working, conscientious contractor needs proper recognition from that less-than-perfect hard-working design consultant.

What is the solution?   If owner, architect, and contractor can agree that problems will arise relating to the quality of construction documents, then the first hurdle in resolving these problems will have been broached. It would appear that the best solution to some of the controversies that arise over change-order work require that the owner, architect, and contractor be *reasonable and responsive* in their approach to these problems of extra cost.

Is it reasonable to assume that it is the contractor's responsibility to ensure that the mechanical and electrical drawings have been properly coordinated with each other, as well as with the architectural drawings? If during the coordination process when construction is underway, it's determined that the ductwork, sprinklers, and ceiling-mounted recessed light fixtures can't fit in the space above the proposed ceiling, and the ceiling can't be lowered, is this not the basis for a *reasonable* change-order request? On the other hand, if additional window blocking is required for proper and secure window installation and this blocking detail is either missing or not sufficiently shown on the drawings, is this a legitimate claim for an extra from the contractor?

The answer to the first question is "Yes." The contractor should not be held responsible for the coordination of multidisciplinary drawings when a *reasonable* amount of above-ceiling space has not been allotted to accommodate all utilities.

The answer to the second question is "No." It does not seem *reasonable* for an architect to precisely detail window blocking, or roof blocking or other similar types of blocking since different suppliers have different requirements and an experienced and competent general contractor ought to be familiar with the required blocking details. Similar questions regarding responsibility should be viewed with an eye to what is *reasonable*—and when *reasonable* parties come together, some of the difficulties in implementing and approving change-order work may be eliminated.

Can trade-offs help?   The project manager should be sensitive to the fact that the architect does not relish advising their client that extra costs may be involved because of design errors, omissions, or inconsistencies in the drawings or specifications. Compromise solutions may resolve some of these problems.

One approach that an architect is often willing to accept is the "trade-off" solution. If the contractor is able to substitute another material or product of similar quality, but that material or product is somewhat less expensive and the savings would offset the added costs in the proposed change order, is the architect amenable to this solution? It is important that the contractor present these proposed changes or trade-offs honestly. If, in fact, the substitution is of lesser cost, but still retains the quality level the project requires, the contractor could possibly offer a slight credit using the balance of the savings to offset the added costs.

But sometimes when a reasonable approach does not prevail, the project manager must adopt a different tack when a proposed change order meets stiff resistance from the owner and architect. Using legal concepts in a nonconfrontational approach may be helpful.

Two legal terms might be offered up when an architect displays reluctance regarding the added cost of change-order work. The first is *quantum meruit*, and the second is *unjust enrichment*. Both present the project manager with two added approaches to resolving change orders.

Quantum meruit. The Latin term *quantum meruit* (pronounced "quantum mare-o-it") is also referred to as a "quasicontract" method of recovery of costs associated with change orders. Both terms refer to the fact that this additional change-order work benefits the owner even though the extra work does not meet the *specific contract language* required to deal with the procedures for change-order approval. The contractor must prove that the owner has, in fact, benefited by the work that was completed and accepted by the owner. Recovery of costs is based on the owner's implied promise to pay for the benefit received by the incorporation of work into their project. If there is a dispute between the owner and contractor over change-order work, and whether it was performed within the terms and conditions of the contract, the contractor may be able to base their claim for reimbursement on the quantity and quality of the extra work and the benefits that accrued to the owner after the additional work was completed.

Unjust enrichment. Requesting reimbursement for change-order work utilizing the theory of unjust enrichment is based upon a contractor claim that the owner has been enriched by the work performed by that contractor. The concept of "You can't get something for nothing" is the essence of unjust enrichment. If the project manager proceeds with change-order work based on a good-faith verbal commitment from either the architect or the owner, and this oral agreement for extra work is not confirmed in a letter or memorandum, how can the project manager proceed? Now that the work has been completed, the project manager's request for a change order is met with silence from the architect or owner who doesn't recall authorizing the work or doesn't recall any such high price being quoted. With no written authorization to document the approval of extra work, the project manger is now rightfully concerned about receiving payment for the cost of the work and their fee.

Hypothetically, let's assume that this change involved upgrading the specified birch hollow core doors to oak veneer, solid core doors. If 10 doors were involved, for example, and the additional cost of each upgrade was $200, the total cost of this extra work would be $2000 plus contractor mark-ups. The owner, very clearly, has had his project enriched by this contract modification, and value has obviously been added by the nature of this change in the work. If it appears that no change order will be forthcoming from the owner, the project manager should familiarize them with the concept of *unjust enrichment* while sending them a letter requesting a change order for the work.

The owner may argue the costs associated with this change and require documentation to substantiate every aspect of the $2000 plus change, but the argument that value has been added to the job by virtue of this change cannot be denied or dismissed. This is another approach to be pursued in order to get that written change order issued and approved.

Betterments and enrichments. The two terms, *betterments* and *enrichments*, are somewhat similar to the unjust enrichment concept in the previous paragraph, except that they're likely to appear in some public works contracts. Therefore, the project manager should be familiar with their implication, which may find applicability in private sector work.

The example of the upgrade from birch doors to oak doors would be considered a *betterment*, and thus subject to the acceptance of a change order. However, if the fire marshal or building inspector conducted an inspection of the structure prior to the issuance of a certificate of occupancy and requested that additional exit or emergency lights be installed, along with more fire alarm pull stations or fire or smoke detectors, the general contractor may have difficulty submitting a claim for additional costs for this work in the public sector.

Remember that the contract provision limiting the contractor's responsibility for compliance with local building codes was in the general conditions contract generally attached to private sector contract work. Public sector work may not use this same AIA A201 document, and therefore their contractor responsibilities may not include such provisions.

The public owner may interpret this additional fire protection work to be such that it does not make the project any "better," but was required simply to comply with the local ordinances as interpreted by the local officials. The contractor, on the other hand, will certainly argue that these extra cost items are indeed *betterments*. Without these changes, a Certificate of Occupancy (C of O) would not have been issued, the building would not have been habitable, and would therefore have no value at all. The project would be *enriched* only when the Certificate of Occupancy was issued, and that would occur only after the extra work was completed.

Liquidated Damages and the Change-Order Process

When change orders increase the completion time of the project, extra costs may be warranted for such items as extended general conditions. Therefore, owner recognition of an extended schedule is important. When a liquidated damages (LD) clause is included in a contract with the owner, recognition of an extended schedule is critical to avoid costs imposed by the owner for late delivery of their project.

A liquidated damages clause stipulates that a certain monetary payment will be paid by the general contractor for failure to deliver the construction project on the date of completion set forth in the contract or adjusted by the change order. These payments are generally computed on a daily basis—for example, liquidated damages of $5000 per day will be assessed against the general contractor for every day work goes beyond the date of completion stipulated in the contract or adjusted for by any change orders. Any such extensions of time granted by the architect and owner and included in approved change orders will be taken into account in these calculations.

A typical liquidated damages clause

A liquidated damages clause will usually be inserted into the paragraph in the contract that refers to "contract time." It will likely be similar to the following:

> If the work is not substantially completed in accordance with the drawings and specifications, including any authorized changes, by the date specified above, or by such date to which the contract time may be extended, the contract sum stated in Article (*whichever article contains the contract sum*) shall be reduced by $_____ (*the daily value of liquidated damages*) as liquidated damages for each day of delay until the date of substantial completion.

It is important to note whether the contract defines "day" as a calendar day or a workday.

Although the intent is to have the contractor pay the owner for damages incurred due to the late delivery of the building, the liquidated damages clause states that the total LD amount will, in actuality, be deducted from the final payment, since it has the effect of decreasing the contract sum.

The purpose of the liquidated damages clause. Some contractors view the liquidated damages clause as a penalty clause—a monetary slap on the wrist for late delivery of a construction project. But an LD clause in a contract is not a penalty, and if included as such, it must include a bonus clause for early completion of the project, or else in the eyes of the law, the LD clause is voided. The purpose of the LD clause is to compensate the owner for late delivery of the project without having to pursue a legal recourse to collect damages.

For example, an office building developer may have signed leases affording tenant occupancy on January 1st, but because of late delivery of the project by the contractor, occupancy must be extended to January 21st. Obviously, the landlord has been deprived of income for this three-week period, and the LD will compensate them for this loss. The tenant may have had to extend their old lease and would expect to be reimbursed for those costs.

One of the key elements of the liquidated damages clause is a definition of what is meant by "complete." The typical LD clause defines completion as "substantially complete," a condition certified by the architect. Substantially complete means that the structure is at that point where it meets the purpose for which it has been designed. A certificate of occupancy from the local building department is notification that the building can be occupied, but the structure does not have to be 100-percent complete. The lack of carpet in the lobby or possibly unpainted walls do not affect occupancy nor the definition of substantial completion. So the issuance of a certificate of Substantial Completion or a Certificate of Occupancy, in most cases, is enough to stop the LD calendar while the building is still under construction.

Enforcement of the liquidated damages clause. Liquidated damages clauses will not be enforced by the courts if they are construed as penalties to be levied against the contractor. Even though the contract language states that the liquidated clause is not a penalty clause, that stipulation is not binding without meeting certain criteria.

1. The amount fixed as the daily dollar amount of damages must be a reasonable assessment of the costs the owner will incur if the project is not completed on time.

2. The effects of the breach of contract (inability to finish on time) must be very difficult to establish at the time of contract preparation.

The amount of the liquidated damages must bear a relationship to the real, not imagined, harm to the owner if completion is delayed. In a court case referred to as *Harty* v. *Bye,* 483 P.2d 458 (1971), the Supreme Court of Oregon ruled that, although a contract contained a liquidated damages clause and the contractor did not complete the project on time, the owner was unable to substantiate proof of actual losses due to the late completion. In another court case, known as *Nomellini Construction Company* v. *Department of Water Resources,* 96 Cal. 682 (1971), neither party could establish the amount of damages which were incurred as a result of the other party's delays. The court said that if it is not able to discern a clear-cut distinction between the damages caused by the parties, liquidated damages cannot be assessed.

A further illustration of unenforceability is the case of *Mosler Safe Company* v. *Maiden Lane Safe Deposit Co.,* 199 N.Y. 479 (1910). There was a liquidated damages clause in the construction contract and when the contractor completed the work later than the date specified in the contract, final payment was requested. The owner, however, was determined to invoke liquidated damages. During the court action, the contractor stated that they had been delayed because the architect required the work to be installed in a manner different from that indicated in the contract drawings. The contractor also claimed that the architect delayed the job by not acting promptly on shop drawings submitted for review. The court ruled that when delays are caused by both parties, it will attempt to apportion the damages and will refuse to enforce the liquidated damages provision in the contract.

When the contract does not include an LD clause. Just because a construction contract does not include a liquidated damages clause, the contractor may still be subject to "actual" damages, which in some cases may exceed those of the liquidated damages. Delays in completing a project where LDs are not included may subject the contractor to the actual damages the owner incurs due to failure to meet the contract completion date. So what's the difference?

When a liquidated damages clause is to be inserted in the owner's contract, the owner calculates the added costs they will incur if the project is late. Costs such as interest on construction loans, loss of revenue from rental income if an office building is involved, or possibly loss of income from operations for a hotel or manufacturing plant are calculated. These costs, assembled as accurately as possible, may fluctuate during the length of construction as interest costs increase or economic conditions change. Under the LD concept, the owner can only collect the damages in the per diem amount indicated in the contract even if their actual costs exceed that per diem. The concept of "actual damages" does not limit the owner to a specific amount, but they can collect as much as they can prove in a court of law. As a general rule of law, damages do not have to be precisely calculated or even documented, but they must stand the scrutiny of what is *reasonable.*

Change Orders Reflecting Costs Due to Job Delays

Some delays encountered by the contractor allow them to extend the completion time of the project and be compensated accordingly via change orders. Some delays do not.

Delays occurring in construction projects can be categorized as follows:

- *Excusable.* The contractor is granted a time extension, but no monetary compensation.

- *Concurrent.* Delays occur, but neither the contractor nor the owner can collect monies for damages due to these delays.
- *Compensable.* Delays for which either the owner or the contractor is entitled to additional monies.

Excusable delays

Excusable delays allow the contractor to extend the completion date of the contract but may not allow them to recover costs associated with the delays. This list of excusable delays includes

- Acts of God
- Fires or other significant accidents
- Illness or death of one or more of the contractors
- Transportation delays over which the contractor has no control
- Labor strikes or disputes
- Unusually severe weather

Concurrent delays

Concurrent delays happen when two or more delays occur within the same time frame, both or all of which impact the project's completion date. These types of delays are also called overlapping delays, for obvious reasons. Concurrent delays may be caused by the owner or the contractor, but if it appears that both are responsible and these delays overlap, neither party will be able to recover damages; the owner can't assess liquidated damages and the contractor cannot collect for damages due to the delay.

Compensable delays

Compensable delays, as defined by several court decisions, are those delays for which damages can be claimed. Compensable delays include delays caused by elements beyond the control of the contractor, but within the control of the owner (including their design team). Changes in the work, access to the site, and site conditions differing materially from those specified in the contract are some other examples of compensable delays. Specific court cases established some delays as "compensable." These are listed next:

- Delays in the approval of shop drawings submitted by the contractor [*Specialty Assembling and Packing Co. v. United States, 274 Ct.Cl. 153 (166)*]

- Delays caused by the owner's improper inspection procedures [*Gannon Company* v. *United States*, 189 Ct.Cl. 328 (1969)]

- Inadequate or defective drawings or specifications [*United States v. Spearin*, 248 U.S. 132 (1918) and *J.D. Hedin Construction Co.* v. *United States*, 171 Ct.Cl. 70 (1965)]

- Contract changes when the nature of the work changed affects the original or unchanged work and causes an extension of time [*Conduit and Foundation Corp* v. *State of New York*, 425 N.Y.S. 2d 874 (App.Div. 1980)]

- Owner work-force interference. The owner, complying with the provisions of the contract is allowed to award separate contracts for work on the project and this may affect the general contractor's work. Such a situation could arise on a union construction worksite when an owner engages nonunion subcontractors and jurisdictional disputes lead to work stoppages. [*Bateson Construction Co.* v. *United States*. 319 F.2d 135 (Ct.Cl.19630)]

Documentation required for excusable delays.   With respect to excusable delays (those delays that extend the contract time but won't support a claim for additional monies), the project manager should submit a written request for a project extension whenever it occurs, even if it appears at a time when the job is ahead of schedule. If unforeseen events create delays of any nature as the project progresses, this early request may prove the necessary cushion to offset potentially costly penalties assessed by an owner.

When a request for a project completion extension is submitted, there may not be an immediate response from the architect or owner. If a response has not been received after a week or so, it is prudent to write another letter referring to the first request and stating that if there is no response to the second request within, say, five (5) days after receipt, it will be assumed that the delays have been accepted and therefore the contract completion time will be extended by the requested number of days. This will surely trigger a fast response.

Weather delay documentation.   Severe or adverse weather delay procedures may be defined in the construction contract but if no mention is made, actual weather delays are generally measured against the 10-year average for the specific time of the year and the location of the project. If the number of actual delay days is not unusual for the area at that time of the year, the contractor may not be entitled to a delay no matter how severe.

It is important that the job superintendent or project manager diligently document weather conditions on the daily time sheet or daily diary, and they should be reported at the start of the workday, at noon, and at mid-afternoon. If inclement weather delayed the start of work or canceled the entire day's work, the report should contain an entry such as: *No work day today due to extremely heavy rains from 6:30 A.M. to 10:30 A.M.* If any workers had reported for work awaiting a decision on whether the day would be considered a "no-work" day, that should also be noted in the report identifying these workers by name. A visit to the local weather bureau's Web site to determine the amount of rainfall during that period is also advisable, and a download can be attached to the daily log.

Pitfalls to Avoid when Preparing Change Orders

The initial submission of a proposed change order should be complete in every respect. Enough documentation should be attached to describe the changed condition, sufficient and detailed cost data to allow the architect to analyze the change request and any other explanatory information. It is important to avoid a series of questions raised by the architect regarding scope or price. The back and forth Q&A procedure that often accompanies the contractor's proposed change order is what prolongs its resolution.

The following pitfalls-to-avoid list may be helpful in that regard:

- To avoid questions regarding the billable costs for labor and equipment, prepare a detailed hourly rate for each type of labor that may be potentially incorporated in change-order work, and a schedule of hourly rates for equipment. Submit this list to the architect in the earliest stages of the project to obtain acceptance of these rates, which will help prevent future disputes over labor and equipment costs.

- Hourly rental rates continue even though the equipment is idle. If work is delayed pending certain decisions from the architect or owner and the equipment is idle, these costs should be included in the cost of the work. If the operator is also on standby, their hourly rate should be included just as though the piece of equipment was being operated. Costs billed as "idle" or "equipment downtime" should be submitted with a brief explanation: Backhoe—Active rate: $65.00/hour exclusive of operator; Idle rate: $50.00 reflects no fuel or maintenance costs.

- Consider whether the cost of work will require premium-time work. It may be beneficial to the owner to proceed with overtime work to avoid delay claims from other trades dependent upon this work in order to proceed on schedule the following day.

- If overtime is required, particularly if extended periods of premium-time are involved, consider the effect of lower worker productivity and increase the number of hours and the total cost of labor proportionately.

- Lump-sum proposals from subcontractors must be broken down into labor, materials, and equipment, with overhead and profit added as a separate line item. Will an owner or architect be able to understand the nature and extent of all costs for which reimbursement is requested? Could *you* if you were an owner?

- Be specific about the reasons for requesting change-order work. Include copies of portions of the drawings or specification section—ASI, RFI, and SK—to clarify the claim and include them as additional back-up.

- What about costs to distribute materials or equipment from the delivery truck to those areas of the building where they are to be installed? Are they included?

- Review all potential general conditions costs such as added utility costs, equipment rentals, small tools, scaffolding (up and down) winter condition, additional copies of plans, and/or specifications required for distribution of subs and vendors.

- Review all scope issues to confirm that the proposed change does or does not involve deletion of any work that may create a credit to be applied against the added costs. How often have you heard "Where is my credit to delete 3 cubic yards of concrete for those sidewalk changes?"

The Change-Order Cost Checklist

Three categories of costs should be considered when preparing a change order: direct costs, indirect costs, and impact costs.

- **Direct Costs** Hard dollar costs required to complete the work, such as:
  - Bond premiums
  - Coordination of trades
  - Equipment—whether active or idle; with or without an operator
  - Estimating costs
  - Insurance premiums
  - Labor, fringe benefits, payroll taxes (*Tip*: When figuring the hourly rate for a salaried employee, don't forget they only work 50 weeks per year (assuming a 2-week vacation) so divide their yearly salary by 50 weeks and again by 40 hours to get the hourly rate)
  - Material rehandling costs
  - Materials, taxes, delivery costs
  - Phone calls
  - Photographs
  - Postage, express deliveries
  - Rental equipment or contractor-owned equipment equivalent rates
  - Restocking charges
  - Safety equipment
  - Subcontractor costs

- Temporary heat
- Temporary protection
- Travel—autos, tolls, parking
- Utility costs for the building under construction
- Winter conditions
- **Indirect Costs** Costs not allocated to any specific item of direct work. These fall into two basic categories:
  - *Field Overhead:*
    - Project management staff
    - Project engineer
    - Project superintendent
    - Field office and field office supplies
    - Temporary utilities (light, heat, power, data communication, etc.)
  - *Home Office Overhead:*
    - Accounting and payroll costs
    - Corporate management
    - Change-order preparation, research, negotiations
    - Computers and office equipment
    - Insurance
- **Impact Costs** Costs associated with changes that impact project performance, such as:
  - Loss of productivity due to trade stacking, or other inefficiencies
  - Idle equipment and idle equipment maintenance
  - Under absorbed corporate overhead
  - Lack of availability of skilled tradesmen
  - Cost of disruption to the orderly flow of work; working out-of-sequence
  - Cost of extended warranties of equipment being installed during the project

Effective Change-Order Control

The following are some guidelines for effective change-order control:

- Alert all project participants (the project superintendent, foreman, and others with a role in the project) to be sensitive to *all* suspected changes in scope and report them to the project manager immediately. The project manager can then determine whether or not to pursue the change-order route.

- Instruct the project superintendent to identify and document any potential changes in their daily report or daily log. Better to err in regards to adding questionable events than to not create the written record.

- When a change in scope affecting the contract sum is identified, notify the architect and owner ASAP, even if the exact cost of the work is not ascertained.

- Any changes in scope of work or schedules reported from the field must be documented by referring to the drawing number and/or detail, finish schedule, specification section or name and position of person issuing verbal instructions to affect the change.

- Don't let change-order proposals stack up, issue them as soon as they are prepared, and in most cases don't combine unrelated changes in one change order. In case one is disputed, the entire change-order will be delayed.

- Don't wait until the end of the job to submit the change orders.

- Don't forget to note the impact on the schedule. Will completion time increase, decrease, or remain the same?

- Determine whether any escalation costs ought to be included in the change order if the work will extend the project's completion date, or the work will be performed at a time when increased labor and/or material costs may be incurred.

- Schedule a postconstruction meeting and review all issued and all missed change orders. Request that estimating, purchasing, and field supervision attend to review the extent and nature of those change orders for which the owner was billed, those change orders in which there was no owner reimbursement and those conditions where change orders could have been issued but weren't. All attendees should benefit from what went right and what went wrong.

10

Quality Control and Quality Assurance

There is a difference between quality control (QC) and quality assurance (QA). Quality control can be defined as the standard(s) to which the construction or assembly of a building component has been incorporated into the project's design, while quality assurance is the process that verifies these standards have been met.

When the architect indicates compliance with American Concrete Institute (ACI) Specification 301 for cast-in-place columns, the contractor is directed to install the column plumb to a tolerance of plus or minus 1/4 inch in any 10 feet—that is quality control. The contractor, in verifying that this tolerance has been met, has performed quality assurance.

This issue of quality in construction has gained more momentum recently as owners are adding quality to the old demands of competitive pricing and schedule adherence, and contractors look at quality issues to not only enhance their reputation but to control costs. "Do it right—the first time" is a mantra more contractors are embracing, recognizing that rework is not only disruptive to the project's schedule but also adds to their cost of construction.

Quality Begins with Quality Design

Respondents to the FMI/CMAA's *Fifth Annual Survey of Owners in 2004* had several things to say about the quality of project documents. About 98 percent of owners surveyed said that architects need to be held more accountable for completing a quality design, one that can be built without numerous change orders and RFIs.

About 60 percent of those questioned said that the quality of design documents had declined to the point where frequently electrical and mechanical subcontractors were actually completing the design using their shop-drawing submissions.

But owners share some responsibility for these problems, too, which was also pointed out in the FMI/CMAA survey:

- The process of computer-assisted design has not only greatly increased productivity but has also resulted in less attention to single elements within the design process.

- Owners, demanding compressed design time, compound the lack of attention to details and problems associated with coordination issues.

- Tremendous pressure on design fees leads to less-experienced individuals working on design details.

- Mechanical and electrical subcontractors and their manufacturer representatives often are more aware of technological advances than the design engineers.

- The advances in materials and systems make designs more complex; therefore, shop drawings are more accurate in detail than the designs created by the architect or engineer.

On one project, the author of this book was negotiating with an owner who kept pressing the mechanical engineer to complete the drawings by a certain deadline. The owner persisted and the engineer finally said, "Do you want expediency or accuracy? At this point, you can't have both." The owner quickly backed down and said that, of course, he wanted accuracy. We need more design consultants like that.

Total Quality Management

Total quality management (TQM) encompasses elements that form the basis of a companywide focus on quality—elements that extend from the office to the field. Preparation of the monthly requisition by the Accounting department to ensure that all required attachments have been included, and making sure the requisition is mailed out on time is a function of TQM. Preparing change orders where sufficient documentation and cost information is attached so an owner can reasonably understand and review the change order is also a function of TQM, just as the QC/QA functions in the field fit the TQM mold.

Over the years, the QC/QA movement has spawned a number of processes focusing on quality. Understanding and implementing many of these processes or procedures requires professional instruction and training, and although several are tailored for the manufacturing industry, some can be applied in construction.

Benchmarking

Benchmarking is a procedure where a company seeks out and studies the best practices in order to produce a superior performance. Benchmarking can take several forms: internal benchmarking that investigates the practices and performance within the company; and external benchmarking, which compares the company's practices and performance with world-class performers.

Internal benchmarking can initially isolate the performance of a specific component of construction for analysis. If the company self-performs carpentry or concrete work, a look at a project where costs were lowest and quality levels were highest would be followed by an investigation into the practices that created this quality, cost-effective job. If high productivity coupled with strong supervision produced these results, this could be a benchmark standard that is reviewed with all field and office personnel with the intent to have these procedures established as company standards. If the concrete work was subcontracted and a particular subcontractor performed flawlessly and was competitive in their price, then the selection process that produced this quality subcontractor might be reviewed and implemented as a benchmark against which future selections could be made.

External benchmarking attempts to match the company's operations against superior firms in their field. The Construction Financial Management Association (CFMA) in Princeton, New Jersey publishes financial data compiled from their members annually. CFMA then collates this information to create Best in Class statistics. So, a contractor in the heavy and highway business can see that Best in Class companies maintain gross margins of 9.3 percent to 9.9 percent, and net profit margins of 2.2 to 3.0 percent. But how do they stack up? This is called external benchmarking.

The International Standards Organization

The International Standards Organization (ISO) was founded in 1946 to promote voluntary manufacturing and trade communication standards. This organization is based in Geneva, Switzerland and the ISO acronym actually refers to the Greek work "isos" meaning equal, not the acronym for International Standards Organization.

ISO 9000 and its variants ISO 9001, 2, 3, and 4 relate to construction and involve basic quality management principles, including:

- Management responsibilities
- Quality systems principles
- Internal audits
- Contract review

- Design control
- Purchasing
- Process control
- Product identification and traceability
- Inspections and testings
- Corrective actions
- Handling, storage, packaging, and delivery
- After-sales servicing
- Document control
- Quality records
- Training
- Purchaser-supplied products

Several construction companies have sought ISO certification to indicate that they have instituted the basic quality management principles as outlined by ISO. In 1980, only two of the largest 100 construction firms in the United States were ISO-certified. By 1993, this list had increased to 22, and continues to grow. The U.S. Department of Defense has adopted the use of ISO 9000 standards in lieu of their MIL-Q-9858A spec where practical.

One of the more practical aspects of ISO involvement in the design and construction industry are their efforts to work with other world organizations to create a common computer language enabling vendors and manufacturers to communicate clearly with architects and engineers worldwide. Combined with the efforts of various trade organizations to develop software that can speak to all participants, a process called Interoperability (discussed further in Chap. 16) was created. This is a worthwhile exercise that should benefit all contractors—large and small.

The Six Sigma Approach to Quality

Motorola Corporation was the originator of the Six Sigma approach to quality control. In the 1970s, Motorola sold their Quasar television set manufacturing business to a Japanese company located in America. Japanese management quickly made changes to the factory's operations and, to Motorola's surprise, began producing TV sets with $1/20^{th}$ the number of defects than previous owners experienced. Motorola's investigations showed that the Japanese managers used the same labor force, and the same technology and designs. Thus, they had to admit that their quality control and quality assurance programs compared to the Japanese were pretty poor. Over the next 10 years, Motorola devoted considerable time and effort to improving their manufacturing process, and thus Six Sigma was born—the letter for sigma in the Greek alphabet being used by statisticians to measure the variability in any process. To achieve Six Sigma, the process cannot produce more than 3.4 defects per million *opportunities* (opportunity being defined as a chance for nonconformance, or not meeting required expectations).

The Six Sigma approach is to train a small cadre of in-house leaders in management techniques required to attain high levels of proficiency in the manufacturing process. These full-time highly trained people have received 160 hours of classroom training, and are known as Black Belts. Lesser-trained project leaders are referred to as Green Belts, and to reach this stage an employee must have five days of classroom study and training that includes project management, quality management tools, problem solving, and data analysis. These Green Belts act as assistants to the Black Belts, who also form teams and manage Six Sigma projects from concept to completion.

At the head of the Six Sigma hierarchy is the Master Black Belt, one who has achieved the highest level of technical and organizational proficiency. These Master Black Belts are the ones that are responsible for the technical leadership within the Six Sigma program.

The process of improving a particular operation is known as DMAIC, where each letter represents an operation, as explained next:

D: Define the goals.

M: Measure the existing system to establish a metric by which progress towards the goals can be tracked (for example, create a baseline against which to measure progress).

A: Analyze the system to identify ways to eliminate the gap between existing performance and the desired goal.

I: Improve the system.

C: Control the new system.

Although Six Sigma is basically a manufacturing process system, the DMAIC approach can also be applied to the construction industry.

From Theory to Practice

QA/QC in the construction business begins at home. The quality of the estimating department's product, the quality of the purchasing function, and the quality of the accounting department all impact the concept of total quality management like a mirror, reflecting the company image back into the community. Quality is more a way of life than a process since its practice is not merely confined to the workplace.

A project manager and a project superintendent don't need ISO or Six Sigma or any other combination of Greek alphabets to tell them that a concrete slab is substandard or a wall is obviously out of plumb. The question remains, will this be handled with an "I guess its okay unless the architect rejects it," or will an investigation follow to find out why this substandard work was allowed and what should be done to prevent it in the future.

Quality starts with a thorough review of the contract documents

The control of quality construction begins with an intensive review of the plans and specifications. As stated previously, there may be a few (or more than a few) deficiencies in the plans and specifications, some of which may have been uncovered during the estimating process, while others were unearthed by the subcontractors and vendors. The project superintendent may have unearthed constructability issues during his or her review. The need for all parties to thoroughly examine the documents under their control cannot be overemphasized. When problems are uncovered early in the game, the impact and associated costs are considerably less.

Quality Control and the Specifications

Various specification sections are loaded with abbreviations of trade and professional organization quality standards, but how many contractors have the full set of American Concrete Institute manuals at $750, or a $299 set of the American Institute of Steel Construction manuals, or a set of ASTM Building Product books for $850?

A typical cast-in-place concrete specification section refers to both ACI and ASTM, and the project manager must often rely on the integrity of their subcontractors and suppliers to insure that these specifications have been met.

Without some guide to ASTM E-1155 (the F-number system) for measuring the flatness of concrete slabs, how does the project superintendent or project manager verify that the quality assurance levels have been met even when testing is performed? How will curling of the slab be dealt with (tests must be performed within 72 hours of installation), and when should elevated slabs be measured? (Answer: Before the removal of shores.) Architects and engineers inserting product and procedure quality levels in their specifications should become more aware of the contractor's need for specific quality standard information instead of simply asserting "Refer to ACI 301-R."

Using trade organizations to learn about quality standards

Several trade organizations offer low-cost or no-cost quality standard information. The following lists but a few:

Steel Door Institute. Provides hollow metal frame and door installation tolerances

Southern Pine Inspection Bureau. Offers specifications on southern pine framing lumber

Western Wood Products Association. Offers specifications and tolerances on fir, spruce, hemlock, hem-fir, various types of pine, and other species of softwood

Steel Stud Manufacturers Association. Presents the section properties of steel studs, floor joist allowable spans, fastener data, and typical details

The Preconstruction Conference—QC and QA

Some specifications include a provision for a preconstruction conference as part of a quality program to deal with specific construction components in order to ensure that both product and installation procedures are understood by the installer prior to the commencement of work. These preconstruction conferences bring together the general contractor, subcontractor, architect, and sometimes the engineer to review and discuss the installation procedures outlined in the specifications. An experienced installer may offer suggestions to improve quality or will listen to tips from the architect on both good and bad past experiences with this particular system or component. The preconstruction conference affords other related trades involved in the component to listen and comment as well.

Even if not required by their contract, many project managers wisely hold their own preconstruction conferences to ensure that the critical components of construction proceed in such a way that high-quality results are achieved.

For example, let's look at a structure that will have the following exterior wall components:

- Steel stud with exterior grade gypsum board or fiberglass reinforced gypsum board applied to the outside face of the studs, structural or otherwise

- Cavity wall construction with an application of some form of moisture protecting membrane on the face of the gypsum board, wall ties of some sort screwed through the sheathing into the stud framing

- Framed rough openings for subsequent window installation
- Flashings
- Brick veneer

Depending upon how various subcontracts were awarded, several different trades will be involved in this operation: drywall, metal framer, mason, moisture protection, and a preconstruction conference affords each subcontractor an opportunity to review their part in the process and possibly offer suggestions to improve the quality of the assembly. With the project manager orchestrating the conference, everyone's part is clearly delineated and comments are heard, discussed, and resolved. Detailed minutes documenting the procedures to be followed should prevent any disputes or disagreements over the upcoming procedures, as well as the role of each participant in the meeting.

Preconstruction conferences or meetings are especially useful when working with any of the following construction operations:

- Sedimentation control and maintenance, alerting all subcontractors to the process and the care they must exhibit in order to avoid damaging the installation
- Cast-in-place concrete work
- Structural steel and metal deck work
- Waterproofing—foundations, above grade walls, roof
- Exterior wall assemblies such as stucco, EIFS, masonry, precast concrete, metal or vinyl siding and metal panels
- Windows, vents, louvers, other types of wall openings that must be flashed into the surrounding wall system
- Roofing, skylights, hatches, roof penetrations, flashings, curbs, gutters
- Joint sealers—caulking
- Gypsum board and related attachments requiring in-wall blocking and reinforcement
- Kitchen and bath cabinets, requiring coordination with MEP trades and in-wall blocking installers
- Finish wall items—signage, window treatments, artwork requiring lighting, and in-wall reinforcement to insure that sufficient blocking is installed for the installation

Sample panels and mock-ups

Sample panels are frequently required by the architect so they can comment on, accept, or reject the quality of the masonry wall construction.

Sample panels can also be used to incorporate related work such as flashings or items to be embedded in the wall panel by other trades.

Mock-ups, even if not a contract requirement, can be important portals to better quality levels. Curtain wall construction often begins with the creation of a mock-up that clearly displays structural attachments and weatherproofing systems. Masonry exterior walls with complex window or glazed openings, several different methods of attachment, assorted flashings, and drainage details are certain candidates for a mock-up. Even interior wall treatments with splays, soffits, or articulated design may also provide a visual display of details that can't be shown otherwise.

The value of a mock-up to the general contractor, and, more particularly, to all concerned subcontractors, is the ability to study and change complicated details before production work begins so as not to impede the progress.

The punch list and QC/QA

Does the punch list qualify as a QC/QA exercise? It sure does! In a perfect world, there would be no need for a punch list—all trades would inspect their work and correct any incomplete or defective items before leaving the job. The project superintendent would inspect the entire project, issue a series of "to do" items to each subcontractor and when the architect performed their inspection at the closing portion of the project, no formal punch list would be produced.

In a previous chapter, we discussed the importance of prompt and complete project close-outs and the poor impression that a lingering punch list conveys to an owner. Neither the project manager nor the project superintendent can be everywhere on the site checking on the work of each subcontractor, but some rules established upfront and enforced religiously throughout the life of the project may get closer to that Zero Punch List goal.

Quality means a clean site.   A clean site, one free from accumulated debris inside and outside the building improves the quality of work. With trash cans placed strategically around the site, a worker will think twice before balling up that leftover lunch and tossing it on the ground. Conversely, a sloppy, trash-strewn site and building sends the message that order is not a priority on this project. Each project meeting should stress the need to keep the various work areas free from accumulated debris and offenders should be identified and put on notice to clean their area or expect a significant backcharge.

The field office should be kept in a clean and orderly condition so as to send a positive message to anyone attending project meetings that cleanliness and orderliness are important to the office.

Weekly quality reminders. The weekly subcontractor meeting is the perfect forum to discuss and review quality issues. Any "shortcuts" that result in marginally acceptable work should be pointed out and accompanied by a statement that any such work that fails to meet the quality standards of the project manager or superintendent will be rejected and must be replaced within the same workday it was discovered—no matter how long that work day is.

Obviously, the watchdog activities of the project superintendent must be apparent to all workers on the site, too. No unacceptable work should go unnoticed.

Developing a Company QC/QA Program

Developing a QC/QA plan (more commonly known as a Construction Quality Control Plan (CQCP)) involves not only formulating a set of policies and procedures, but requires a qualified individual to assume the duties and responsibilities of a CQCP Manager. This manager, once the plan is put in place, will be the individual responsible for the implementation and monitoring of the program, much in the way that a Director of Safety is responsible for the company's safety program.

Let's discuss the preparation of a CQCP, which should include, at a minimum, the following:

- An organization chart showing where the CQCP manager fits into the current management hierarchy
- Proposed personnel and support staff
- Designated authorities, and their duties and responsibilities
- Operating procedures—inspections, tests, and related documentation
- The development of a subcontractor and vendor rating system whereby supervisors can evaluate and report on subcontractor/vendor quality and performance
- Assist in selecting competent, quality-oriented subcontractors and vendors acting in liaison with the Purchasing department
- Training sessions to familiarize field personnel with quality issues
- The development of field inspection reports to assist field personnel in their inspection procedures

The inspection checklist method

The CQCP manager, as part of their responsibilities can develop a series of forms to assist field supervisors in their weekly inspection of the work of various trades. These reports not only serve as a guide, but also provide the inspecting supervisor with some basic knowledge of the quality points associated with each trade. Figures 10.1 thru 10.5 are some examples of these inspection reports:

Inspection Checklist

Concrete Placement   **Project No.**_____

1. Shop drawings approved and on site. _____
2. Verify correct psi ordered from plant. _____
3. Chutes, elephant trunks required? _____
4. Verify approval of forms and rebar prior to pour. _____
5. Requirements for testing, mix design, ingredients. _____
6. Test lab notified and tests required. _____
       Slump _____
       Number of cylinders _____
       Temperature/truck waiting time _____
7. Testing required at plant. _____
8. Vibrators to be used during pour. _____
9. Temporary form openings O.K.? _____
10. Arrange for specified curing and saw cut joints. _____
11. Arrange for cold weather protection. _____
      -or
12. Arrange for hot weather protection. _____
13. Embeds available for insertion in pour. _____
14. Box-out properly installed in form work. _____
15. Verify finishes – smooth troweled, broom. _____
16. No troweling while bleed water is on surface. _____
17. Slopes to drain properly designated. _____
18. Wet spray or curing compound adequately performed. _____
19. Traffic over area controlled. _____
20. Preparations for repairs at hand. _____

General Notes:_____
_____
_____
_____
_____

Inspected by:_____Date:_____

FIGURE 10.1   A concrete placement checklist. (With permission from McGraw-Hill, New York.)

Inspection Checklist

Concrete Reinforcement   **Project No.**_____

1. Shop drawings are approved and on site._____
2. Grade of steel delivered as required. _____
3. Spacing coordinated to suit masonry/concrete units. _____
4. Required clearance of steel from forms provided. _____
5. Length of splices and staggered splices as required. _____
6. Bends within radii and tolerance are uniformly made. _____
7. Additional bars at intersections, openings, and corners provided. _____
8. Bars cleaned of materials that affect bond. _____
9. Dowels for marginal bars at openings. _____
10. Bars tied and supported to avoid displacement. _____
11. Spacers, tie wires, chairs as required. _____
12. Conduit is separated by 3 conduit diameters minimum. _____
13. No conduit or pipe placed below rebar material except where approved. _____
14. No contact of bars is made with dissimilar metals. _____
15. Bar not near surface which may cause rusting. _____
16. Adequate clearance provided for deposit of concrete. _____
17. Verify that contractor has resolved conflicts with embeds. _____
18. Verify that contractor has coordinated for anchors, piping, sleeves. _____
19. Special coating as required. _____
20. No bent bars and tension members installed except where approved. _____
21. Unless approved, boxing out is not approved for subsequent grouting out. _____
24. Rules for bar splices: For 24d lap – multiply bar size by 3 = lap in inches
                                          For 32d lap – multiply bar size by 4 = lap in inches
                                          For 40d alp – multiply bar size by 5 = lap in inches
25. Agency/Engineer inspection is performed, if required. _____

General Notes: _____

_____

_____

_____

_____

Inspected by:_____Date:_____

FIGURE 10.2 A concrete reinforcement report (which also contains bar splice lap tolerances). (With permission from McGraw-Hill, New York.)

Inspection Checklist

Metal Framing - Gypsum Drywall   **Project No.**_____

Metal Framing

1. All submittals, samples, shop drawings are approved and on site. _____
2. Material is stored in a dry location. _____
3. Material galvanized as required. _____
4. Studs are doubled up at jambs, unless otherwise required. _____
5. Structural and/or heavy gauge studs as required. _____
6. Studs allow for movement, slab deflection. _____
7. Studs securely anchored to walls, columns, and floors. _____
8. Soundproofing provided at floor and walls as required. _____
9. Observe location, layout, plumbness. _____
10. Channel stiffeners are provided as required. _____
11. Special fastening and connections are observed. _____
12. Anchor blocking, plates, other equipment provided. _____
13. Cut studs for openings are properly framed. _____
14. Observe size, gauge of runner, and furring channels. _____
15. Hangers are saddle tied, bolted, or clipped as required. _____
16. Tie wire for channels to runners properly tied. _____
17. Elevation and layout of furring is understood. _____
18. Observe that surfaces are plumb and level. _____
19. Observe that long single lengths are used. _____
20. Control joints are installed per contract requirements. _____
21. Requirements for adjoining surfaces of different materials are accommodated. _____
22. Seating provided for sound or thermal insulation. _____
23. Spacing and construction are as specified. _____
24. Observe location of all blocking, bracing, and nailers. _____
25. Type, thickness, length, and edges are as required. _____
26. Type fastener, length, and spacing as required. _____
27. Installation complies with manufacturer's requirements. _____
28. Special type suited for damp locations if required. _____
29. Special lengths are provided as required. _____
30. Verify if horizontal or vertical application is required. _____

FIGURE 10.3   A metal framing and drywall checklist. (With permission from McGraw-Hill, New York.)

Simple checklists like these can be prepared for all components of construction from sitework to low-voltage electrical work. With the help of an engineer or highly qualified subcontractor, checklists can be prepared for various HVAC components, plumbing installations, fire protection, and electrical work.

31. Wall board is installed with staggered application. _____

32. Internal and external metal/plastic corners as required. _____

33. Number of coats of compound required is provided. _____

    Level 1 - All joints and interior angles have tape embedded in
        one layer of joint compound. _____
    Level 2 - All joints and interior angles taped & receive two
        coats of taping compound. _____
    Level 3 - All joints and interior angles taped & receive three
        coats of taping compound. _____
    Level 4 - Apply Level 3 plus skim coat of joint compound
        over entire surface. _____

34. Sanding between coats is performed. _____

35. Feathering is out 12" to 16". _____

36. Provide air circulation with adequate dry heat. _____

37. If fire rated, recesses over 16" are boxed in. _____

38. Penetrations tight and sealed as required by code. _____

39. Verify contractor has coordinated cut-outs and outlet boxes correctly to avoid patching.
_____

40. Wall board is held up from floor 3/8" minimum. _____

41. Vertical joints are aligned with door jambs. _____

42. Damaged sheets are not used and are removed. _____

43. Observe minimum piecing or joining. _____

44. Non-metallic cable, plastic, or copper pipe is not damaged. _____

45. Check for bubbles and dimples. _____

46. Curing time is adequate for subsequent finishes. _____

General Notes:_____
_____
_____
_____
_____
_____
_____

Inspected by:_____ Date:_____

FIGURE 10.3   (*Continued.*)

Focused inspections that use these checklists not only serve as a means of checking the work of a particular trade, but of also sending out a message that the quality of the project is an important issue with the general contractor. Subcontractors can be given copies of the checklist after their work has been inspected in order to document compliance with "good practices" or to alert them to deficiencies in their work.

Inspection Checklist

Masonry   Project No._____

1. Approved shop drawings on site. _____
2. Approved samples on site or evidenced. _____
3. Sample panel, if required, constructed and approved. _____
4. Materials stored off ground and covered. _____
5. Correct type/color mortar. _____
6. Concrete masonry units are not wet. _____
7. Reinforcement: type, size on site, spacing specified. _____
8. Excessive bending of rebar not allowed. _____
9. Pipes, sleeves, boxes located. _____
10. No shovel measures for job-mixed grout. _____
11. Climatic and temperature controls are acceptable. _____
12. Adequate lighting available for good workmanship. _____
13. Joint size, type, tooling method as required. _____
14. Bonding is as required. _____
15. Observe full head and bed joints. _____
16. Joints tooled to provide dense surface. _____
17. Bond beams in place, properly reinforced and grouted. _____
18. Wythes or cavities kept free of excess mortar. _____
19. Check anchors & ties for materials, sizes, etc. _____
20. Bucks and anchors - secured, plumb, and level. _____
21. Provisions for flashings, cut-outs, & later items. _____
22. Provisions for parging, if required. _____
23. Expansion and control joints are located. _____
24. Structural members have suitable attachments. _____
25. Debris is removed periodically, not piled. _____
26. Protect work from freezing for at least 48 hours. _____
27. Clean off splatters. _____
28. Observe bond beam filling. _____
29. Hollow metal frames fully grouted, if required. _____
30. Backfilling after proper securing & support. _____
31. Proper support of high walls. _____

General Notes:_____

_____

_____

_____

_____

Inspected by:_____Date:_____

FIGURE 10.4   A masonry checklist. (With permission from McGraw-Hill, New York.)

Inspection Checklist

Membrane Roofing   **Project No.**_____

1. All shop drawings, product data, samples as required and approved are on site. _____

2. Attend or conduct preconstruction meeting. Review construction sequence, any deviations from approved submittals, field problems, warranty issues. _____

3. Before roofing contractor commences work, observe the following:

    a. Surfaces are clean and free of debris. _____

    b. Excess mortar or concrete is removed. _____

    c. All holes, joints, and cracks are pointed. _____

    d. All rough or high spots are ground smooth on concrete deck. _____

    e. Wood nailers or other attachment conditions are adequate. _____

    f. Surfaces are dry as required by manufacturer. _____

    g. Concrete deck tested for dampness if necessary. _____

    h. Slope is as required. _____

    i. Pipes, conduits, and other roof penetrations are in place and ready to receive flashings. _____

    j. All sheet metal and roof accessories are in place or on hand to be installed as the roof installation commences. _____

4. Materials of types required are provided. Materials are identifiable and comply with ASTM standards. _____

5. If rolled roofing, stand rolls on end and keep free of moisture. _____

6. Nails and fasteners are of length, shank, head, and coating required. _____

7. Surface to receive roofing is primed, if required. _____

8. Observe lap, nailing, and quantity of adhesive applied. _____

9. See that membrane is laid so that it is free of air pockets, wrinkles, and buckles. If present, rolling may be required. _____

10. All surfaces to be kept free of moisture. Under no condition allow exposure of insulation or incomplete install membrane to remain overnight without protection. Protect stored material from moisture. _____

11. Observe installation of cant strips, vertical surfaces, reglets, and penetrations. _____

12. Observe sealing of roof membrane envelopes where the use of envelope is permitted. _____

13. All concrete walls to receive roofing are primed. _____

14. Observe welds; do not allow any skips or unwelded joints. _____

15. Avoid plugging of drains and weeps and do not damage adjacent surfaces. _____

FIGURE 10.5   A membrane roofing checklist. (With permission from McGraw-Hill, New York.)

Quality inspection tips.   The following are a few tips to help you with any quality inspection tour:

- Inspect for quality on any walkthrough, and at a bare minimum check on quality issues once a week.

16. Roofing membrane is fully set into clamping ring. Lead collar flashing is installed and stripped in, if required. _____

17. Roofing is protected against damage by other trades. _____

18. Observe cut samples, if required. Observe that patching of cut samples is properly performed where samples are cut. _____

19. Clean-up provided after installation; drains cleared. Debris removed from site. _____

General Notes: _____
_____
_____
_____
_____
_____
_____
_____
_____
_____

Inspected by: _____ Date: _____

FIGURE 10.5   (*Continued.*)

- Promptly notify a subcontractor or vendor of any unacceptable work, materials, or equipment. Set a date for any replacement or rework, and later check to see if the date has been met.

- If substandard work or a repeated lack of quality continues, call the subcontractor's owner or manufacturer/vendor representative to the site to have them observe the poor quality work and obtain a verbal commitment to change and improve. Follow up with a written memorandum of the meeting. If a change in crew supervision is deemed necessary, make the request.

- Prior to a subcontract demobilizing, conduct a walkthrough to create a punch list and e-mail a copy to the subcontractor's office. Advise that all punch list items and unacceptable work must be completed before demobilization. Let them know that further payments may be jeopardized if the work is not done. Also mention that this list does not absolve the subcontractor of any future punch-list work issued by the architect.

Remember that the quest for quality is unrelenting and requires commitment from the top down. Once achieved, however, it can be a rewarding experience.

11

Project Documentation

The proper documentation of a construction project involves the creation of sufficient records to affect a history of the construction process. Proper documentation does not mean creating gigabytes of electronic data or reams of paper to fill row upon row of filing cabinets. It *does* mean making certain that when important events are about to happen, or are happening, or have happened in the not-too-distant past, they are accurately and promptly recorded. The key words are "important" with regard to *events*, and "accurately" concerning *recorded*.

That smooth running project of yesterday may turn into a quagmire today, and without the documentation to support this turn-of-events, grave consequences can result; something which could apply to the general contractor as well if they're asleep at the switch. It is not by chance that this chapter on documentation precedes the one on claims and disputes later in the book.

The Documentation Process

With e-mail used in lieu of letter writing these days, those back and forth messages should periodically be downloaded, printed, and filed to provide easier access to documents than by scrolling through hundreds of messages looking for the right one. Throughout this chapter, "e-mail" can be substituted for "letters" in almost all cases.

Facts rather than opinions are important in every facet of documentation, although some opinions gathered from remarks made by attendees at a project meeting ought to be recorded separately to shed some light on the intent of the comment.

Documentation of an event should consider

- What happened
- When it happened
- How it happened
- When it was discovered
- Whether responsibility can be assigned with certainty
- Who was notified
- To whom and when was notification given
- What the impact of the event is
- What immediate action is required, if any, and by whom
- What longer-term action is required, if any, and by whom
- Which party or parties will be responsible for resolving the problem

Project documentation accomplishes the following:

- Creates a history of the project which can be referred to when similar jobs with comparable problems are encountered
- Provides enough information so that, if project reassignments are made within the organization, a new project manager can trace the job history to date and continue the administration of the project will little or no difficulty
- Provides more than mere reliance on one's memory to reconstruct various segments of a project's activity long after it has been completed
- Reduces the possibility of future misunderstandings, disagreements, or disputes by committing important events or verbal communications into writing
- Makes details available in the event that mediation, arbitration, or litigation is pursued

Recordkeeping is important since it relates to the project manager's relations with

- The owner
- The architect and engineer
- Government authorities
- Subcontractors and suppliers
- Field operations
- Office staff

Documentation to the Owner

The form and content of any owner documentation will vary with the general contractor's contractual relationship with that owner. When the architect is the owner's agent or representative, most owner-related communications will pass through the architect. Some owners may also request copies of all documents transmitted to the architect. Occasionally, some owners will bypass the architect entirely and direct instructions to the general contractor instead. When this occurs, the project manager must alert the architect to the situation with a request to ask the owner to transmit future documents to the contractor via the architect's office.

The unique provisions of a GMP contract with respect to documentation

A cost-plus-not-to-exceed-a-GMP (guaranteed maximum price) contract requires communication between the owner, architect, and contractor for several reasons. All parties want to be kept apprised of any significant cost increases or decreases, and any decisions that may affect the cost of the project.

Subcontractor selection. When a cost-plus or GMP contract is administered, the final selection of subcontractors may be subject to approval by the owner. If so, copies of the subcontractor quotes along with the project manager's recommendation will be forwarded to the owner. The letter accompanying the proposals should include a tabulation of scope and prices so that each quoted price represents equal scope, or is adjusted accordingly.

Included in the letter to the owner is a time frame required for a response—within 5 working days, 72 hours, or whatever seems appropriate to maintain job progress. This notice must be monitored so that if no response is received within the requested time frame, another written request should be sent indicating that there is some urgency attached to a timely response.

As is the case in all forms of notice, strict monitoring of the requested response to that notice must be maintained, and when no response is forthcoming, a second more urgent request must be sent.

Documentation to an owner from the construction manager

A construction manager (CM) contract creates another dimension in owner-contractor relations. The CM is now the owner's agent and consultant in matters related to construction. Because of that relationship, the documentation takes on a different perspective. The CM's daily activities as the owner's agent will require the CM to prepare reports, schedules, and various cost and estimate analyses, recommendations pertaining to contract awards, general contractor and subcontractor requests for extras, and interrelationships with the architect and engineer, so correspondence that would have previously been transmitted by the general contractor to the architect/owner will then flow instead from the CM.

A review of the Construction Management Association of America's A-1 document lists the basic services required of a CM, each of which requires some form of documentation.

- Time management—preparing a master schedule or milestone schedule
- Cost management—surveying the construction market to provide the availability of local construction services, and project and construction budgets
- Operating a management information system—preparing various reports such as cost reports, cash flow reports, and scheduling updates
- Project management—working with prequalifying bidders, placing bid notifications, arranging prebid conferences, conducting bid analyses and recommendations, organizing postbid conferences, and preparing construction contracts
- Setting up project conferences
- Arranging progress meetings
- Reviewing contract documents
- Acquiring approval from regulatory agencies
- Managing change orders
- Monitoring contractor safety programs
- Controlling documents

The owner's responsibility to the contractor.   The owner must not only pay the contractor's monthly requisitions, but pay the contractor on time. It's also the owner's responsibility to make all the necessary arrangements for financing prior to the start of construction. The owner's obligation to a general contractor generally commences after a construction contract has been signed. Formal bidding instructions require general contractors to submit bids to the owner that must remain valid for periods ranging from 30 to 60 days. The owner must honor those commitments and notify the contractor of the final decision in writing; if the owner doesn't, the contractor is under no obligation to accept an award if one is offered. Why would a contractor want to reject a potential contract? Well, suppose that in reviewing their bid the general contractor discovered that a major item of work had either been inadvertently omitted from the estimate, or seriously undervalued. As a result, the contractor may wish to withdraw their bid; however, if a bid bond was required, that would mean forfeiting that bond.

If the owner fails to abide by the bid notification procedure though, and the contractor declines to accept the contract when offered (and a bid bond is involved), written notification of nonacceptance to the owner along with a request to return the bid bond would be in order.

The owner also has an obligation, in most cases, to provide certain surveys and easement information to the contractor if applicable. If the required surveys have not been submitted in a timely manner, a letter to the architect/owner needs to be sent, along with any follow-up.

Documentation to the Architect and Engineer

Correspondence with the architect, their engineers, and consultants generally makes up the most voluminous part of the entire general contractor's recordkeeping process. Most of the following categories of items require the project manager's attention. When "architect" is mentioned, the word "engineer" may also apply.

Shop drawing submittals, review, returns and logs

The prompt receipt, review, transmission, distribution, and tracking of shop drawings is one of the most critical elements of project administration. Documentation of this process must be complete, accurate, and timely. Architects frequently require the general contractor to affix their stamp to the shop drawing certifying that it has been reviewed and that it complies with the contract requirements. Often, hundreds of shop drawings are generated and pass through the project manager's hands, thus a proper shop drawing log is required to monitor the flow of shop drawings. This log should include, at a bare minimum:

- A tracking number
- The name of the sender
- The date received
- A brief description of the content
- To whom it should be sent
- The date it was sent
- The date due back

- Days received—plus or minus the date requested
- The action taken—approved, approved as noted, rejected, resubmitted, and so on

Figure 11.1 shows a standard format shop drawing log like that included in most project management software packages.

The log must be reviewed periodically at the owner's meeting. This is important in order to track the flow to and from the architect and to note any drawings that are overdue and when they can be expected to be reviewed at the subcontractor's meetings, to track the flow of shop drawings from their vendors, and also to respond to requests for resubmission of some questionable shop drawings. Finally, this should be done so as to notify a subcontractor of any late submissions and the consequences thereof.

To ensure a successful shop drawing management system:

- Create a tracking log
- Monitor the flow of documents, minimally on a weekly basis
- Provide written notification to the delaying party
- When delays continue, issue a letter warning of the consequences, job delays, impact on other subcontractors, out-of-sequence work, and so on

Requests for Clarification and Requests for Information

Requests for Clarification (RFCs) and Requests for Information (RFIs) are used during the bidding process to obtain clarifications or information affecting the cost of the project and, of course, after the contract award when an in-depth analysis of the plans and specifications is made by contractors. Requests may have originated from the general contractor or from their subcontractors and suppliers.

Because it is not uncommon for RFIs and RFCs to be composed of several hundred documents, a tracking and monitoring device is essential. The RFI log answers this need.

The RFI log (Figure 11.2) should contain the following elements:

- A tracking number
- The date when the RFI was created by the general contractor, or when received from their subcontractor (since all RFIs are sent to the architect, there's no need for an entry of this type)
- The date sent to the architect
- The date when a response is required

Job No: 000				Submittals by Package					Date: 2/6/2006		
Project No:									Page: 8 of 45		

Package	Submittal	Rev.	Title	Status	Required		Latest Dates				BIC
					Start	Finish	Rcvd.	Sent	Return	Forward	
03300			Cast In Place Concrete								
	03300-072	001	Shp Dwg: Level 3 Bottom Rebar	AAN		4/25/2005	4/29/2005	4/4/2005	4/22/2005	4/25/2005	
	03300-073	001	Shp Dwg: Level 3 - Beams	AAN		4/29/2005	3/29/2005	4/4/2005	4/22/2005	4/25/2005	
	03300-074	001	Shp Dwg: Level 3 - Top Reinf.	AAN		4/29/2005	3/29/2005	4/4/2005	4/22/2005	4/25/2005	
	03300-075	001	Shp: SP Columns 3-7	AAN		4/29/2005	4/5/2005	4/14/2005	5/3/2005	5/3/2005	
	03300-076	001	Stair #3 Landing Detail	APP		4/29/2005	4/14/2005	4/15/2005	5/23/2005	5/25/2005	
	03300-077	001	Shp: SP Stair #3 Shearwalls L 6-R	AAN		4/29/2005	4/5/2005	4/15/2005	5/3/2005	5/3/2005	
	03300-078	001	Shp: SP Shearwalls L6-R	AAN		4/29/2005	4/5/2005	4/15/2005	5/3/2005	5/3/2005	
	03300-079	001	Shp: Column Schedule 7-R	AAN		4/29/2005	4/8/2005	4/15/2005	5/3/2005	5/3/2005	
	03300-080	001	Shp: SP Slab on Grade	APP		4/29/2005	4/8/2005	4/15/2005	5/3/2005	5/3/2005	
	03300-081	001	Shp: SP Level 4 Slab Bottom	AAN		4/29/2005	4/13/2005	4/18/2005	5/10/2005	5/10/2005	
	03300-082	001	Shp Dwg SP-Level 4 Beams	AAN		4/29/2005	4/13/2005	4/18/2005	5/10/2005	5/10/2005	
	03300-083	001	Shp Dwg:SP Level 4 Slab Top	AAN		4/29/2005	4/13/2005	4/18/2005	5/10/2005	5/10/2005	
	03300-084	001	Shp Dwgs: CB- Columns to 3rd Flr	AAN		4/29/2005	4/13/2005	4/20/2005	5/20/2005	5/20/2005	
	03300-085	001	Shp Dwg:SP Stair 4 Shearwalls	AAN		4/29/2005	4/7/2005	4/20/2005	5/3/2005	5/3/2005	
	03300-086	001	Shp Dwg: SP Stair 4 Shearwalls 6-R	AAN		4/29/2005	4/7/2005	4/20/2005	5/3/2005	5/3/2005	
	03300-087	001	Shp Dwg: SP Stair 4 Shearwalls 6-R	AAN		4/29/2005	4/7/2005	4/20/2005	5/3/2005	5/3/2005	
	03300-088	001	Shp Dwg:SP Lvl 5, 7, 9 Btm Rebar	AAN		4/29/2005	4/21/2005	4/25/2005	5/23/2005	5/25/2005	
	03300-089	001	Shp Dwg: SP Lvl 5,7, 9 North Beams	AAN		4/29/2005	4/21/2005	4/25/2005	5/23/2005	5/25/2005	
	03300-090	001	Shp Dwg: Lvl 5,7,9 South Beams	AAN		4/29/2005	4/21/2005	4/25/2005	5/23/2005	5/25/2005	
	03300-091	001	Shp Dwg: SP Level 5, 7, 9 Top Rebar	AAN		4/29/2005	4/21/2005	4/25/2005	5/23/2005	5/25/2005	
	03300-092	001	Shp Dwg: CB- Stair 2 Shear A-A, B-B	AAN		4/29/2005	4/25/2005	5/3/2005	5/23/2005	5/25/2005	
	03300-093	001	ShDwg: CB Stair 2 Shear C-C, D-D, E	AAN		4/29/2005	4/25/2005	5/3/2005	5/23/2005	5/25/2005	
	03300-094	001	Shp Dwg: CB Stair #2 Shear 4-Roof	AAN		4/29/2005	4/25/2005	5/3/2005	5/23/2005	5/25/2005	
	03300-095	001	Shp Dwg:Stair 2 Shear 4-R Schedule	AAN		4/29/2005	4/25/2005	5/3/2005	5/23/2005	5/25/2005	
	03300-096	002	Shop: Site Retaining Wall	AAN		4/29/2005	5/12/2005	5/12/2005	5/17/2005	7/6/2005	
	03300-097	001	Shp Dwg: SP Rebar 6, 8 & 10 Fl.	AAN		4/29/2005	5/23/2005	5/31/2005	6/10/2005	6/13/2005	
	03300-098	001	Charles Bldg. Reiforecment	AAN		4/29/2005	5/17/2005	6/2/2005	6/14/2005	6/15/2005	
	03300-099A	001	Charles Bldg. Reinforcement	AAN		4/29/2005	5/17/2005	6/2/2005	6/27/2005	6/29/2005	
	03300-099B	001	Charles Bldg. Reinforcement	APP		4/29/2005	5/17/2005	6/2/2005	6/27/2005	6/29/2005	
	03300-099C	001	Charles Bldg. Reinforcement	AAN		4/29/2005	5/17/2005	6/2/2005	6/27/2005	6/29/2005	
	03300-100	001	Charles bldg. Reinforcement	AAN		4/29/2005	5/23/2005	6/3/2005	6/14/2005	6/15/2005	
	03300-101	001	Charles bldg. Reinforcement	APP		4/29/2005	5/23/2005	6/3/2005	6/14/2005	6/15/2005	
	03300-102A	001	Charles bldg. Reinforcement	ANR		4/29/2005	5/23/2005	6/3/2005	6/27/2005	6/29/2005	MILLER
	03300-102B	001	Charles bldg. Reinforcement	AAN		4/29/2005	5/16/2005	6/3/2005	6/27/2005	6/29/2005	
	03300-102C	001	Charles bldg. Reinforcement	AAN		4/29/2005	5/23/2005	6/3/2005	6/27/2005	6/29/2005	
	03300-102D	001	Charles bldg. Reinforcement	APP		4/29/2005	5/23/2005	6/3/2005	6/27/2005	6/29/2005	
	03300-103	001	St Paul 3rd fl Construction Joint	AAN		4/29/2005	6/14/2005	6/14/2005	6/16/2005	6/17/2005	
	03300-104A	001	Design Mixes: 3500 AEA	APP		4/29/2005	6/23/2005	6/28/2005	7/7/2005	7/8/2005	
	03300-104B	001	Design Mixes: 3500 NAEA	APP		4/29/2005	6/23/2005	6/28/2005	7/7/2005	7/8/2005	
	03300-104C	001	Design Mixes: 4000 AEA .50	APP		4/29/2005	6/23/2005	6/28/2005	7/7/2005	7/8/2005	
	03300-104D	001	Design Mixes: 4000 NAEA .50	APP		4/29/2005	6/23/2005	6/28/2005	7/7/2005	7/8/2005	

FIGURE 11.1   A standard shop drawing log.

RFI #	Subject	Date Created	Date Req'd	Date Resp	Question	Answer
047	Penthouse Wall Finish Clarification	1/26/2005	2/2/2005		Drawing A.6:9 Detail 6: Should the penthouse wall finish be just dense glass or does it require any waterproofing (we could bring the rubber roof membrane all the way to the cant strip since it seems not to be the case).	
048	Concrete beam reinforcement	1/26/2005	2/2/2005	1/31/2005	Drawings JBS-1.3 and JBS-1.4 opening between L1&M and C&D lines Revision 6 is deleting the concrete beams around the opening. Please confirm.	Yes, this is confirmed.
049	2 1/2" Chilled Water Supply and Return Valves and Capped for Future	1/31/2005	2/7/2005		Reference Drawings JB.H.1.1 This line is shown penetrating the Royal Room Wall around the C line. Do you want us to penetrate the Room Wall or leave it on the Tenant Side. Please Advise	
050	Magnetic Starters	1/31/2005	2/7/2005		Revision #6 is changing the magnetic starters for roof mounted equipment from indoor to outdoor. The durability is usually higher when installed inside. Please confirm the change.	

Total Number of RFIs for this project: 50

FIGURE 11.2 The RFI log.

- A brief description of the question (the detailed question will have been included in the RFI itself)

- An answer or response from the architect (the detailed answer will be stated in full in the architect's response, accompanying the return of the RFI)

A periodic review of outstanding RFIs/RFCs should take place at the weekly owner's meeting, where their status should be documented accordingly. There again, the monitoring process is important. Many of these RFIs will, when answered, impact either cost or time, or both, and as such will generate a change-order proposal, starting another process that will require close monitoring.

Field Conditions Documentation

When a field condition occurs that requires a construction detail, procedure, or dimension to be changed, a Request for Clarification often takes the form of a field information memo (FIM). FIMs are also used when a contractor wishes to change a construction detail and needs to create a sketch or narrative in order to convey this information to the architect. Conversely, when an architect during a site visit requests a change, this data can be transmitted to the contractor's field office via a FIM.

Another similar document is the ASI (architect's supplementary instruction), which may also be issued after a visit to the field. This directive may or may not have cost and time implications. Some contractors include FIMs and ASIs on the RFI log because they are somewhat similar in nature.

The affect on "as-builts"

RFIs, RFCs, FIMs, and ASIs impact another part of the project: the "as-built" drawings. Subcontractors should be provided with copies of all of these documents as they come forth from the architect, not only to determine if their scope of work is impacted but to alert them to actual changes in the work that have no cost impact. Even a no-cost charge to the plans may affect the as-built drawing(s) for which a specific subcontract has the responsibility to produce. This is a further reason to distribute all such documents to the field and to interested subcontractors. Some general contractors make it a practice to send copies of all RFIs, ASIs, and FIMs to all subcontractors just to ensure that all the interested parties have received the information they need.

Field inspections required before close-in.   A subcontractor or general contractor may, at times, request an inspection by the architect/engineer

of an item of work before it is covered or enclosed. This will be in addition to the inspections required by local building officials during the course of construction. There may be a specific structural detail or mechanical piping installation that the project manager would like to have inspected, or that may be required by the contract to be inspected before being enclosed. An e-mail requesting the inspection, a verification that the inspection was made, and a report of the results of the inspection should be documented. The general conditions of the contract, if you recall, state that when a portion of work has been covered without being inspected as requested by the architect/engineer, the contractor may have to uncover the area in question for inspection. The cost to uncover and recover thus may be borne by the general contractor.

The Coordination Process

Coordination problems occur frequently and they reflect the complexity of today's construction project. Making things fit in their designated space requires a great deal of time and effort and a section of the specifications generally includes a directive that the general contractor is to instruct that all systems are to be properly coordinated. A typical specification will read as follows:

> General contractor shall circulate coordination drawings to the following subcontractors (usually MEP, fire protection, electrical) and any other installers whose work may conflict with other work. Each of these subcontractors shall accurately and neatly show the actual size and location of the respective equipment and work. Each subcontractor shall note apparent conflicts, suggest alternate solutions and return drawings to the General Contractor.

Several pass-throughs are required before a "coordinated" set of drawings is produced and transmitted to the architect for review and comment. Not infrequently, it becomes apparent that some mechanical or electrical components will not fit in their allotted space above a ceiling or within partitions and the project manager must present the facts, as they are known, to the architect so that a resolution can be obtained. If a solution can be offered, it should be sent along with the drawings— and if additional costs are involved, they should be included.

This process often does not go smoothly and costs may be called into question by the architect, or alternate schemes may be passed back and forth to avoid increased costs. This entire process, if it becomes bogged down must be documented until a mutually agreed upon solution is found. But the problems are not over yet!

Once a coordinated set of shop drawings has been prepared and approved, each subcontractor participating in the process should "sign-off" on all the drawings, either directly on the drawings themselves or in a letter accepting the drawings. Too often, a subcontractor will agree with their portion of the coordination process only to find that their initial review was not correct and their work will not fit as planned. By "signing off" and accepting the drawings, any future costs associated with making work fit will be charged to their account.

Other Important Documents

Cost proposal or cost estimate requests

All cost proposals or estimates for extra work, if presented verbally, must be confirmed in writing so that there is no misunderstanding regarding the amount of the quote or the scope of the work involved. All such proposed change orders (PCOs) should be numbered sequentially for ease of identification and for ease of tracking.

Two response times should be monitored, both of which are critical:

- The response from the vendor or subcontractor involved in the pricing of the PCO
- The response from the architect prior to their acceptance or rejection of the PCO

When either an architect or owner takes an exceptionally long time to review a particular proposed change order and implementation affects job progress, retrofitting or rework may be required at additional cost to the general contractor, who in turn should pass these costs onto the owner. Without tracking and notification that a quick response is required, when these added costs are presented to the owner, the answer will assuredly be "Why didn't you tell me you needed approval in a hurry? I certainly would have responded promptly."

It is therefore a good idea to insert the following caveat in every PCO:

> If this proposal is not acted upon within 10 days after receipt, at the option of _____ (*the general contractor*), it may be modified or withdrawn.

Conditions that impact completion time

A contract containing a liquidated damages clause requires that all justifiable delays be documented as they occur whether or not it appears that they may ultimately affect the contract completion time. Even when the contract does not contain the LD clause, delays due to conditions beyond the contractor's control should be documented in some fashion, both in the superintendent's log book and in the project meeting minutes.

When work stoppages due to strikes or other types of job actions occur, the architect should be notified in writing the day the strike or dispute happens. This notification can state that at such and such a time the duration and effect of the strike or dispute was established; another notice should be forwarded later reflecting the effect, if any, on the construction schedule.

Alternates and allowance reconciliation.   Many contracts contain "alternates" for additional work that the owner can incorporate into their project. The alternate concept allows an owner to get a fixed price for additional work that, depending upon budget and other considerations, they may decide to incorporate in the project as construction progresses. However, some of these alternates must be accepted or rejected at specific stages in the construction cycle so as not to disrupt or delay progress. An alternate for water and waste lines in an employee lounge, for example, must be accepted or rejected while mechanical rough-ins are in progress in that area in order to maintain the contract price for this work. The project manager, preferably at the start of construction, should submit a list of each alternate to the architect with a date by which it must be accepted or rejected in order to maintain the price structure included in the contract. *In fact, this time frame for acceptance/rejection should have been included in the contract, probably in the Allowance/Alternate exhibit.*

If there is no response to these deadlines from the architect, the project manager can reference their initial letter and stipulate that the "alternate" in question cannot be incorporated into the project at the "contract" price, and if the "alternate" is to be elected, the additional costs will be required.

With regard to "allowance" items, the project manager has to review the contract to determine how they are to be reconciled. Will the architect be soliciting bids for that item? If so, select a vendor or subcontractor so that the general contractor can reconcile costs on that basis and issue an add or deduct change order. Or is the project manager to solicit bids when the work is defined and then present them to the architect for approval? Depending upon their nature, some of the allowance items might be time-sensitive.

Finish hardware is often an "allowance" item and since an approved hardware schedule is required in order to purchase doors and frames, it becomes an allowance item that needs resolution early in the project.

Certain floor treatments, although not needed until later in the project, may require resolution early in the schedule if they require depressions in the cast-in-place concrete slab.

Documentation of all allowances and alternates acceptance/rejection/reconciliation efforts will defuse any "Why didn't you tell me you needed a quick response?" comments.

Unforeseen subsurface or unusual conditions. Whenever conditions encountered below the surface or above ground, in the opinion of the project manager, are unusual, or at variance with the norm, or of a peculiar nature that could have a cost or time impact, a letter ought to be sent to the architect outlining these conditions. If groundwater is observed during excavation and it does not appear to be significant at the time, it is best to document its existence. Even if its presence was noted on a geotech report, if tests were made during a dry spell only a trickle may have been observed, but after a significant rainfall there could be a river. Document!

The topic of unforeseen or "differing" site conditions is dealt with in more detail in the following chapter on claims and disputes.

Disputes, claims, or requests for arbitration. Today's congenial relationship with an architect, engineer, owner, or subcontractor can deteriorate rapidly for any number of reasons—usually triggered by problems relating to money. If conditions arise that have a potential for a future dispute, document them even though the relations between all parties to the contract are, at that time, favorable. Documentation might be extremely important in the future if that initial congenial atmosphere deteriorates. And when that happens, everyone's memory of past events tends to be selective, recalling only those facts that support their position, and forgetting others that don't.

Monthly requisitions. Will the schedule of values as submitted prior to, or shortly after, the contract award be acceptable, or will more detail be required of the contractor? How are stored materials, both onsite and offsite to be handled in a requisition?

The project manager should obtain an agreement on the requisition format at the inception of the project. Some architects prefer to review a preliminary or "pencil copy" requisition with the project manager a few days in advance of the formal submission. The project manager should then promptly submit the monthly requisitions with a letter of transmittal so that there is a record of the date when each was submitted. Then, any late payments, if they do occur, cannot be attributable to the "late" submission by the general contractor.

Of course, all necessary documentation of costs, if a GMP-type contract is in force, must be submitted with the requisition so the architect can reconcile actual costs incurred during the period with invoices, bills, subcontract requisition forms, and so forth.

Documentation of close-out requirements

Proper compliance with the contract closeout procedures, and the proper documentation of those procedures, are necessary to trigger the contractor's final payment and start the clock ticking on any retainage requirements. There will be requirements for as-built drawings, owner's operating and maintenance manuals (O&Ms), and warranties and guarantees, most of which must be supplied by subcontractors and vendors. Requests to subcontractors and vendors for these documents should be made by letter or transmittal well in advance of the completion of the project.

When subcontractors or vendors don't respond to repeated requests for this information, remind them that final payment for all subcontractors will not be forthcoming until all of this documentation is received and approved by the architect. Submission of all lien waivers and other certificates to the architect as part of the closing documents package, should be transmitted to them with a cover letter and through a delivery service where receipt can be verified. And, of course, one copy should always be retained for the files.

If the contract stipulates that extra materials for maintenance or replacement are required (attic stock), such as floor tile, carpet scraps over a certain size, ceiling tiles, or extra cans of paint, a signed receipt should be obtained from the owner's representative accepting delivery of these materials so there is a clear record of the occurrence.

Sign-offs on acceptance by the architect/engineer of equipment such as boilers, hot-water heaters, DX units, cooling towers, and emergency generators are also necessary in order to establish the date of official acceptance, and, most importantly, the commencement of the warranty period.

Documentation to the Subcontractors

Most general contractors today subcontract the greater portion of work to specialty contractors, and the project manager assumes the responsibility for the administration of these contracts. Monitoring subcontractor performance will occupy the major portion of the project manager's time as construction demands and schedules accelerate.

The formal subcontract agreement should contain the terms and conditions expected from both parties in the contract, and is usually amended with a series of exhibits to cover specific obligations. Unless the project manager and subcontractor are familiar with the provisions in these agreements, how else can they be managed and monitored?

Although this appears to be a basic principle, many project managers and subcontractors have *never completely read nor fully understood* all of the provisions in their subcontract agreement.

These lengthy legal documents vary from contractor to contractor, but they all seem to contain more or less the same provisions to protect the contractor from poor subcontractor performance. Proper documentation of subcontractor performance, as it relates to contract language, is an important element of the project manager's administrative responsibilities because a subcontractor's performance, or lack thereof, can severely impact the work of other subcontractors and the project as a whole.

The standard subcontractor agreement contains provisions for progress payments, the scheduling of the work, the scope of work, and the cost of that work, as well as insurance requirements, and compliance with local, state, and federal laws and labor practices, along with a long list of work practices. The standard boilerplate language is supplemented by specific job requirements such as:

- The precise scope of work to be included in the contract, often in the form of "Inclusion" and "Exclusion" lists, preceded by a list of specific plans and specifications included in the agreement.

- The time frame in which the work is to take place.

- The name of the owner of the project, the date of the agreement between the owner and general contractor, the name of the architect, and the name and address of the project. (This is important when the subcontract agreement has a "pass-through" provision.)

- The contract sum, requisition period, and retainage to be applied to each payment request.

- The applicable addenda, alternates, allowances, and other documents included in the bid proposal.

- A tax-exempt status, if applicable, and the tax-exempt number or certificate.

Is the scope of the work fully understood by all?

Has the subcontractor acknowledged that their price reflects the scope of the work as outlined in the latest drawings as referenced in the subcontract agreement? All too frequently, a subcontractor will sign a subcontract agreement, start to perform their work, and somewhere in the early stages of construction indicate that they had not really received all of the drawings referenced in the subcontract agreement. Now that they have seen all of these drawings and reviewed all of the specification requirements, they find that the cost of their work has increased and will require an adjustment in price. Situations like this can be avoided if a subcontractor interview form described in the Chap. 8 on buy-outs had been used during the negotiations, thereby documenting the specific scope of work discussed and agreed upon during negotiations.

The project manager, if and when the subcontractor disputes their scope of work, can quite simply ask "But why did you sign our subcontract agreement that referenced the proper contract drawings and contract sum when you did not have the latest drawings?" The answers will range from "I guess I didn't notice the difference in drawing dates" to "I never received a set of the latest drawings" to "I don't know." No matter what the answer, the project manager should attempt to resolve the problem amicably rather than delay resolution. Thus, it might be best to possibly allow a slight adjustment in the contract sum, while making a mental note to keep an eye on this subcontractor who signs documents without reading them.

Avoiding problems related to subcontractor misunderstandings

One way to avoid misunderstandings is to transmit whatever plans and/or specifications there are to a subcontractor via a transmittal that identifies the specific documents being sent, the date of each drawing enclosed, and the date of the specifications or any addendums or bulletins.

This will provide a record to confirm or deny the allegation that the most recent drawings, specifications, or addendums were never received and that the subcontractor's quoted price did or did not include the scope of work contained in these documents. Transmittals used during the bidding process can serve another purpose: to keep track of the distribution of the drawings so they can be retrieved from nonsuccessful bidders and redistributed to subcontractors who will be performing the work.

Addressing questionable items in the agreement.   Any potential for misunderstandings of scope should be clearly written and included in the subcontract agreement. For instance, if the general contractor requires an electrician be available during *working hours* to turn the temporary electric power on or off, this requires more specific information in the contract. What is the normal workday? If the electrician's workday ends at 3:30 P.M., but other trades normally work until 4:00 P.M., agree on whose "normal workday" will be used. If it is the intention to have the electrician stand by until 4:00 P.M. every day to turn off the temporary power, state these hours of operation in the subcontract agreement.

The time frame for subcontracted work—the anticipated start and completion—is an important part of the contract. Inserting dates avoids any misunderstanding that the price quoted anticipates a specific starting date and that any increased labor rates for work performed during the period have been taken into account. The cost of materials and equipment necessary for the work will take into account the time frame required for the construction period.

There is, however, a danger in inserting specific start and finish dates. If the project is delayed and the time frame for that subcontractor's work is delayed, they can use these specific dates to document their claim for extended general conditions and possibly other costs associated with delays. So, before inserting specific start-finish dates in the subcontract agreement, carefully weigh the consequences of doing so.

Linking the subcontract agreement with the owner's contract

Most subcontract agreements include a provision that links that agreement with the terms and conditions of the general contractor's contract with the owner. In fact, Article 5.3 of the 1997 edition of AIA Document A201, the General Conditions, states that "where legally required for validity" the general contractor will require each subcontractor, "to the extent of the work to be performed by the Subcontractor," to be bound to the general contractor by the terms of the contract documents and assume toward the general contractor all the obligations and responsibilities that the general contractor assumes toward the owner and the architect.

The subcontractor's payment schedule will then be tied to the payment schedule from the owner to the general contractor. The general contractor's philosophy is "We can't pay you on the date agreed upon in the subcontract agreement if we have not yet received payment from the owner." The subcontractor's answer may well be that they have a contract with the general contractor and not the owner, and unless they are made aware of the fact that their payment schedule is indeed tied directly to the owner's, future disagreements will certainly occur.

Several states have enacted laws that reject the "pay when paid" clause in public works as not in the best interests of the public. In fact, various subcontractor organizations have petitioned the contractor associations to strike the "pay when paid" clause. Project managers need to keep abreast of new developments in the "pay when paid" contract clause in both public and private work to ensure that they are on firm ground in keeping this provision in their subcontract agreements.

Subcontractor performance—the major concern

One of the more restrictive clauses inserted into most subcontract agreements has to do with the remedies to correct a subcontractor's poor or otherwise unacceptable performance. The general contractor must be able to control the performance of a lagging subcontractor so that the entire project's progress is not severely impacted. One of three often-used clauses concerning performance needs to be included in every subcontract agreement.

Contract Clause Option 1.   Should the subcontractor be adjudged bankrupt or insolvent or repeatedly fail to prosecute the work hereunder with promptness and diligence in keeping with the then-existing work schedule, the contractor may take possession of all materials, equipment, tools, construction equipment, and machinery of the subcontractor after serving three (3) days notice to that effect and may through itself or others provide labor, equipment, and materials to prosecute and finish the work hereunder.

Contract Clause Option 2.   Should the subcontractor fail to prosecute the work or any part thereof with promptness and diligence, or fail to supply a sufficiency of properly skilled workers or materials of proper quality, or fail in any other aspect to comply with the contract documents, the contractor shall be at liberty, after seventy-two (72) hours written notice to the subcontractor, to provide such labor and materials as may be necessary to complete the work and to deduct the cost and expense thereof from any money then due or thereafter to become due to the subcontractor.

Contract Clause Option 3.   Should the subcontractor be adjudged bankrupt or should the subcontractor at any time refuse or neglect to supply a sufficient number of skilled workmen or sufficient materials of the proper quality, or fail in any respect to prosecute the work with promptness and diligence in keeping with the project schedule, or allow a lien to be filed against the building or cause by any action the stoppage of, or interference with, the work of other trades, or fail in the opinion of the contractor in the performance of any of the agreements contained herein, or fail to comply with any order given to him by the contractor or architect in accordance with the provisions of this subcontract, the contractor shall be entitled to provide for the account of the subcontractor and without terminating this subcontract, any such labor and materials, after 24 hours notice to the effect, and to deduct the cost thereof from any money then due or thereafter to become due, or the contractor at his option at any time may terminate this subcontract after 24 hours notice to that effect.

The Associated General Contractors of America (AGC) has published a subcontract agreement in conjunction with the American Subcontractors Association Inc. (ASA) and the Associated Specialty Contractors (ASC) known as AGC Document No.640/ASA Document No.4100/ASC Form No.52, 1994 edition and is intended to be used in conjunction with AIA Document A201. These contract forms are available from any local AGC chapter at a nominal cost.

Article 16 of the AGC contract entitled "Recourse by Contractor" includes somewhat the same provision for failure of performance as the three options listed above, giving the subcontractor three (3) working days to perform—or else!

These restrictions require written notification to the subcontractor in order to be enforceable, and without a document trail it will be difficult to institute corrective action.

Danger signs and how to interpret them

There are easily recognizable danger signs that portend trouble with subcontractors. Watch out for any of the following danger signs and document them:

Lack of adequate manpower.   There may always be disagreements between the general contractor and the subcontractor about what constitutes an adequate workforce. But we are not talking here about needing 12 workers on the job instead of the 10 currently working, we're talking about *two* workers on the job for a week when there was an obvious need for five times that number. A situation like this, if allowed to continue, will no doubt affect other trades; thus, something must be done and done quickly. When a subcontractor is experiencing financial difficulties, it will be most noticeable in the size of their work crew. Workers must be paid weekly; no pay, no work. Material or equipment suppliers invoice on a monthly basis and may agree to extend their credit terms to 45, 60, or even 90 days, but this flexibility does not extend to the weekly payroll.

When subcontractor work crews are insufficient to maintain job progress, notify the subcontractor in writing. If the situation persists, invoke the subcontract agreement notification provisions regarding performance. This letter starts the "lack of performance" clock, and at this point you should have a heart-to-heart talk with the subcontractor to find out exactly what the problem is. If the problem is a temporary shortage of funds, consider advancing monies for weekly payroll until the problem is resolved, *but only if the subcontractor has put sufficient work in place to justify an advance.*

Ask the question: "Are there sufficient funds remaining in that subcontractor's account so if they default on their obligation, another subcontractor can complete the work and still stay within budget?"

Delays in submitting shop drawings.   A subcontractor, after being awarded a contract, will purchase materials and equipment and in so doing may often try to "package" various materials and equipment with one or two vendors in order to obtain the best possible price for the larger value of that package.

Under normal circumstances, this can take several weeks for a subcontractor to test the market for the most competitive price, issue their purchase order, and request the necessary shop drawings. Vendors do not issue shop drawings until they receive a firm order from a subcontractor, so the issue of prompt purchasing by a subcontractor becomes a matter of concern for the project manager. Prolonged negotiations between vendor and subcontractor can be a reflection of the subcontractor's inefficient purchasing department or a desperate need to get rock-bottom prices to compensate for a too-low bid. Or worse, the subcontractor may be searching for a supply house or vendor to accept their order because they have such a poor credit rating.

Shop drawings that are not being submitted in a reasonable, timely manner should prompt a letter to be generated advising the subcontractor that any further delays in the submission of a particular shop drawing (or group of drawings) will seriously affect the progress of the job, and that their submission must be received by a certain date. If the drawings are not received within that time frame, another letter should be sent invoking the appropriate paragraph in the subcontract agreement pertaining to nonperformance and delays.

Inability to provide day-to-day working materials.   If, along with reduced manpower, adequate day-to-day working materials are not readily accessible on the job when there is adequate manpower to do the work, this could portend big problems. We are talking about, for example, a plumbing contractor having trouble keeping enough small-diameter copper pipe and/or fittings on the job even though the field supervisor calls the office daily requesting these materials. This could signal a lack of credit at the local supply house. After sending the subcontractor written notice of poor performance, a telephone call to the subcontractor's supplier might be helpful in determining or confirming the problem connected with the shortage of supplies.

During one of this author's projects years ago, he overheard the electrical crew supervisor on a telephone call to the supply house indicate that he would use his personal credit card to purchase some conduit and couplings. This was a red flag I couldn't ignore.

Requests for joint checks.   Requests from subcontractors to have joint checks issued are not necessarily danger signs. A valued subcontractor might be dealing with a new supplier who feels more comfortable using joint checks for the payment of supplies in their first transaction with that subcontractor. A subcontractor might be involved in that first big job which normally would be beyond the company's present credit line and their supplier may need the added assurance of a joint check to accept the order.

Joint checks are desirable from the general contractor's viewpoint because it affords them added insurance that monies they disperse are being passed on to the appropriate supplier or vendor. The danger flag is raised when a subcontractor with a good track record suddenly asks that joint checks be issued on the new project. Does this mean that they cannot obtain adequate credit any other way? The subcontractor should explain their sudden request for joint checks to determine if there is a problem. Another danger sign is raised when a subcontractor, halfway through the job, requests a joint check for a supplier who had been supplying materials to the project without that requirement. When initiating a joint check policy, a joint check agreement should be prepared for signature by a subcontractor's officer and placed in their contract file.

Delinquent payroll deduction notices. When a subcontractor is party to a union collective bargaining agreement and the required payroll contributions are not being made to the appropriate union office, the project manager had better find out why these payments are not being made. Of course, this would be true of monies owed to local, state, or federal agencies for various taxes as well.

The subcontractor should be requested to explain the circumstances surrounding all nonpayment issues and submit a plan indicating how they expect to meet these obligations. Since the general contractor may be liable for all such funds if the subcontractor defaults, a properly written letter outlining the problems should be sent to the subcontractor. The letter should state that unless the general contractor receives a satisfactory repayment schedule, they may withhold funds sufficient to satisfy the payroll contributions or taxes should the need arise.

Requests from the subcontractor or their supplier for immediate payment. If the subcontractor has been routinely submitting monthly requisition requests and receiving payment within the normal pay period but suddenly asks for accelerated payment schedules, find out the reasons for this change. It could be a temporary problem caused by late payments from other general contractors or the beginning of a bigger problem.

**That low subcontract bid—are problems
waiting to surface?**

When a subcontract agreement has been awarded to a company and their competitive bid was substantially lower than their competitors, that initial "buy-out" may quickly disappear.

We are not talking about the subcontractor who submitted a bid of $95,000 when the other four subcontractors' quotes ranged from $98,500 to $105,000. This variation of 6.5 percent between high and low bidders is a fair spread. What should be of concern is one subcontractor's bid of $75,000 that is 24-percent lower than the *lowest* bid. If that $75,000 bid is accepted, the project manager's danger antenna should be raised all the way.

General contractors, when faced with a decision whether to accept or reject a substantially lower bid, will look at the situation differently.

One general contractor will reason that if this very low-bid subcontractor is selected, the potential savings look tempting, but the subcontractor probably omitted some item of work and will not be able to complete the job. The additional costs to complete the work, assuming the subcontractor defaults, will probably be greater than the second bidder's price and will also impact the work progress of the entire project; therefore, this low bidder should be disqualified.

Another general contractor will reason that maybe this unrealistically low-priced subcontractor does not know what they are doing and could possibly complete the work before they find out that they have lost money. And if this subcontractor defaults near the end of the project and another subcontractor is engaged to complete the work, the total cost may still be below the second bidder's price. So this subcontractor is brought on board and fingers are crossed. After all, isn't this a risky business?

If a low bid subcontractor is awarded a contract, against the project manager's best judgment, but not that of his boss, the project manager should accumulate a list of all material and equipment suppliers, as well as second and third-tier subcontractors, so that if the subcontractor defaults, it will be easier to uncover those lower-tier subs and suppliers that may not have been paid.

This is a Las Vegas–type decision that once made demands that steps be taken to live with that decision, and to be prepared for the inevitable. The odds can be reduced substantially if the very low bidder is called into the office before an award is made so that their bid can be carefully scrutinized to determine why it is significantly lower than their competitors. If a major portion of the work has been inadvertently omitted from the subcontractor's estimate, then the general contractor's decision may be somewhat affected.

But when things begin to go wrong, document everything, every day, including written memorandums of telephone conversations, because every bit of documentation will surely be needed.

Documentation When Major Drawing Revisions Are Made

There may come a time when an architect makes series revisions to the contract drawings because of a major change requested by the owner. These changes may encompass a substantial number of architectural, mechanical, and electrical drawings. And this usually occurs when work at the site is progressing smoothly and rapidly.

An interior designer hired by the owner to provide furniture layouts and decorating assistance in an office project may develop furniture drawings to locate desks, workstations, possibly task lighting and power, and data and voice communication terminals that impact the location of electrical and HVAC devices. Demountable partitions, of either partial or full height may require relocation of ceiling lighting fixtures, sprinkler heads, and VAV terminal devices.

The first priority is to get these drawings to all affected subcontractors as quickly as possible via transmittals with a note to review them and respond as to the nature of any changes and their cost implications within X number of days. The subcontractors should be instructed to identify changes that either add or delete scope, and all such changes having a cost impact must be clearly defined and accompanied by a detailed labor and material cost breakdown to facilitate review and comment from the architect.

If the subcontractors can be requested to delay work in the affected areas for several days at no impact to their work schedule, so much the better. The architect should be advised that all subcontractors whose work is affected by these potential changes have been directed to refrain from working in those areas for X days and that a prompt review and authorization to proceed is required to avoid added costs. It is important to note in this letter that if no response is received from the architect within the required period of time, all affected subcontractors will be permitted to commence work in the areas where changes are being considered, and any costs to retrofit will be added to the cost proposal previously submitted.

A meeting with the architect and owner a day or two after submitting the cost proposal in this type of situation may facilitate a quick decision. Multiple revisions to a wide range of plans are often accomplished by an architect via a series of small sketches, not only for time expediency but to save printing costs, but frequently this process creates other problems.

What to do with all of those 8^1/$_2$ × 11s

A multitude of 8^1/$_2$ × 11 drawing revisions or sketches can create havoc when major plan changes are to be initiated. First of all, there is the problem of gathering them together for submission to all the interested parties—subcontractors and vendors. If one or more critical sketches become lost along the way, the consequences are easy to imagine.

And as stated earlier, architects may have objections to the reissuance of a full set of revised drawings for both time and money reasons. The project superintendent and various subcontractor crew supervisors will be seen busily pasting these sheets over the applicable sections of the drawings in order to insure that all changes are properly noted.

But there are hidden dangers that lay ahead in situations like this:

- Some sketches may have gotten lost in the process.
- Some sketches many not have been distributed to all subcontractors.
- Some field supervisor may have placed these sketches in a loose-leaf binder but failed to refer to them when required and therefore will not be incorporated in the work.

Architects may be reluctant to reissue full-size drawings incorporating all of the changes, but the project manager should request that this blizzard of small sketches be incorporated into an updated set of drawings. In the letter to the architect, the following disclaimer might be considered:

> Due to the nature and extent of the changes reflected in the (*number of—20, 30, and so on*) sketches generated by your office during the period (*time*), we request that the appropriate full-size drawings be reissued no later than (*date*) reflecting these changes, or else we cannot be held responsible for their incorporation into the work.

Even if the architect refuses to reissue the drawings, with such a letter on file, if correction of some missed details is required, the project manager would have gone on record with their concerns.

Documentation Required When Contracting with Public Agencies

When entering into a construction contract with local, state, and federal entities, the project manager should note that these agencies have their own contract formats with specific requirements, and sometimes unusual requirements, for submission of shop drawings, change orders, requisitions, schedules, releases, affidavits, and various work rules that vary from those in private sector work Additionally, requirements to comply with the Davis-Bacon Act, and various mandated executive orders and laws relating to equal opportunity, minority hiring, and environmental issues will need to be addressed.

There may even be an extensive list of requirements before the first requisition is submitted. The close-out procedures for these types of projects oftentimes contain numerous documents to be filed with the agency that involve subcontractor obligations as well.

The project manager should carefully read all of the general, special, and supplementary conditions that accompany the contract, highlighting all requirements to be followed before, during, and at the close of construction. Subcontractors should be given copies of any requirements affecting their work, and at the first project meeting these unusual or special requirements should be reviewed with them and documented in the project meeting minutes. On the assumption that most contractors and subcontractors don't thoroughly read the contract "boiler plate," this extra effort by the project manager may make everyone's job a little easier.

The Davis-Bacon Act

The Davis-Bacon Act (DBA) requires the payment of prevailing wages on federal government construction projects in excess of $2000. Local and state public works projects, where federal funds are used, also require compliance with Davis-Bacon.

The Davis-Bacon Act became law during President Herbert Hoover's administration, when this country was in the midst of the Great Depression. The act required that all laborers and mechanics employed on the site of a federally funded construction project in excess of $5000 (later amended to $2000), must be paid rates determined to be prevailing in that area.

The U.S. Department of Housing and Urban Development (HUD) requires compliance with Davis-Bacon because of a labor provision contained in one of HUD's "related acts"—the U.S. National Housing Act of 1937, the Housing and Community Development Act of 1974, the National Affordable Housing Act of 1990, or the Native American Housing Assistance and Self-Determination Act of 1996.

Davis-Bacon requires that worker wages meet the highest prevailing wage in the area as stipulated by the Secretary of Labor. Today, the prevailing wage scale is regional, but may contain different wage rates for similar tradesmen working in various part of the same area. A laborer working in one part of the state may have a different wage rate from a laborer working in another part of the same state. These prevailing wages, more or less parallel union wage scales in the region.

The requirement to comply with Davis-Bacon and pay prevailing wages also requires certification that those wages were actually paid. A Statement of Compliance form, WH-347 and WH-348, requires the contractor to list each worker, their job classification and the hourly wage paid during the previous pay period.

Gross wages will be reported, and deductions for Social Security and other fringes are to be included, and then the net pay tabulated. The only workers that can be paid less than prevailing wages are apprentices and trainees who are registered in approved apprenticeship or training programs. The project manager should verify that the apprenticeship programs are "approved" because some may not be, meaning the request for payment may be rejected.

Workers on projects requiring compliance with DBA that have been hired as "piece workers" have weekly earnings calculated according to how much work they actually completed during the pay period. Employers reporting wages of piece workers must certify that weekly earnings are sufficient to satisfy the wage requirement based upon the prevailing wage rate for that period, including any overtime if incurred. If the weekly piece-rate earnings are not sufficient to meet this DBA standard, the employer must recompute weekly earnings based upon the actual hours worked times the rate of the prevailing wage rate and pay the difference to the affected employee.

The prime contractor is ultimately responsible and will be held liable for any wage restitution due the government by improper or false reporting, including workers employed by their subcontractors. Falsifying payroll records will be cause for further government legal action, and willful violation of the labor law is cause for disbarment from government projects for periods of up to three years. Falsifying government documents, meanwhile, is a criminal offense.

Complying with other government requirements

To ensure compliance with all documents required when a government project commences, it is wise to review the project specifications thoroughly before construction begins.

Make a checklist of documentation required by the General Conditions and when each item will be needed. A typical list could include the following requirements:

1. Contract signing
   a. Within 10 days, submit a schedule of value for approval for the purpose of the requisition format.
   b. An approved project sign should be installed before the first requisition is submitted (both the design and contents of the sign need approval).
   c. The field office should contain a separate space for the inspector's office, as well as a phone and computer terminal for their use. (Some contracts require a separate trailer for the inspector(s), the copier, file cabinet, and computer access.)
   d. Prevailing wage scales should be posted prominently with Executive Orders 3, 17, and Public Act 79-606—Notice of Non Segregated Facilities outside the office trailer.
2. To be submitted before construction starts
   a. An estimated progress schedule (sometimes an "S" curve is required)
   b. The list of subcontractors to date (update as required)
   c. Surety bonds
   d. A schedule of values
   e. Insurance certificates
   f. A site logistics plan
   g. A project organizational chart
3. During construction
   a. Weekly payroll certification
   b. Monthly manpower utilization reports
   c. Requisitions to be submitted by the 20th of month that projects work that will be completed by the end of the month (or other set dates).
   d. A list of other forms required during construction
4. Due before Substantial Completion can be obtained
   a. A certificate of compliance

b. An architect/engineer certificate of substantial completion, along with a list of escrow items and a punch list
c. Product warranties, guarantees, and operating and maintenance manuals (O&Ms)
d. A list of other closing documents such as a final waiver of liens, a Consent of Surety, and so on

This list should be prominently displayed in the project manager's work area so it is a constant reminder of the documentation and data required at specific project milestones.

Project Documentation from the Field

The project superintendent must be kept apprised of what the project manager is doing as it relates to his or her area of responsibility. A good rule-of-thumb for the project manager is to ask themselves: "If I were the super in this job, is this something I would need to know in order to run my job effectively?" If the answer is "Yes," send the information to the field. Sketches or revised drawings issued by the architect/engineer should be sent to the jobsite promptly and the transmittal should indicate what action is required—FYI (For Your Information), For File, Review with Subcontractor, and so on. All *approved* shop drawings and equipment catalog sheets will be sent to the job with accompanying transmittals, and the project manager, from time to time, ought to ensure that they are properly filed and not scattered around the field office where they can easily be lost.

Note: The word approved *is italicized. There may be occasions when it is necessary to send unapproved shop drawings to the field so that the superintendent can verify certain dimensions, or review and comment on installation details, but these unapproved shop drawings should be filed away or thrown away after being reviewed. There have been too many times when unapproved shop drawings are left on the plan table in the field office and an inquiring subcontractor will come in when no one else is around, obtain information from that unapproved shop drawing and proceed to follow directions that may prove to be totally different from the approved set.*

Verbal commitments between the project manager and the architect/engineer or a subcontractor that affect the ongoing work at the site should be e-mailed to the project superintendent.

The superintendent's record of daily activities

Every superintendent must keep a daily record of job activity, if for no other reason than to keep track of their own workers on the job and the hours they have worked for payroll purposes. Superintendents keep these job records in the form of either a bound daily diary or daily reporting sheets either handwritten or computer-generated. The diary, with entries made on a daily basis, fulfills the legal definition of a *business record* and may be introduced in court. A bound volume is required to meet this criteria and often individual daily reports filed in a loose-leaf binder may not meet these standards. Figure 11.3*a* is a page from a typical bound daily log book, while Fig. 11.3*b* is a typical computer-generated version.

The whole discussion of whether certain computer-generated documents meet the criteria of acceptable legal documents is ongoing, and a review of these types of documents with the company attorney is a good idea.

No matter what form the daily diary or daily log takes, the following information must be reported, at a minimum:

- Month, day, date, and year—Every day, whether work is performed or not, requires an entry. When no work takes place due to holidays or weekends, enter "No Work."

- Weather conditions and temperature—Should preferably be done at the start of work, mid-day, and at the end of the workday. If inclement weather occurs (rain, snow, sleet) include the amount (light, heavy, and so on).

- List of subcontractors on the site that day, the number of workers in their crew, the work performed, and the location within the building or onsite.

- The number of the company's workers onsite and the operations they were performing.

- Visitors to the project and the purpose of their visit. Should include any of the company's office staff, the owner, and so on.

- The list of inspections that took place, either by local building inspectors, testing labs, or the architect/engineer.

- Briefly describe the type of work performed that day and, if possible, the location within the building or site where it took place.

- Record deliveries of materials, equipment and refer to delivery ticket numbers if possible.

- Record any unusual events, occurrences, and work stoppages.

- Any accidents, whether reportable by OSHA or the insurance company or not.

Look at the daily diary as a reference book of project history. If required to recall events at the jobsite on a particular day a year or two

◄◖═► DAILY LOG ◄═◗►

Day: _____

August 24, 20____   Weather: 8:00 A.M. ___ Noon: ___ 4:00 P.M. ___

WORK FORCE   **SUBCONTRACTORS ON PROJECT & THEIR ACTIVITY**

Work Force	
Supt.	___
Foreman	___
Carpenters	___
Laborers	___
Masons	___
Oper, Engrs.	___
Iron Wrkers	___
Electricians	___
Plumbers	___
Steam Fittrs	___
Sheet Metal	___
Glazers	___
Roofers	___
Sprinkler	___
Painters	___
Tile Setters	___
Carpet Lyrs	___
Controls	___
_____	___
_____	___
_____	___
_____	___

DAILY ACTIVITY - VISITORS - INSPECTIONS

Supervisor's Signature: _____

FIGURE 11.3a   Two types of daily log book pages. (With permission from McGraw-Hill, New York.)

DAILY REPORT

Date:		
Superintendent		
Weath Cold, windy		
Visitors		

Project:

Personnel Sub-contractors	Number	Hours	Men	Remarks	Description of work
Piles Donaldson					
Sandblasting NER					
Excavation Scully	2 machines	8	2 lab 1 Fore 2 ops		Loading out conc. w/ trailer
Demo Deprizio	Brokk w/ op	our ticket			chopping, removing conc
Concrete S & F					
Marble/Ceramic					
Masonry Pizzotti					
Steel/Iron Marr					
Millwork					
Roofer					
Glass					
Ceilings					
Floor covering Ceramic					Change orders/Backcharges/Extra work
Painter					
Drywall					
Sprinkler					
Plumber					
Pipe fitter					
Duct work					
Electrician Sully					
Misc. Trade					Verbal discussions/Instructions

SDC:

Superintendent

FIGURE 11.3b (*Continued.*)

from now, will the entry in this daily diary allow the recall of events with some clarity?

Photographs: important documentary components

The old saying "a picture is worth a thousand words" is certainly true in the construction industry when there is a need to document some events.

With the digital camera or video camera, lots of photographic documentation can be achieved at a small expense and with little effort. A 250MB memory card in a digital camera will hold 1000 photographs. Onsite 24 hours cameras are becoming popular, particularly on urban sites, affording a complete progression of events both day and night, some of it for security reasons.

Photographs can be a valuable adjunct to the documentation of a project, for reasons such as the following:

- To further document the job progress, or lack of it
- To record the uncovering of unusual conditions or document conditions that differ "materially" from those normally encountered
- To act as further substantiation for a change-order request
- To record a complex construction process or detail for future use by others

Many general contractors use some special photographs for in-house training sessions or in future sales and marketing presentations, as well as for lobby-wall decorations.

Photos to document lack of progress.   In the administration of contracts with liquidated damages, it may be important to document the lack of progress regarding conditions that are beyond the contractor's control, such as encountering unanticipated site conditions, severe weather, labor disputes, or owner-directed changes. If the project is in the excavation and foundation stage, a torrential rainfall can cause more damage than just the loss of one workday. Photographs will vividly depict the aftermath of severe weather. During work stoppages due to labor disputes or strikes, photographs can serve several purposes.

- They can document the effect of the strike, not only on the disputed trade but on trades that may have joined the dispute in sympathy, which may constitute an illegal secondary boycott. The photos can show how progress of seemingly unrelated work was affected.

- When tempers flare, altercations can erupt, equipment or suppliers may be vandalized, and photos will be helpful in any claims against proven offenders or for insurance claims.

Photographs during rehabilitation or renovation work.   Photos can be especially helpful in documenting conditions uncovered during the demolition stage and reconstruction of a rehab or renovation project. Contract drawings showing the location of a nonbearing wall that is actually determined to be a bearing wall can provide documentation of this changed condition and assist in the approval of a change order, if one is required. Photos are useful in recording structural cracks or structural failures which may have existed prior to being uncovered by the demolition of existing plaster or drywall finishes.

When taking close-up detailed photos, it is sometimes difficult to determine the size of the item being photographed and its relationship to surrounding areas. Proper scale can be displayed by inserting a carpenter's rule or metal tape rule into the photograph.

Unsuitable subsurface conditions are other areas that can be effectively documented with photographs. Subsurface rock formations, underground water, and buried trash are all conditions that warrant documenting with photos even it if appears that they will not be required within the terms of the contract.

Remember.   Documentation (recordkeeping) is an essential component of the construction process. It is just as important as the correct placement of concrete foundations or the superstructure on which it rests. Disagreements and differing interpretations of contract responsibilities and obligations abound in this business.

To reduce or avoid future misunderstandings during the construction process, remember three important things:

1. Document
2. Document
3. Document

12

Claims, Disputes, Arbitration, and Mediation

At one time or another, chances are that a project manager will have to deal with a dispute that results in a claim. The problem may arise at any point during the construction process—before the process even begins or after it has been completed. Although it may be the general contractor's intent to avoid it at all costs, it may happen that litigation is forced upon them. A project manager who is doing a proper job of documentation and who is somewhat familiar with legal terminology, previous court decisions, and past industry practices will be better prepared to deal with that inevitable dispute.

The practice of law is best left to the lawyers, but as the name "contractor" implies, we deal with contracts (legal papers), and therefore need to become familiar with the legal implications and responsibilities that go hand-in-hand with the administration of all of these documents.

What Triggers Claims and Disputes?

The principal reasons for misunderstandings leading to disputes and claims are as follows:

- Plans and specifications containing errors, omissions, and ambiguities, or which lack the proper degree of coordination

- Incomplete or inaccurate responses or nonresponses to questions—or resolutions of problems—presented by one party in the contract to another party in the contract

- The inadequate administration of responsibilities by the owner, architect/engineer, contractor, subcontractors, or vendors

- An unwillingness or inability to comply with the intent of the contract or to adhere to industry standards in the performance of work

- Site conditions which differ materially from those described in the contract documents

- Unforeseen subsurface conditions

- The uncovering of existing building conditions which differ materially from those indicated in the contract drawings—situations that occur primarily during rehabilitation or renovation work

- Extra work or change-order work

- Breaches of contract by either party in the contract

- Disruptions, delays, or acceleration to the work that creates any deviation from the initial baseline schedule

- Inadequate financial strength on the part of the owner, contractor, or subcontractor

The key to dispute resolution is promptness in addressing the issue(s) and conducting negotiations with the goal of early settlement. Experience has shown that the longer a dispute lingers, the more entrenched each party's position becomes, and the more difficult it is to resolve. Rapid resolution can be achieved if there is complete, accurate, and indisputable documentation regarding the events surrounding the disagreement.

Only when efforts at resolution fail is there a need to consult a claims consultant and/or the company attorney, and again, if this is done at an early stage, the strengths and weaknesses in both party's claim may become apparent and the decision to pursue or dismiss the claim made somewhat easier.

The Bid Proposal Process and the Potential for Disputes

Disputes often arise even before a contract is awarded. In fact, a dispute can occur even before a sealed bid proposal is submitted at a formal bid opening. In public sector work, and frequently in private industry, a formal bid procedure is established: bids are submitted on a preissued bid proposal form, the form is completed and signed by an officer of the construction company, sealed in an envelope, and presented to the owner's representative at a predetermined place, time, and date. If bid bonds are required, a certified check or letter of credit may also be acceptable.

The bid bond offers assurance to the owner that if the contractor's bid is accepted, the owner will be protected if that low bidder is unable or unwilling to accept a contract when offered. Other requirements may also accompany the bid proposal and, per the bid instructions, any deviations may be cause for rejection.

In the public sector, strict compliance with all aspects of the bid proposal must be adhered to, otherwise formal protests may be lodged by other bidders protesting acceptance of a bid that fails to meet all the required criteria.

In the private sector, compliance is not so strict, and the owner may waive any or all requirements in their selection of an acceptable bidder if it appears to be in their best interest. In a way, so can public officials.

Do late bids count?

What's described next is very familiar to project managers as they dash from their office with an incomplete bid form, several pens, and a cell phone on their way to the bid opening. The scenario generally is as follows: the project manager arrives at the bid opening, looks for a spot away from the crowd and calls the office for final instructions and late-breaking price adjustments. With only a few minutes to bid closing, they find themselves looking at a blank bid form that must be filled in with a page or two of alternates, allowances, unit prices, and who knows what. Time is running out and only a minute or two is left, but many blanks remain to be filled in before that mad dash to the bid office. A cold sweat begins to form on the project manager's brow: "What will happen if I make a mistake on this form or enter the wrong number in the wrong place—or worse yet, if the battery in my cell phone dies before I'm finished consulting with the office?"

Don't despair! Complete the form as quickly and as accurately as possible and proceed, posthaste, to the office where the bids are to be submitted. Turn the bid into the clerk and have it date/time stamped. Just because a bid is received after the exact specified time on a public work project, does not mean it will automatically be disqualified.

In private bid situations, the owner is free to waive any prebid qualifications, but in public bid openings, the courts have ruled that the public bidding requirements are there for the public's benefit. If a bid is received a few minutes late and there is no evidence of fraud, collusion, or intent to deceive, all is not lost. If the local authority *refuses* to accept the late bid, an immediate protest must be filed and can be voiced in the presence of the official refusing to accept the bid, as well as in the presence of witnesses.

In the meantime, the project manager should remain during the entire bid opening process and keep detailed notes of the other bidders' proposals, noting their competitive bids and any exceptions that they may have taken to the bidding instructions. As quickly as possible, upon return to the office, a written protest should be filed citing the circumstances involved in the late submission (all other portions of the bid having been met with strict compliance) and stating that the bid was delivered at, say, 2:05 P.M. instead of 2:00 P.M. as required.

If your late bid happens to be the low bid, but could be rejected because it was submitted five minutes after the deadline, it may well be worthwhile to lodge a formal, legal protest and challenge. It is possible that the public officials—or the court, if it gets that far (which it probably won't)—will rule that it is in the *public interest* to accept the low bid even though the late submission must be considered a violation of the bidding procedure. In fact, if other "minor" deviations are noted during the bid opening process and another contractor requests that your bid be disqualified, or the public official opening the bids indicates your bid may be disqualified, continue to take notes relating to all other bids being opened, then return to the office and discuss lodging a protest.

When a "nonconforming" bid is significantly lower than the second bid, the public agency may well decide to interview your firm to ascertain that all other aspects of the bidding procedure have been met, that your company fully understands the scope of the work, has included costs for the full scope of work, is a reputable builder, and is ready, willing, and able to enter into a contract. The agency may then rule that it is in the best interests of the public to accept this bid even though there was a "minor" discrepancy in the bidding process. Emphasis must be placed on the word "minor."

Who is the low bidder?   Unless the bid documents state otherwise, the low bidder on a lump-sum contract is the "apparent" low bidder, subject to a review of compliance with all other provisions of the bidding instructions. When a bid contains not only a lump-sum price but several alternates, as additions or deletions to the base bid, the determination of who is actually the "low" bidder becomes less clear. In the absence of any language to the contrary, usually the owner may select the alternates that will be accepted. Therefore, the low bidder may not be apparent until all accepted alternates are included with the lump-sum bid. It is also possible that the "apparent low bidder" wishes they were not the lowest bidder, which brings up another point.

Withdrawing a bid.   In the rush to assemble a competitive bid, it is not unheard of to make a mistake, either in the extension of unit prices, the omission of a few cost items, or to just plain make a mathematical error. If these items are minor and if the general contractor has been designated "apparent low bidder," they will usually accept the contract and chalk up the mistakes to experience—afterward having a few words with the person that put the bid together.

However, the situation might take on a different twist if a bid bond or bid deposit check was required and, upon the opening of the sealed bids, the contractor finds to their dismay that they were low bidder by such a substantial margin that their bid must have been defective in some way. If a contractor discovers that their bid was, say, $400,000 lower than the next bidder's price of $2 million, and the third and fourth bidders' quotes were 5- to 10-percent higher than the second bidder, there may be reason for alarm. The contractor's thoughts immediately go to the bid bond submitted with their bid. If they would elect not to accept a contract if offered, the likelihood of having to forfeit their bid bond is real, but the other option is to accept a contract and, more than likely, anticipate a substantial loss in the process. Neither choice is very palatable, but does the contractor have any other?

In the case of *M. F. Kemper Construction Co.* v. *City of Los Angeles,* 376 Cal.2d 696,235 P.2d 7 (1951), the contractor inadvertently omitted a $300,000 item from their estimate, causing them to be $250,000 lower than the second bidder. The contractor contacted the Public Works Department a few hours after discovering their mistake and requested that their bid be withdrawn and their bid bond returned. The City of Los Angeles decided to award the contract to the second bidder and directed the "low" bidder to forfeit their bid bond. A lawsuit followed. The court made a distinction between errors in judgement and mathematical or clerical errors. On the basis of the evidence submitted, the court judged the $300,000 error to be a justifiable one and allowed the contractor to withdraw their bid and not forfeit their bid bond. It was the contractor's responsibility, however, to prove that a clerical error had, in fact, been made that accounted for the erroneously low bid.

Based upon this decision, it becomes critically important that all adding machine tapes, scrap paper with notes, calculations, bid tabulation sheets, confirmation of telephone bids, and other related matter be kept until well after a contract award has been made and accepted by the contractor. A contractor should not assume that they only have two choices upon discovery of a defective bid: forfeit the bid bond or accept the unwanted contract. It may be possible to withdraw the bid with no penalties.

Verbal subcontractor quotations.   During the preparation of a hard bid, several subcontractor proposals may be submitted over the telephone and recorded by the general contractor on a telephone bid form. This form may include the subcontractor's name, address, date, the project name, the sub's phone and fax number, and the name of the person phoning in the bid. As these telephone bids are received, occasionally the full scope of work being quoted is not reviewed or recorded, because bids are coming in fast and furious and the estimate must be completed quickly. But when the bidding process is over, these telephone bids will be reviewed for completeness and any missing information filled in by contacting the subcontractor. Each subcontractor or vendor should then be requested to submit a written confirmation of their verbal bid.

A general contractor who receives a verbal bid from a subcontractor and incorporates it in their "winning" proposal may subsequently be awarded a contract, at which time that competitive subcontractor will be requested to submit their formal, written quotation if they have not already done so. But suppose one of these subcontractors with that competitive bid responds that, due to an error in their bid or because they have been awarded another big job in the interim, they wish to withdraw their previous quotation and will not agree to enter into a contract with the general contractor. Without a written proposal from that subcontractor, can anything be done? In some cases, yes, most assuredly.

The principle of *promissory estoppel* may apply and render the oral contract enforceable. The word *estoppel* is defined in the dictionary as "an impediment that prevents a person from doing something contrary to his own previous assertion to do so." This appears to be a long-winded definition of "to stop."

Under this doctrine, the offerer (subcontractor) may be held liable for any damages incurred by the general contractor. In other words, the subcontractor may be required to pay the general contractor the difference between their price and the next lowest quote. In the case of *Bridgeport Pipe Engineering* v. *DeMatteo Construction Company*, 159 Conn. 242, 244 (1970), the subcontractor submitted a verbal bid by telephone to furnish labor and materials for plumbing, heating, and ventilating work for a housing project, but later declined to enter into a contract. The Connecticut Supreme Court ruled that the telephone bid was an oral offer and the general contractor accepted the offer when it became the successful bidder on the housing project.

Although this Connecticut case involved a subcontractor supplying both labor and materials, several other state courts have ruled that bids received representing the sale of goods only also constitute a contract, and the acceptance of that contract is complete when the general contractor receives a contract from the client. Even if the subcontractor attempts to withdraw the bid before the general contractor is notified of an award by the owner, the principle of promissory estoppel may apply to prevent the subcontractor from withdrawing their bid without incurring a penalty.

In another court case, *H. W. Stanfield Construction Corp* v. *Robert McMuller and Son, Inc.,* App.92 Cal.Rptr.669 (1971), the court ruled that the subcontractor was liable for the bid submitted to the general contractor assembling a bid for the U.S. Navy. The general contractor had obtained prices from a number of painting contractors and the low bid was submitted by a company whose first bid was 50-percent below the next highest one. After being advised of this low bid, the subcontractor resubmitted their proposal which H. W. Stanfield used in their bid. The general contractor advised the painting contractor that they had incorporated their bid, and after receiving the Navy contract, the painting contractor would be awarded a contract for their portion of the work. As a result, the subcontractor refused to sign the subcontract agreement when it was offered. The court held that the general contractor had relied on the promise of the subcontractor to do the work for the stipulated sum and would be damaged by the refusal of the painting contractor to do that work. The decision was upheld by the California Supreme Court.

In the practical world of contracting, however, working with a subcontractor who has been directed to accept a subcontract agreement or face litigation might present such hurdles and problems that this arrangement would, in most cases, not be worth the savings in costs. But being aware of one's rights under the law may be used to advantage if the situation permits.

Other Potentials for Disputes

Oral contracts

Oral contracts are often recognized as creating the same obligation between two parties as a written contract. A case tried in a North Carolina court was based upon an architect performing work for an owner without benefit of a written contract. The owner, a developer, requested the architect to complete some preliminary design work on a condominium project. The architect, according to the owner, had prepared such a complex contract for the design work that it was rejected. The owner indicated that they would be willing to sign a simpler contract if presented. The architect's fees, at that time, were not a subject of disagreement, and the owner directed the architect to proceed with the design work. Months later, the owner advised the architect that another designer had been hired to work on the project, but by that time the original architect had completed a major portion of the work and so submitted an invoice for it. The owner refused to pay the invoice and the architect filed suit, claiming breach of contract.

The North Carolina trial court dismissed the case due to the absence of a written contract between the owner and architect, but the North Carolina Court of Appeals disagreed with this decision. The appeals court ruled that even though a written contract did not exist, the architect had an enforceable oral agreement. Therefore, the court determined that a jury trial was in order and went further in stating that if the jury were to agree that an oral contract did not exist, the architect could still pursue the recovery of actual value of the services performed in connection with the project design [*Willis* v. *Russell*, 315 S.S. 2nd 91 (North Carolina Appellate Court, 1984].

Disputes regarding contract interpretation

Ambiguities in the contract documents are usually resolved by reasonable parties taking a reasonable approach. When reasonableness does not prevail and one party pursues a hard line, what to do? There are no tried and true procedures for contract interpretation and each case seems to stand on its own. What documents have priority over others? Do the plans or specifications take precedence? Will the specifications take precedence over full-scale drawings? Some contract requirements specifically establish an order of the precedence such as: contract requirements—first precedence; schedules—second; drawings—third; and so forth. Typical contract language, when there is a conflict between the plans and specifications, will be similar to the following clause:

> In the event of conflicts or discrepancies among contract documents, interpretations will be based on the following priorities:
> 1. Contract
> 2. Addenda, with those of a later date having precedence over those of an earlier date
> 3. Supplementary conditions
> 4. General conditions of the contract for construction
> 5. Drawings and specifications with schedules and large-scale details having precedence over small details
> 6. If drawings and specifications are not in concurrence regarding quality or quantity, the contractor shall request interpretation from the architect.

Another standard contract phrase regarding this same topic is

> In a case of conflict between drawings and specifications as to the extent of work, or location of materials and/or work, the following order of precedence will govern:
> 1. Large-scale drawings
> 2. Small-scale drawings
> 3. Schedules (door, finish, equipment, and so on)
> 4. Technical specifications

In case of conflict regarding the quality of materials, the specifications will govern. However, the courts are not consistent in their interpretations. Regarding a specific statement taking precedence over a general statement, the court may look for the purpose *intended* by a specification section relating to the drawing requirements before arriving at a decision. The point is that there are no hard-and-fast rules that apply to the interpretation of which document has priority over another, unless, as previously stated, a contract Order of Precedence prevails.

Errors and omissions concerns.   Another provision that architects often include in the contract requires the contractor to advise the architect of a discrepancy, error, or omission before submitting a change order relating to that error or omission. This provision may appear in the bid documents or most certainly in the contract for construction.

Article 3.2 of the A201 General Conditions document requires the contractor to notify the architect *at once* of any errors, inconsistencies, or omissions discovered. If the contractor recognized such an error, inconsistency, or omission and *knowingly* fails to report it to the architect, the contractor may be held liable for any damages resulting from them.

This unilateral edict by the design consultants can be frustrating to the contractor who rarely has sufficient time to thoroughly scrutinize all of the bid documents during that hectic process known as "bidding." In most cases, a contractor may uncover only the most glaring drawing mistakes at bid time and won't find any minor discrepancies—or major discrepancies—until construction is underway. Upon discovery and notification of the architect, the response might be, "Well, the bid documents required that you notify us, in writing, when these errors were discovered prior to submitting your bid, otherwise you are considered fully responsible," which is a response that will raise the hackles on most project managers. Is there any way to deal with these kinds of situations? I think so.

Dealing with problem drawings.   With barely enough time to completely scrutinize the drawings and specifications before and during the hectic competitive bidding process, some things will undoubtedly slip through the cracks. Last-minute calls from subcontractors, and suppliers e-mailing their bids along with estimate adjustments by the boss as the clock ticks away, leave little time to concentrate on potential drawing errors and omissions. Doesn't the contractor have a right to assume that the plans and specifications submitted by the owner's architect/engineer are reasonably complete for bidding purposes? If there are any major discrepancies, shouldn't they be able to rely on everyone's good faith to find an equitable solution to correct major errors?

In a court case identified as *John McShain* v. *United States,* 412F.2d 1218 (1969), the contractor stated that the true condition of the drawings was not known at the time of bidding and that, after being awarded the contract for construction, they found that several drawings which were illegible at bid time had not been replaced with legible ones. The addenda drawings, furthermore, did not correct many of the coordination errors in the bid documents. The general contractor instituted legal action to recover damages incurred by their company and their subcontractors because of the inadequate drawings. The U.S. Court of Claims said that although the plans furnished by the owner need not be perfect, they must be adequate *for the purpose for which they were intended.* The court went on to state that the contractor was under no legal or contractual obligation to inspect the drawings to determine their adequacy for construction prior to a contract award. Furthermore, the court ruled that the documents were to be used for estimating purposes only, and it had not been proven that McShain knew or should have known how defective the drawings were.

Owners have a responsibility to present the general contractor with drawings and specifications that are adequate and reasonably accurate, and the GC has a right to expect that. If there are considerable problems relating to deficiencies in the documents, the general contractor should be afforded consideration in their request for additional compensation for any delays the substandard or deficient drawings might have caused them.

In the case of *J. D. Construction Co., Inc.* v. *United States,* 171 Ct., Cl.70 (1965), the court ruled that if faulty specifications prevent or delay completion of the contract, the contractor is entitled to recover delay damages from the defendant's breach of implied warranty. This breach cannot be cured, said the court, by the simple expedient of merely extending the contract time and performance.

The contractor's guarantee regarding design.   When an architect specifies a certain component design and the installation results in poor performance, who is responsible? In a case brought before the courts in the State of Washington, an architect had modified a curtain wall design. The contract specifications contained a standard clause requiring the general contractor (GC) to notify the architect if any materials, methods of construction, or workmanship changes were needed to ensure compliance with the contract documents. The GC did not notify the architect of any changes they felt were necessary relating to this curtain wall design and installation.

When the curtain wall was installed and the building completed and signed off by the design consultants, a series of leaks appeared in the system. The GC denied any responsibility to correct the leaks and the owner sued. In this case, *Teufel* v. *Wiener,* 68 Wash.2d 31,411 P.2d 151 (1966), the court concluded that the leaks were caused by design error. The owner had claimed that the specifications called for the curtain walls to be fabricated and installed by a manufacturer regularly engaged in the manufacture of this type of system and that the work be first-class and performed in a manner that did not allow any weather infiltration. However, the court's ruling was that the curtain wall was modified by the architect and was not suited to its use, and that the leaks were not caused by faulty materials or poor workmanship but were the result of a design defect.

The Spearin doctrine.   A landmark court decision rendered in 1918 is still applicable today: the Spearin case—sometimes known as the Spearin Doctrine.

Spearin, a contractor, bid on a U.S. Navy drydock project that included replacing a 6-foot section of storm sewer pipe, which they did. The replacement sewer line proved to be inadequate to carry the volume of water run-off and it broke due to internal pressure. The Navy held Spearin responsible and told them to replace it. Spearin refused and the lawsuit progressed all the way to the Supreme Court.

The resulting "Spearin Doctrine" stated that:

> If the contractor is bound to build according to plans and specifications prepared by the owner, the contractor will not be responsible for the consequences of the defects in the plans and specifications.

The court continued, saying:

> The responsibility of the owner is not overcome by the usual clauses requiring bidders to visit the site, to check the plans, and to inform themselves of the requirements of the work.

Although today, the 1997 edition of the AIA Document A201-General Conditions in their Article 3 recognizes that the "contractor's review (of the plans and specifications) is made in the contractor's capacity as a contractor and not as a licensed design professional." Knowledge of the provisions of the Spearin Doctrine may come in handy when a situation involving defects in the plans and specifications is being reviewed.

Subsurface, changed, and differing conditions

A significant number of disputes involve the sitework phase of a project. Even with numerous test borings and other geotechnical site investigations, conditions uncovered during excavation may be at variance with the conditions assumed by the information in the geotechnical survey. Test borings accurately display the subsurface soil strata in the *exact location* where they have been taken, but another boring drilled just yards away may reflect totally different subsurface conditions. In fact, a standard clause in most geotechnical reports recognizes this condition and state.

Regardless of the thoroughness of a geotechnical engineering exploration, there is always the possibility that conditions will vary from those encountered in the test borings, or that conditions are not as anticipated by the designers.

Another disclaimer often included in the geotechnical report reads as follows:

The analyses and recommendations submitted in this report are based upon information revealed by this exploration. This report does not reflect any variations which may occur beyond the locations of the test borings and test pits. Since the nature and extent of variations may not become evident until during the course of construction, an allowance should be established to account for possible additional costs that may be required to construct the foundations as recommended herein.

Using geotechnical disclaimers to advantage

Although geotechnical disclaimers may be considered by contractors as limiting their claim for additional costs incurred during excavation work, they can also be used to advantage.

Even the geotech recognizes that actual conditions will not be known until mass excavation or trench excavation is underway. If the contractor uncovers conditions that are materially different from those reflected in the geotechnical documents, wouldn't that substantiate the contractor's claim for added costs due to these variations?

Furthermore, the geotechnical report is generally prepared at the owner's request and paid for by the owner, thus it can be stated that the reference in the disclaimer to an allowance—a contingency—was actually inserted for the owner's benefit, not the contractor's.

The contractor must thoroughly document their claim of encountering conditions that differ materially from those anticipated by the information available in the geotechnical site investigation report.

The court and differing site conditions

The case *Randa/Madison Joint Venture III* v. *Dahlberg* [239 F.3d 1264, 2000 U.S. App/(U.S. Fed. Cir. Feb 7, 2001)] dealt with differing site conditions in another way. Randa/Madison, the contractor, had a contract to dewater the excavation for a pump-house foundation, but found that it had grossly underestimated the extent of dewatering and so filed a claim with the government to recover these additional costs. The government contract contained two clauses: that the contractor satisfied itself to the character and quantity of work and that the bid documents addressed the physical data. The contract documents also stated that *soil test results and soil and rock samples were available for inspection* but that these tests results were not included in the documents.

In their appeal, the contractor argued that they had no duty to review this information and that the government had an affirmative duty to disclose the additional information beyond just making it available. The review court disagreed and said that the government placed the contractor on notice that this added information existed and the contractor was presumed to have reviewed it.

Rock excavation. Suppose that in the course of excavating, rock was uncovered where no rock was indicated in the borings because those borings happened to straddle the rock formation. Does the contractor have a legitimate claim for an extra? One answer to that question relates to whether or not the site is denoted to be "Unclassified" or "Classified." A "classified" site includes the specific subsurface conditions to which a contractor is bound. An "unclassified" site is one in which the contractor "owns," or is responsible for, costs relating to all subsurface conditions necessary to complete the sitework. If unsuitable soils, rock, debris, underground structures, and so on are encountered, the contractor is obliged to remove them and replace them with suitable materials at no cost to the owner. A typical unclassified site section in the specifications will state the following:

> Excavation shall be unclassified and shall comprise and include the satisfactory removal and disposal of all materials encountered regardless of the nature of the materials and shall be understood to include rock, shale, earth, hardpan, fill, foundations, pavements, curbs, piping, and debris.

But is that position of denying responsibility for all such costs defendable in all cases? The answer is No!

The author's company entered into a contract with an owner to build two office buildings in a campus setting with a total square footage of approximately 500,000 ft². The site was deemed "unclassified." Accompanying the plans and specifications were a series of test borings, and the isometrics of various soil and rock strata along with a test boring location plan superimposed over an outline of the building footprint (see Figs. 12.1 through 12.3) Excavation proceeded in the area of test boring B-23, and rock was discovered at Elevation 162.8. Both the isometric and the test boring log clearly indicated the presence of rock at Elevation 152. Continued excavation in areas represented by other test boring data uncovered rock in areas at lower and higher elevations than those indicated by the borings. The architect and owner were made aware of these disparities and even though the contractor prepared a written request for an extra, excavation continued, uncovering more rock requiring blasting. When the site work was complete and the contractor submitted a bill for $288,000 it was dismissed out of hand by the

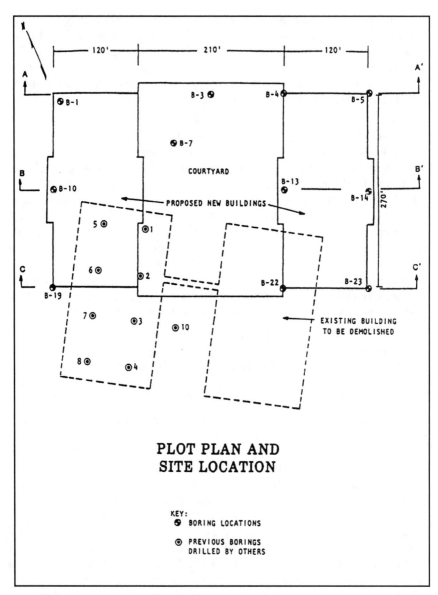

FIGURE 12.1 Test boring location plan, locating test boring B-23.

architect and owner claiming that the site was unclassified and the contractor "owned" all conditions, most notably, rock. The contractor was of the opinion that the owner/architect decision was grossly unfair.

The architect and owner stood by their interpretation of the contract obligation relating to the unclassified nature of the site after weeks of

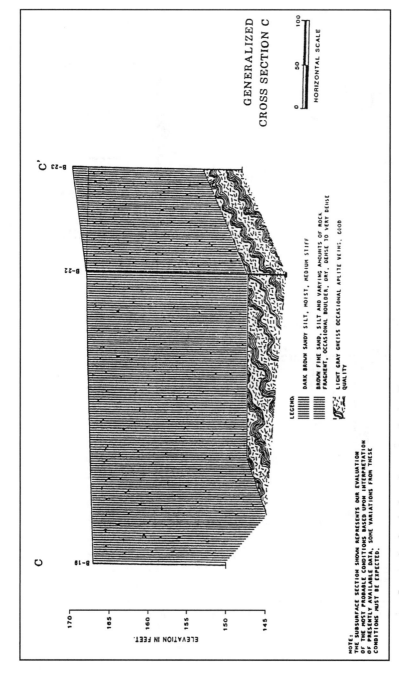

FIGURE 12.2 Isometric of rock elevation at test boring B-23.

275

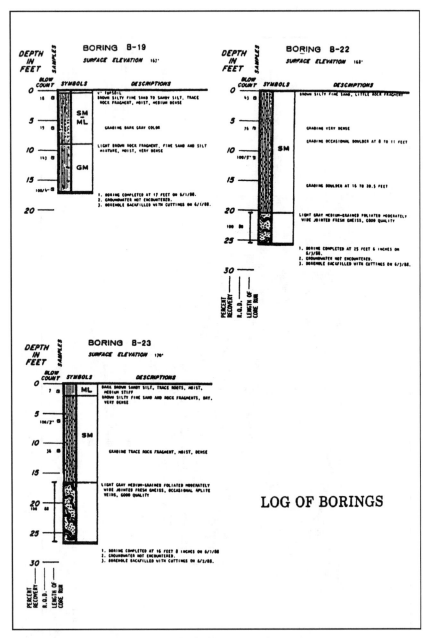

FIGURE 12.3 Test boring data relating to test boring B-23.

discussion, some very heated. There seemed to be a wide disparity between the geotechnical report and the actual field conditions.

The author then made a transparent overlay of the test boring location plan and began to shift it, first to the east and then to the west of the building footprint outline. Suddenly it became evident what had occurred. The geotechnical engineer had inadvertently shifted the entire test boring location plan approximately 50 ft to the west of where it should have been in relation to the building's footprint. When the overlay was shifted 50 feet to the east, all test boring data matched the conditions actually encountered.

When presented to the owner, they paid the $288,000.

Another site war story.   On a different occasion, the author of this book, while administering another contract with an "unclassified" site encountered considerable unsuitable soils below the footing and foundation "payline" (the design level for those footings and foundations). A proposed change order was presented to the owner who dismissed it, claiming that the contractor was responsible for all unsuitable soils, no matter where uncovered, because, again, the site was unclassified. As the contractor, we argued that we "owned" all soils *above* the payline in an unclassified site, but that did not mean that the contractor "owns" *all* unsuitable soils *under* the payline. This argument was again dismissed by the owner. At the next meeting to review this issue, I drew a round circle meant to represent the earth and placed the project in question on top, drawing a deep "V" meant to represent excavation to the bottom of the circle (China) and said, "So you're telling me that we are obligated to dig all the way to China if necessary?" The owner's representative said, "Well, no that's unreasonable." The concept of degree of "ownership" of responsibility to remove unsuitable soils was thereby established and the claim was settled by agreeing to limit the amount of soil the contractor was contractually responsible to remove, at no cost, to a depth somewhere between the payline and China.

This was a compromise, but one that settled a contentious matter, and it established an atmosphere of "reasonableness" that pervaded the relationship between owner and contractor during the balance of the project.

Two important points illustrated in the preceding experiences are the following:

- The project manager should not give up their claim for compensation for unforeseen subsurface conditions, *even in the face of restrictive language in the contract.* If something appears to be unfair, it probably is and the project manager just needs to continue to investigate the situation until they can develop enough information and documentation to refute the unfairness of the architect/owner's ruling.

- Equally as important: Attack any dispute early on and persist in its resolution promptly and fairly.

Dealing with the exculpatory language in the contract. Most contracts include statements to defuse claims by a contractor for unforeseen site conditions. Look for the following types of clauses.

1. The contractor may be required to visit the site prior to submitting their bid and any condition visually observed or reasonably assumed by this prebid visit will be interpreted by the owner as being "disclosed."

    This means that a close and careful inspection of a site is necessary to document not only that all existing conditions were observed, but the extent to which other conditions were not apparent, or hidden, and therefore could not have been anticipated.

2. A "no damages for delay" clause prevents the contractor from claiming additional costs beyond direct costs if unforeseen conditions are encountered, resulting in excusable delays to the project's completion.

    If this clause cannot be stricken from the contract, it must be passed through to all subcontractors, thereby reducing the exposure for subcontractor delay claims if the project is delayed after unforeseen subsurface conditions are encountered.

3. A clause that requires the contractor to examine the contract documents and report, in writing, any obvious errors, omissions, and ambiguities within a specific time period, otherwise a claim will not be considered. Re-read Article 3 of AIA Document A201 which provides a clear presentation of the contractor's responsibility with respect to the review of contract documents.

4. Geotechnical reports included in the bid documents often state that they "are not to be relied on" since the contractor must make their own conclusions as to the representation of the information contained in these geotechnical reports.

The general contractor must rely on the information presented by the geotech in order to prepare their sitework estimate which will include a contingency of some sort for unknown or unanticipated conditions. As discussed earlier in this chapter, if the contractor can demonstrate that this information was misleading or insufficient, and that it affected the amount of a *reasonably assumed* contingency, the contractor may be able to overcome this exculpatory contract statement.

If there are sufficient concerns about the validity of the geotech, the contractor can also request permission to perform additional site exploration. (This is plausible in a negotiated contract scenario, but what about a hard-bid situation? Will further exploration once disseminated to other bidders reduce the competitive edge or is the contractor willing to forego that advantage to ensure that their own bid will be more responsible?)

Differing or changed conditions

Article 4.3.4 of the 1997 edition of AIA Document A201 includes the definition of changed or differing conditions, and the procedure to follow when presenting a claim. In order to prepare a viable change order for additional costs utilizing "differing conditions" as the basis for that claim, a contractor must show that:

- The contract documents reflect certain conditions that formed the basis for the contractor's estimate.
- The contractor's interpretation of the documents was reasonable and based upon previous experience in such matters.
- The subsurface conditions actually encountered differed "materially" from those represented in the documents.
- The actual conditions encountered could not have been "reasonably" anticipated.
- The costs claimed must be solely attributable to the materially differing conditions—for instance, the difference in cost between the *assumed* material or condition and the cost of the *actual* material or condition.

If a claim is to be prepared, the contractor should include the following:

- A clear statement as to the *usual* conditions a contractor would have expected to encounter on that site.
- What conditions were actually encountered.
- How these conditions differed "materially" from the known and usual.
- What costs and delays were incurred because of the encountered conditions.

These types of situations often result in delays to the construction process and the impact of these delays, in most cases, are much the same as other significant interruptions to the planned sequence of work.

The impact of differing or changed conditions resulting in delays will incorporate the following costs:

- *Direct costs:* Labor, materials, and equipment employed in dealing with the work.
- *Weather-related costs:* Weather-sensitive work tasks established in the baseline schedule may become shifted to another part of the year

where, for example, concrete foundation work initially scheduled to be completed in mild weather now must be completed in cold weather. Winter conditions, in those parts of the country where cold weather occurs, will add considerable costs to labor, materials, and equipment such as curing blankets, temporary heat, and enclosure materials.

- *Acceleration costs:* Costs required to accelerate the progress of work, as directed by the owner may entail shift work, extended work hours, additional materials and more equipment. Studies have proven that extended periods of overtime reduce worker productivity and this reduction in productivity needs to be taken into account when submitting an estimate to accelerate the pace of the job.

- *Idle equipment:* Rental equipment on the project remaining idle during the delay period will continue to generate costs whether it is company owned or leased. When a backhoe is rented on a weekly basis, the contractor must pay the weekly rate whether the equipment is active or idle; only fuel and operator costs are eliminated. The type of equipment, its idle and active periods during the delay must be documented so that all related costs can be isolated for reimbursement.

- *Field office expenses:* Some field expenses are time related. Since the project will be delayed, added costs for field office rentals and associated utility costs, temporary toilets, and field supervision will certainly accrue.

- *Indirect costs:* Costs incurred but not allocated to any specific item of work.

- *Home office expenses:* Expenses at the home office assigned to administer the project currently being delayed will still accrue.

- *Loss of productivity:* This cost is a difficult one to quantify, but the interruption of a continuous sequence of work results in a loss of productivity. Also, extended periods of overtime work causes worker productivity to drop drastically and it basically takes more man hours to complete the work.

Differing conditions when unit price contracts apply.   In public works projects, where unit prices apply, the general contractor is provided with approximate quantities upon which to base their unit price(s). If site-work is involved, the agency may advise the contractor to base their unit price on a quantity of, say, 10,000 cubic yards of structural fill borrow, 3500 cubic yards of unsuitable materials to be hauled offsite, and so forth. If actual quantities either exceed these thresholds, or are less than these stated requirements, a *differing condition* may have occurred and the contractor may be permitted to review and adjust their unit price(s). If, as an example, only 3000 cubic yards of structural borrow are required and only 1000 cubic yards of unsuitable soils are to be hauled offsite, the contractor's initial bid anticipated higher quantities and therefore actual conditions *differ materially* from those estimated. Although many public agencies may have different threshold criteria, a standard yardstick allows the differing conditions concept to be employed if actual quantities exceed bid quantities by 20 percent or more. In dealing with private owners, the same criteria can be used, citing acceptable public sector principles as the basis for requesting the differing conditions change order.

Claims due to scheduling problems.   The AIA's General Conditions document, relating to the contractor's schedule has changed over the years. The 1970 edition required the contractor to submit their schedule for the architect's *approval*; the 1976 version required submission for the architect's *information*, while the 1987 edition of A201 required the contractor to prepare and submit the schedule *promptly*. The current 1997 version of A201 requires the contractor to submit a schedule that "complies with the contract completion date."

Construction schedules are dynamic and subject to change, and with the latest AIA General Conditions requirement, the general contractor has more flexibility in modifying the baseline schedule as long as the completion date is not extended by a change order. The use of the critical path method (CPM) schedule with its ability to display the relationship of one work task to another has come to occupy a key role in any potential delay claim being considered by a contractor. The baseline schedule—the initial schedule formulated at the beginning of the project—would have been prepared with input from subcontractors and vendors, the project manager acting as coordinator for this critical task.

Be wary of subcontractor input

The invitation to subcontractors to participate in the preparation of a CPM schedule is proper, but must be monitored closely. A subcontractor may not be accurate in determining the length or duration of an operation due to inexperience or not spending sufficient time to properly estimate durations. A subcontractor may have more than enough "float" time to cover any potential contingencies to insure that they can absorb some glitches that will occur and still maintain their schedule. By not advising the project manager of the added days, this subcontractor will have significantly overstated the duration of their task(s) and, as a result, impact subsequent trades. So subcontractor input, while valuable, should be carefully reviewed and not just accepted at face value.

Once published, a baseline schedule becomes the "official" project roadmap, and any changes to the sequence or time allotment for selected activities should be noted to determine their impact. Conditions that are cause for an extension of a baseline completion date must be promptly communicated to the architect and owner—either through a Request for Time extension, or in the case of a nonexcusable delay, with a recovery schedule.

Remember that change orders are issued to affect a change on contract sum or contract time, requests for time extensions will actually take the form of a Proposed Change Order (PCO) where the facts documenting the delay and related costs are included.

The CPM schedule, accepted by all parties, turns into a two-edged sword. It becomes the official time frame for the completion of the entire project. This baseline "accepted" schedule can now (1) document deviations in that schedule due to owner directed changes and (2) document delays created by the general contractor. A subcontractor can also point its finger at the general contractor who has delayed the start or completion of their work.

Beware of subcontractor claims for delay.   General contractors need to be especially sensitive to delays that may impact subcontractors who have not contributed to that delay. A delay early in the project, say, in the erection of the structural system may push the start of drywall and drywall taping operations into winter months where work may be less productive, and thus require added costs for temporary heat. When delays early in the project occur, the general contractor would be wise to get all subcontractors to agree, in writing, that the delay upfront, will or will not impact their operations.

Delays created by one subcontractor may affect other subcontractors who will certainly look to the general contractor for assistance, and possibly additional compensation, in order to accelerate their work or be compensated for lost productivity. If the project manager is planning to prepare a delay claim for submission to the owner, a well-documented series of CPM schedules graphically displaying how one delay or a series of delays, has impacted the overall completion date, will be strong evidence to support such a claim.

A word about lost productivity

Added costs due to lost productivity are probably one of the more difficult claims to present or dispute. The contractor, or subcontractor, is basically stating that interruption to their normal daily operation, created by work stoppage because of others, or significant changes in the sequencing of work, has caused them to be less productive. They state that in order to recapture this loss of productivity, they must work longer hours, which generally equates to overtime. These interruptions may result in trade stacking, where, say, MEP subcontractors are working in such close proximity to metal framing and drywall subcontractors that each trade's ability to work productively is denied. Working out of sequence, having to go back and complete one section of wall left open for another trade certainly deprives a subcontractor of the ability to work productively. Claims for loss of efficiency employ a simple calculation, but the basis for establishing an inefficiency factor does not.

Total labor incurred during the period in question	$1,000,000
Inefficiency % (whatever has been established, say, 7.5%)	7.5%
Inefficiency loss	$75,000

Both premises, the value of lost time and the cost to recoup these losses are, in the main, subjective evaluations, not objective ones—even though studies by various groups have been published.

Even the concrete subcontractor's use of the "measured mile," a standard cost to form and place a cubic yard of concrete under normal conditions, is not purely objective. Although this measured mile cost of, say, $400 per yard, is culled from thousands of yards placed in the past year, and even adjusted for inflation, can it be applied as a standard when so many variables have gone into this averaging of work?

Similarly, productivity itself is a subjective term. Does each worker produce the same amount of work as their co-crew members? Although an average of all worker productivity may serve as some form of measured mile, suppose one crew is exceptionally productive while another isn't. If so, which one is used to support the claim, and which one can be used to deny the claim? Thus, loss of productivity is also a subjective matter.

Numerous studies have been made to try and quantify loss of productivity due to overtime, such as the following.

Bureau of Labor Statistics.   The oldest study dating back to 1940. Based on a study of 2455 men and 1060 women working in a variety of manufacturing industries, the BLS found that productivity for a 50-hour week was 92 percent, and for a 60-hour week it was 82 percent compared to a 40-hour work week.

The Business Roundtable.   In 1980, the Business Roundtable issued a task-force report after studying working conditions at the Proctor & Gamble factory over a 10-year period. It was discovered that prolonged periods of extended work days like 50-hour weeks, dramatically reduced productivity, but just one 50-hour week reduced productivity from a baseline of 1.0 to 0.7; two 50-hour weeks, back to back, reduced productivity to nearly half.

The Construction Institute (CII).   In 1988, CII collected data from seven heavy industry projects over a four-year period. They had some difficulty defining overtime in efficiency, but reached the following conclusions:

- Previous studies by BLS and the Business Roundtable were not consistent predictors of productivity losses during overtime time.

- Even on the same project, individual crew productivity rates were not consistent when working in an overtime environment.

- Productivity does not necessarily decrease with an overtime schedule.

- Absenteeism and accidents do not necessarily increase under overtime conditions.

Construction Industry Institute source document 98—effects of scheduled overtime.   Labor Productivity—based on 151 weeks of data collected from 1989 to 1992 from four industrial construction projects—focused on piping and electrical crews only.

Conclusions:

- Short-term overtime can cause a loss of productivity in the 15-percent range; however, they can vary from 0 to 25 percent.

- As overtime efficiency decreases, resource availability (the inability to provide materials) was the root cause for loss of efficiency.

Although CII stated that the data collection and analysis methodologies were sound, the variation of efficient losses from 0 to 25 percent hardly make this a landmark study.

The National Electrical Contractors Association.   The NECA published results of a 1962 study based on a small survey of their members and concluded that:

- Five 9-hour days did not affect productivity.

- Five 10-hour days decreased productivity to 98 percent.

- Five 11-hour days reduced productivity to 95 percent.

- Five 12-hour days reduced productivity to 92 percent.

This issue of productivity is further complicated by a U.S. Bureau of Labor Statistics study published in February 2006 which shows that worker productivity, in general, fell by 0.6 percent during the last quarter of 2005, the first decline since 2001.

The results of all of these studies don't appear to help the case for or against loss of productivity, but there is no tried-and-true method to accurately determine the actual cost of lost productivity and it appears that the choice of which study to use depends upon whether one is the claimant or claimee!

The courts and lost productivity issues

In the *Appeal of Clark Construction Group Inc.* (2000WL 37542) VABCA No.5674,00-1 BCA para 30,870, the Board of Contract Appeals stated that "Quantification of loss of efficiency or impact claims is a particularly vexing and complete problem. We have recognized that maintaining cost records identifying and separating inefficiency costs to be both impractical and essentially impossible."

The Mechanical Contractors Association of America (MCAA) publishes *Change Orders, Productivity, Overtime—A Primer for the Construction Industry* available to members and nonmembers alike. The factors contained in this manual have been supported by several court decisions.

In *Appeal of P.J. Dick, Inc.* (2001 WL 1219552) VABCA No.5597,01-2 BCA, para 31,647, the contractor's expert stated that their client's labor productivity was impacted because of continuous revisions to design. This same expert said that while the "measured mile" is the generally preferred method of dealing with these costs, there was no period during which the work was *not* affected by design problems or acceleration and that damages should be based upon the MCAA manual on change orders, productivity, and overtime. The board found the contractor's quantification of loss based upon this method as reasonable.

In *Hensel Phelps Construction Company* v. *General Services Administration* (2001 WL 43961) GSBCA 01-1 BCA Par 31,249, the board accepted the claimant's use of the factors affecting labor productivity as set forth in the MCAA manual. Although the "measured mille" approach is often used to assess labor productivity issues, the MCAA analysis produced a more accurate valuation of the claim. It should be noted, however, that this claim was made by a mechanical contractor.

In *S. Leo Harmonay, Inc.* v. *Binks Mfg. Co* (S.D.N.Y.1984) 597 Supp.1014 (Harmonay), the U.S. District Court upheld the MCAA manual labor inefficiency factors in dealing with a loss of productivity claim. The measured mile approach was used by the claimant, but it was backed up by testimony prepared from factors gleaned from the MCAA manual. The court awarded damages based upon Harmonay's claim that they had incurred a 30-percent loss in productivity due to: (1) excessive work hours; (2) overly crowded conditions; (3) the unavailability of tools, materials, and storage; (4) the defendant's delay in supplying drawings and equipment; and (5) the constant revision of the contract drawings—all of which combined to create confusion and frequent interruptions in the progression of work.

Claims against professionals

Because a contractor usually does not have a direct contractual relationship with an architect, it has been difficult for a contractor to make a claim against them for increased costs or loss of profit resulting from errors and omissions in the project documents. A January 19, 2005 decision by the Pennsylvania Supreme Court changed that concept by adopting the legal principal that by negligently supplying information for the guidance of others, a party may be held responsible for the results of their negligence. This case may open the floodgates for similar actions. It deals with *Bilt-Rite Contractors Inc* v. *The Architectural Studio* case. Bilt-Rite, the contractor, submitted a bid based upon the plans and specifications prepared by The Architectural Studio. The design of an aluminum curtain wall system on the project would not work as designed, so the builder had to incur substantial increased costs to make it work. As a result, they sued the architect, intending to recover their damages. Initially Bilt-Rite's claim was dismissed by the trial court and the Superior Court, both of which were overturned by the state's Supreme Court. This decision thus paves the way for other contractors to recover damages from design professionals who heretofore may have been protected from such claims.

Delay in processing shop drawings.   One lawsuit that comes to mind arose when the architect failed to process shop drawings within a reasonable period of time.

In *Peter Kiewits Sons Co.* v. *Iowa Utility Co* [355 f.Supp.376,392, S.S.Iowa (1973)], a claim was made to collect damages because the contractor suffered losses due to an unjustified delay in the architect's processing of shop drawings.

Errors and omissions.   The standard professional liability for malpractice insurance refers to "errors of commission" and "errors of omission." Employing the old expression "nobody's perfect," it is certainly understandable that final construction documents may be lacking somewhat in minor details, a few dimensions, or 100-percent coordination.

Most of the problems that involve disputes between architects, engineers, and contractors seem to be due to the following drawing deficiencies:

- Drawings that are not coordinated properly among the mechanical, electrical, fire protection, structural, and architectural trades.

- Conflicts between small and large details, and between written specifications and graphic drawings.

- The failure to apply project-specific requirements in the drawings or in the specifications. These statements or details are often "boiler plate" lifted from other projects and don't apply to the project at hand (like the TV bracket spec for a K–12 school, requiring the TV to be turned to be viewed *by the patient*).

- Lack of communication between the various design consultants so that a change made by one designer is not relayed to the others to determine whether it has any impact on their design.

- Insufficient consultation with owners to afford them the opportunity to participate in decisions that ultimately affect the way in which the design will be finalized.

- Insufficient time for proper and thorough review of all contract documents by design consultants when an owner demands an unrealistically compressed time frame for drawing production and submission for bidding purposes.

Incidentally, this list was excised from one prepared by a *design professional* who was expressing concern over the growing cost of malpractice insurance. The professional was making the point: problems are not being created by what the professionals are *doing*; they are being created by what professionals are *not doing*.

Contractors working in a limited geographic area may be reluctant to file suit against a design professional they are likely to be working with in the future, but if the stakes are high, there may be no alternative.

Acceleration: what it is and how it is used

We previously discussed various costs associated with a claim for differing or changed conditions. One of these costs related to *acceleration*.

The legal term *acceleration* should be a part of every project manager's vocabulary. When an owner recognizes that there have been delays incurred by a construction project, but directs the project manager to maintain the original project completion schedule, these instructions are known as a *demand for acceleration*.

There are two types of acceleration: actual and constructive. Actual acceleration occurs when the owner directs the contractor to complete the project ahead of the date contained in the Baseline or "accepted schedule." Constructive acceleration, on the other hand, occurs when the contractor is delayed by some owner/architect action or inaction, but is requested to maintain the original completion date.

By claiming the condition of constructive acceleration, a contractor can pursue monetary relief from the owner. The legal elements required to establish acceleration in such a case are

- An *excusable delay* has been established that entitles the contractor to a time extension.

■ The contractor submits a written request to the owner for the time extension.

■ The request for the time extension is denied.

■ The owner issues a directive to accelerate performance to complete the project within the original time frame.

■ The contractor proceeds with the work at an accelerated pace and documents costs involved with this speed-up process.

■ The contractor then notifies the owner of the intent to submit a change order for the added costs or to file a claim to recover costs if the owner refuses to recognize the change order.

The project manager should be aware of the steps necessary to document and claim reimbursement of costs created by undue demands on the part of the owner, if the occasion presents itself.

Mechanic's liens.   A mechanic's lien is a charge against the owner's property serving notice that some portion of the labor and/or materials/equipment incorporated into their building has not been paid. It can be filed only against work in connection with a private sector project; the filing of a mechanic's lien against unpaid goods and services on a public project is not allowed by law, which is a law known as the Miller Act.

The Miller Act, passed by the federal government, does not permit liens to be placed against property of the U.S. government. Similar acts have been legislated by most state governments (known as "little Miller Acts") and prevent liens from being placed against state and local government property. In such cases, a general contractor must pursue arbitration (if included in the contract) or litigation to collect a claim.

The bond provided by the general contractor on public works projects protects the owner from claims submitted by subcontractors or vendors who are due monies owed but unpaid. These subcontractors or vendors can file a claim with the appropriate federal/state/local agency and request that the agency "call the bond."

When a mechanic's lien is filed, the title to the property is "clouded" and can't be sold or have its title transferred until the lien is "satisfied." The lien can be "removed" by either paying the disputed sum or bonding the lien, which in effect states that the insurance company will guarantee payment of the claim if it is not resolved between the parties.

The normal expiration date of a lien may vary from state to state, and if the lien remains unsatisfied and no bond has been obtained at the time of expiration, the issuer of the lien can initiate foreclosure proceedings. These proceedings commence after the company attorney files a claim stating that a valid lien has been placed upon the property and that a certain sum is due toward their client. Proof of the amount owed is required, as is proof that the lien was properly filed against the correct owner of the property, the correctly identified property was included and the lien was filed within the time frame established by law.

In theory, when foreclosure takes place, the property is sold and the lien holder gets paid from the proceeds of the sale. In practice, the sale of the property rarely occurs and the lien is usually satisfied. The rules and regulations governing the filing of liens vary with the state, but generally a lien must be filed within 90 or 120 days of the last date on which work was performed on the jobsite in question.

Liens filed by general contractors.   When invoices and requisitions are ignored by private clients and all attempts at collection fail, the general contractor is left with little choice but to file a lien against the owner's property.

When the filing of a lien is being considered, watch for certain pitfalls to avoid:

1. *Make certain the lien is filed within the filing limit time.* If the filing time is 90 days after the last date that work was performed on the job, make sure the filing is made before the deadline. If it isn't, the right to file a lien is lost. However, the court is aware of various tricks used to comply with the 90-day requirement. The date of last work must be the last date of *meaningful work.* If a contractor, realizing that their lien rights will expire, say, tomorrow, sends a mechanic back to the site to replace a filter in HVAC equipment, or adjust a door or replace a broken light fixture lens, the courts will interpret this as a means to circumvent the *intent* of the lien rights and thus the lien will probably be declared invalid.

2. *Make certain the lien is filed against the right property.* Although this might seem rather simple, when urban property is involved, it can get complicated, particularly in subdivisions or projects composed of a number of different parcels. If the wrong property is described, the lien will be invalid.

   The author of this book had been involved with a large senior living community and the concrete subcontractor had not paid their ready-mix concrete supplier (although they signed monthly lien waivers indicating that the supplier had been paid!). The irate owner of the property notified the author's company that the current requisition would not be honored until "the enclosed lien is satisfied and proof of removal of the lien is furnished." The only problem was that the lien was filed improperly against *another* project also owned by that owner for which the author's company was not involved. Therefore, payment of the requisition could not be denied.

3. *Is the lien filed against the proper owner?* The proper owner is usually included on the land records in the tax assessor's office, but quite often defining legal ownership is difficult. When a project involves a limited partnership, a joint venture between corporations or individuals, or a syndication or shell corporation such as a Limited Liability Corporation (LLC), it may be difficult to identify the proper owner of record.

A word about lien waivers submitted by subcontractors.   Not all subcontractors will faithfully and honestly complete their lien waivers. On more that one occasion a subcontractor has been known to falsify a lien waiver. Although the subcontractor may sign the lien waiver indicating payment for all labor and materials placed in the building during the period covered by the waiver, either knowingly or unknowingly the information may not be true. Some subcontractors may not be aware that their lower-tier subs have paid their bills—for example, the mechanical subcontractor who engages an insulation contractor and never requests a lien waiver from that company or never questions whether they have paid their suppliers. There are other subcontractors who just plain lie because they need the money to pay for other nonproject materials and equipment.

So don't assume that the submission of a lien waiver from a subcontractor insures that no liens will be filed in the future. Instead, take steps to lessen the chance that this may occur.

Require each subcontractor to furnish the names of all lower-tier subcontractors, the project manager can request lien waivers from those subcontractors as well.

The project superintendent should maintain a list of all subcontractors working on the project so the project manager is alerted to the need to obtain lien waivers from those companies. Lastly, when a false lien waiver is discovered, along with a threat of legal action, the subcontractor should be advised that from that time on, joint checks will be issued to all of their suppliers and subcontractors.

Arbitration and Mediation

Referring once again to that key document, AIA A201—General Conditions, the Article 4.6, 1997 edition requires mediation and arbitration to resolve disputes prior to resorting to litigation. This clause is actually helpful to the general contractor since it provides alternatives to the costly process of litigation.

A $15,000 claim denied by an architect or owner cannot be realistically pursued by litigation because the costs to do so may exceed the amount claimed. The recent notoriety of a lawyer charging $1000 per hour is testament to the high cost of litigation. Conversely, the low cost to pursue mediation and the somewhat higher costs to arbitrate offer other alternatives.

Article 4.6 of AIA A201 establishes the procedure for commencement of the dispute resolution process and requires mediation (Article 4.6.1) as the first step in that process.

Mediation

Mediation is a nonbinding procedure which means that if initiated as a first step in dispute resolution, either party to the process may decide to withdraw from the proceedings at any time. When the mediation session(s) has been concluded, the parties are under no legal obligation to accept or abide by its conclusions.

This process involves engaging a professional mediator to review the facts of the dispute and attempt to get each party to give a little, or sometimes more than a little, to resolve a dispute. Typically, the mediator will start the proceeding by announcing the steps they plan to take to bring about resolution—discussing the strong and weak points of each party's claim and establishing the fact that there is a genuine desire by both parties to negotiate a settlement. The mediator will act as the go-between to affect that resolution, will separate each party, physically, assigning one group to one room and the other group to a second room. By shuttling back and forth and presenting the mediator's opinion of the strong and weak points of each party, a negotiated settlement will be attempted.

As another old saying goes, "a successful negotiation session is one in which neither party is satisfied with the result," resolving a dispute may mean giving up something. If mediation fails, then each party must ratchet the dispute up one further notch and request arbitration.

The arbitration process

The next step in the dispute resolution process is arbitration. The contract documents will include the proper notification required to demand arbitration and the American Arbitration Association (AAA) can provide all of the details and fees regarding the arbitration process.

Although the arbitration process was initially established to reduce or eliminate the participation by lawyers, nowadays many law firms have developed specific departments specializing in arbitration hearings and, in most cases, attorneys will not only assist in the development of the facts and accumulate the documentation required for the arbitration process but will present the claimant's case at the hearing.

The American Arbitration Association, generally referred to as the Triple A, publishes standard procedures to be followed in requesting and participating in the arbitration process. A list of arbitrators will be submitted to both parties with a request to select a panel that could consist of one or more arbitrators. Each person on the list will have been screened so as to have experience in the field or area to be arbitrated.

Both sides present their case, submit documentation to support their position and all such evidence is accepted and evaluated by the panel. Questions are asked by the panel, but unlike court proceedings, the rules of evidence don't apply. Hearsay evidence is permitted and other legalistic requirements are waived. Depending upon the nature and complexity of the case, these proceedings can be concluded in a day, or can stretch out for a year or more.

When the hearings are completed, the panel will retire, review the evidence, and prepare a written "finding" in which they will determine responsibility and establish a monetary award. In binding arbitration, the predominant form of arbitration, the panel's findings are final and if an award is not forthcoming, the injured party may file a claim in court which will automatically be upheld.

Electronic records as evidence.   The general rule is that computer records are admissible just like any other form of business record. The company attorney can provide the definition of "business record," but the generally accepted meaning is *records that are created and maintained in the company's ordinary course of business.* But with any record—written or otherwise—the manner in which the records were created and maintained is essential. Documents stored in a computer are not regularly or necessarily printed out in hard-copy format since this is one advantage of electronic storage. However, when a stored document is printed in preparation for submission at a trial, the opposing party may object to their admissibility based upon the fact that they were not printed out concurrently with the events stored electronically. The courts may require the original source of the computer program to be delineated (procedures used in inputting the document), to ensure the accuracy and reliability of the material.

Rule 1001(3) of the Federal Rules of Evidence defines an original as:

> An "original" of a writing or recording is the writing or recording itself, or any counterpart intended to have the same effect by a person executing or issuing it. An "original" of a photograph includes the negative or any print there from. If data are stored in a computer or similar device, any printout or other output readable by sight, shown to reflect the data accurately is "an original."

Since most e-mails contain certain identifying material such as the sender's address, the name of the sender, the company name, and the presentation of this information may be enough to satisfy any authentication requirements. Because more and more electronic documents are created, stored, and exchanged and because the courts are still refining their views on acceptability of electronically stored materials, a short session with the company attorney may be in order to avoid any future problems with the legal acceptability of those documents.

In Summation

Just remember that the construction industry is an industry of contracts, and in order to deal with the inevitable conflicts that occur due to misunderstandings of the contract obligations, first and foremost, a thorough understanding of one's rights and obligations under the contract is essential. So, Mr. or Ms. Project Manager, carefully read the contract, the specifications, and review the drawings in detail.

Second, it is important to pay prompt attention to an impending dispute and affect a resolution quickly before positions harden.

Third, not all disagreements can be resolved quickly, but attempts to do so are greatly enhanced when complete and accurate documentation relating to the dispute has been prepared and assembled along the way.

Last, and most importantly, many disputes can be resolved by viewing them from the other party's perspective and approaching resolution with an open mind and an attitude of reasonableness.

Safety in Construction

Accident rates in the U.S. construction industry have had their ups and downs over the last four decades, but thankfully mishaps are currently on the decline.

In the early 1950s, the construction industry in the United States was rife with accidents. Work-related injuries were 54-percent higher than in most other industries, making it one of the most hazardous in the country.

According to a November 2005 report by the Bureau of Labor Statistics, the incidence rate of injuries in the construction industry, however, decreased from 6.8 cases per 100 full-time workers to 6.4. This decline was driven by subcontractors whose cases per 100 full-time workers showed a significant drop from 7.3 to 6.8. Figure 13.1 displays the recordable nonfatal injury rate for all industries and how construction compares with other occupations.

With respect to incidence rates for cases involving days away from work, construction has the second-highest missed workday count of all industries except warehousing (Figure 13.2), something which can have a significant impact on productivity.

The Occupational Safety and Health Act (OSHA)

In 1970, Congress passed the Williams-Steiger Occupational Safety and Health Act, now referred to simply as OSHA. The rules and regulations established by OSHA are governed and enforced by the U.S. Department of Labor's Occupational Safety and Health Administration Division. Over the past two decades, many states have enacted industrial safety legislation that parallels many of the safety regulations contained in the 1970 federal act. In 1990, OSHA announced the establishment of a separate Construction and Engineering Division, and in November of that year

Total recordable nonfatal occupational injury and illness incidence rates by major industry sector, 2003 and 2004

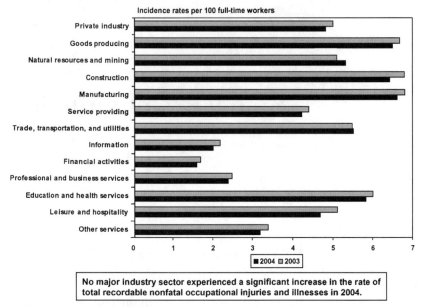

No major industry sector experienced a significant increase in the rate of total recordable nonfatal occupational injuries and illnesses in 2004.

Source: Bureau of Labor Statistics, U.S. Department of Labor
November 2005

FIGURE 13.1 The nonfatal injury rate in various industries.

Incidence rates for cases with days away from work, job transfer, or restriction by case type and selected industry sector, 2004

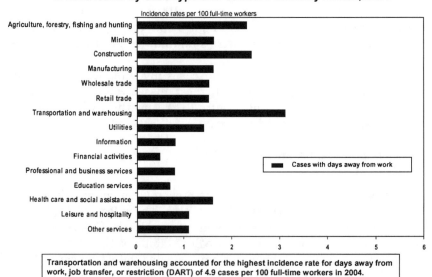

Transportation and warehousing accounted for the highest incidence rate for days away from work, job transfer, or restriction (DART) of 4.9 cases per 100 full-time workers in 2004.

Source: Bureau of Labor Statistics, U.S. Department of Labor
November 2005

FIGURE 13.2 Lost work days due to injuries for the construction industry.

President G.H.W. Bush signed the Omnibus Budget Reconciliation Act, which included, among other provisions, substantially increased penalties for OSHA violations. These increased fines, which took effect on March 21, 1991, were meant to further reduce construction deaths and injuries, adding another violation type—*Willful*—which carried with it a $70,000 maximum fine—and these stiff penalties have been applied.

Factors Responsible for Declining Accident Rates

An increase in training in both quality and quantity is one major factor in the reduction of accident rates.

Some union collective bargaining agreements include provisions for funding their own safety programs. For example, the carpenter's union instituted training safety instruction in 1993 designed to familiarize their members with the new OSHA scaffolding standards. Aided by their various trade organizations, nonunion contractors spent, and are spending, substantial sums of time and money on safety training. Increased safety research has also provided contractors with invaluable information, allowing them to focus their efforts in a more meaningful way.

Studies have shown that most injuries occur during the workers' first months on the job, and most accidents occur in firms employing less than 1000 workers, so particular attention is being paid to those situations.

The accident-prone profile.   OSHA has compiled a whole series of statistics relating to reported accidents, and by culling through these lists, certain patterns emerge. In particular, their study of construction laborers and carpenters is revealing in that they can be interpreted as a profile of the accident-prone worker (see Table 13.1):

OSHA's list of most frequent reporting violations.   Not all OSHA citations are issued for accidents, injuries, and fatalities. Many fines are levied for failure to comply with various reporting and posting violations such as:

- Failure to provide a log and summary of occupational injuries and illnesses

- Failure to adhere to OSHA's General Duty clause (a citation based upon no specific violation, or a citation issued after a previous one had been ignored)

- Failure to report a fatality or multiple hospitalization incidents

- Failure to record occupational injuries and illnesses on the Supplementary Record form

- Failure to record and report occupational injuries and illnesses on the required OSHA form

TABLE 13.1   OSHA Accident Statistics

Construction Laborers—Nature of Injuries	
Sprains, strains	34.0%
Cuts, lacerations, punctures	15.8%
Fractures	11.2%
Soreness, pain	6.3%
Part of body affected:	
Trunk	33.3%
Upper extremities	25.6%
Lower extremities	23.7%
Back	20.7%

Surprisingly, finger injuries represent only 11.8 percent of the injuries, while hand injuries constitute 5.6 percent.

Time of Injury:	
Day: Monday	19.9%
Tuesday	20.4%
Wednesday	20.9%
Thursday	16.7%
Friday	15.3%
Saturday	5.3%
Hours Worked:	
Less than 1 hour	6.5%
2–4 hours	22.4%
4–6 hours	14.5%
6–8 hours	18.4%
8–10 hours	8.9%
10–12 hours	1.6%
Sex, Age:	
Men:	97.9%
Age: 20–24	19.7%
25–34	29.3%
35–44	26.3%
45–54	16%
55–64	4.5%
Length of Service:	
Less than 3 months	25.7%
3 months to 11 months	29.1%
1 year to 5 years	31.1%
More than 5 years	13.2

Carpenters—Nature of Injuries	
Sprains, strains	31.5%
Cuts, lacerations	21.8%
Fractures	11.8%
Bruises, contusions	6.2%

TABLE 13.1   **OSHA Accident Statistics (*Continued.*)**

Carpenters—Nature of Injuries	
Part of body affected:	
Trunk	30.9%
Upper extremities	30.2%
Lower extremities	23.9%
Back	19.8%
Time of Injury:	
Day: Monday	22.7%
Tuesday	21.7%
Wednesday	19.1%
Thursday	18.0%
Friday	13.0%
Saturday	3.5%
Hours Worked:	
Less than 1 hour	20.8%
2–4 hours	27.3%
4–6 hours	15.9%
6–8 hours	16.5%
8–10 hours	9.3%

OSHA listing of most frequently reported safety violations.   For the period ending September 2005, OSHA reported that most citations issued were for violations involving scaffolding, followed by fall protection and electrical-related accidents. These types of accidents have dominated the Most Frequently Cited OSHA Violation list, as shown in Figure 13.3, along with the number of citations issued, and the dollar amount of penalties levied for the period October 2004 through September 2005. Scaffolding violations lead the pack with 912 citations, followed by fall protection and electrical problems.

With respect to fatal accidents, falls (followed by deaths due to powder-actuated tools and abandoned underground storage tanks) ought to be high on the list of any safety inspector.

Figure 13.4 is a list of the top 73 causes for jobsite fatalities, some of which are self-explanatory. Others may not be—such as No. 2, Struck by Nail, which refers to powder-actuated tools and highlights the danger of inadequate training of personnel using those tools.

Fatality No.3—Explosions—relates to hazards in the disposal of abandoned underground storage tanks. With more and more urban renewal projects taking place as cities rediscover the economic value of their downtown areas, more of these abandoned tanks will be uncovered, thus contractors must be aware of them and heed the dangers that await if these tanks are not handled properly.

Standards Cited for SIC 1542; All sizes; Federal

1542 *General Contractors-Nonresidential Buildings, Other than Industrial Buildings and Warehouses*

Listed below are the standards which were cited by **Federal OSHA** for the specified SIC during the period October 2004 through September 2005. Penalties shown reflect current rather than initial amounts. For more information, see definitions.

Standard	#Cited	#Insp	$Penalty	Description
Total	4695	1660	3149686	
19260451	912	382	690495	Scaffolding
19260501	697	595	811520	Fall Protection Scope/Applications/Definitions
19260404	220	190	68388	Electrical, Wiring Design and Protection
19260405	213	151	47109	Elec. Wiring Methods, Components and Equip,Gen'l Use
19261053	180	138	63519	Ladders
19260453	171	148	146995	Manually Propelled Mobile Ladder Stands and Scaffolds
19260502	166	107	110525	Fall Protection Systems Criteria and Practices
19260020	138	120	72985	Construction, General Safety and Health Provisions
19260100	128	127	57108	Head Protection
19261052	119	90	53616	Stairways
19260403	111	99	58261	Electrical, General Requirements
19101200	104	59	9523	Hazard Communication
19260503	96	93	112072	Fall Protection Training Requirements
19261101	76	12	32810	Asbestos
19260454	74	70	21306	
19260651	73	46	104495	Excavations, General Requirements
19260452	72	63	42048	
19260701	72	72	35276	Concrete/Masonry, General Requirements
19260304	67	34	24980	Woodworking Tools
19260760	63	53	92131	
19260021	62	58	31811	Construction, Safety Training and Education
19260150	59	53	11460	Fire Protection
19260652	59	59	116804	Excavations, Requirements for Protective Systems
19260025	56	55	26634	Construction, Housekeeping
19260350	56	39	23459	Gas Welding and Cutting

FIGURE 13.3   25 Most frequently cited OSHA violations.

Better communication between the Department of Labor and the Industry. Over the years, OSHA has become more "user friendly," and on October 1, 1994 they announced their Focused Inspections Initiation plan. The purpose of the Focused Inspection is to further assist responsible contractors and subcontractors who have implemented effective safety programs. Regular OSHA inspections are in-depth and address *all* areas and classes of hazards on a construction site. The Focused Inspection places inspection emphasis on the four leading causes of fatal injuries, which are the source of 90 percent of all construction fatalities:

DISCLAIMER:

The information contained in these Fatal Facts was accurate and correct at the time of issuance. Fatal Facts began in 1984, and since that time, there have been several changes to standards. Please check the current standards (http://www.osha.gov/pls/oshaweb/owastand.display_standard_group? p_toc_level=1&p_part_number=1926) prior to using these documents for training. Many of the addresses, telephone and facsimile numbers, references, and publication numbers may also have changed. You may also want to check these resources for more information on OSHA-related topics.

Government Printing Office now 202-512-1800 or http://bookstore.gpo.gov/

National Technical Information Services now 800-553-6847 or 703-605-600 or http://www.ntis.gov/search.htm

OSHA Office of Training and Education's (OTE) Training Resources, http://www.osha.gov/dcsp/ote/index.html

OSHA Consultation Services, http://www.osha.gov/dcsp/smallbusiness/consult.html

Fatal Facts by Fact Number

No. 01 Fall from Different Level
No. 02 Struck by Nail
No. 03 Explosion
No. 04 Struck by Collapsing Crane Boom
No. 05 Caught in or Between
No. 06 Fall from Elevation
No. 07 Crushed by Falling Wall
No. 08 Struck by Falling Object
No. 09 Trench Cave-in
No. 10 Crushed by Falling Machinery
No. 11 Electrocution
No. 12 Fall from Elevation
No. 13 Collapse of Shoring
No. 14 Fall, Different Level
No. 15 Crushed by Dump Truck Body
No. 16 Fall from Elevation
No. 17 Electrocution
No. 18 Caught by Rotating Part
No. 19 Crushing

FIGURE 13.4 Top causes of jobsite fatalities.

- Falls from elevated areas
- Struck by an object or machine
- Caught in between objects
- Electrical hazards

The Construction Focused Inspections Initiative (see Figure 13.5) and the Construction Focused Inspection Guideline (Figure 13.6) include

No. 20 Fall from Elevation
No. 21 Fall from Roof
No. 22 Cave-in
No. 23 Fall from Tower
No. 24 Fall
No. 25 Fire/explosion
No. 26 Fall through stairwell
No. 27 Fall through Scaffolding
No. 28 Electrocution
No. 29 Fall from Scaffold
No. 30 Electrocution
No. 31 Cave-in
No. 32 Falling from Excavator Bucket
No. 33 Electrocution
No. 34 Caught in Machinery
No. 35 Struck by
No. 36 Asphyxiation
No. 37 Crushed by steel beam
No. 38 Caught in or between
No. 39 Asphyxiation
No. 40 Electrocution
No. 41 Trench Cave-in
No. 42 Fall from elevation
No. 43 Fall from Elevation
No. 44 Electrocution
No. 45 Crushing
No. 46 Fall from Elevation
No. 47 Fall from Elevation
No. 48 Struck by Nail
No. 49 Electrical Shock
No. 50 Caught between Backhoe Superstructure and Concrete Wall
No. 51 Struck By
No. 52 Trench Cave-in
No. 53 Explosion
No. 54 Fall from Roof
No. 55 Trench Cave-in
No. 56 Fall from Scaffold
No. 57 Electrocution
No. 58 Fall and Drowning
No. 59 Struck by Falling Wall
No. 60 Electrocution
No. 61 Trench Collapse
No. 62 Fall
No. 63 There is No Number 63
No. 64 Fall
No. 65 There is No Number 65
No. 66 Fall
No. 67 Asphyxiation
No. 68 Fall (Thrown From)
No. 69 Death due to burns
No. 70 Fall
No. 72 Explosion
No. 73 Struck by/caught between

NOTE: The cases here described were selected as being representative of fatalities caused by improper work practices. No special emphasis or priority is implied nor is the case necessarily a recent occurrence. The legal aspects of the incidents have been resolved, and the cases are closed.

FIGURE 13.4   (*Continued.*)

CONSTRUCTION FOCUSED INSPECTIONS INITIATIVE

Handout for Contractors and Employees

The goal of Focused Inspections is to reduce injuries, illness and fatalities by concentrating OSHA enforcement on those projects that do not have effective safety and health programs/plans and limiting OSHA's time spent on projects with effective programs/plans.

To qualify for a Focused Inspection the project safety and health program/plan will be reviewed and a walkaround will be made of the jobsite to verify that the program/plan is being fully implemented.

During the walkaround the compliance officer will focus on the four leading hazards that cause 90% of deaths and injuries in construction. The leading hazards are:

- Falls (e.g., floors, platforms, roofs).

- Struck by (e.g., falling objects, vehicles).

- Caught in/between (e.g., cave-ins, unguarded machinery, equipment).

- Electrical (e.g., overhead power lines, power tools and cords, outlets, temporary wiring).

The compliance officer will interview employees to determine their knowledge of the safety and health program/plan, their awareness of potential jobsite hazards, their training in hazard recognition and their understanding of applicable OSHA standards.

If the project safety and health program/plan is found to be effectively implemented the compliance officer will terminate the inspection.

If the project does not qualify for a Focused Inspection, the compliance officer will conduct a comprehensive inspection of the entire project.

If you have any questions or concerns related to the inspection or conditions on the project you are encouraged to bring them to the immediate attention of the compliance officer or call the area office at:

_____.

_____ qualified as a FOCUSED PROJECT.

(Project/Site)

_____   _____.

(Date)   （AREA DIRECTOR)

This document should be distributed at the site and given to the Contractor for posting.

FIGURE 13.5 The Construction Focused Inspections initiative.

more information about this program. Your local OSHA office can provide more details on this worthwhile program.

A new process in updating OSHA regulations. Each year, an average of 35 ironworkers die during the erection of structural steel, while another 2,300 are injured. Thus, industry and government have sought ways to prevent this terrible loss of life, and in 1980, government officials began to discuss the possibility of updating antiquated rules relating to ironworkers by meeting with representatives from the ironworker's union

CONSTRUCTION FOCUSED INSPECTION GUIDELINE

This guideline is to assist the professional judgement of the compliance officer to
determine if there is an effective project plan, to qualify for a Focused Inspection.

	YES/NO
PROJECT SAFETY AND HEALTH COORDINATION; are there procedures in place by the general contractor, prime contractor,or other such entity to ensure that all employers provide adequate protection for their employees?	
Is there a DESIGNATED COMPETENT PERSON responsible for the implementation and monitoring of the project safety and health plan who is capable of identifying existing and predictable hazards and has authority to take prompt corrective measures?	
PROJECT SAFETY AND HEALTH PROGRAM/PLAN* that complies with 1926 Subpart C and addresses, based upon the size and complexity of the project, the following:	

_____ Project Safety Analysis at initiation and at critical stages that describes the sequence, procedures, and responsible individuals for safe construction.

_____ Identification of work/activities requiring planning, design, inspection or supervision by an engineer, competent person, or other professional.

_____ Evaluation/monitoring of subcontractors to determine conformance with the Project Plan. (The Project Plan may include, or be utilized by subcontractors.)

_____ Supervisor and employee training according to the Project Plan including recognition, reporting and avoidance of hazards, and applicable standards.

_____ Procedures for controlling hazardous operations such as: cranes, scaffolding, trenches, confined spaces, hot work, explosives, hazardous materials, leading edges, etc.

_____ Documentation of: training, permits, hazard reports, inspections, uncorrected hazards, incidents, and near misses.

_____ Employee involvement in hazard: analysis, prevention, avoidance, correction and reporting.

_____ Project emergency response plan.

* FOR EXAMPLES, SEE OWNER AND CONTRACTOR ASSOCIATION MODEL PROGRAMS, ANSI A10.33, A10.38, ETC.

The walkaround and interviews confirmed that the Plan has been implemented, including:

_____ The four leading hazards are addressed: falls, struck by, caught in/between, electrical.

_____ Hazards are identified and corrected with preventative measures instituted in a timely manner.

_____ Employees and supervisors are knowledgeable of the project safety and health plan, avoidance of hazards, applicable standards, and their rights and responsibilities.

THE PROJECT QUALIFIED FOR A FOCUSED INSPECTION

FIGURE 13.6 The Construction Focused Inspection guideline.

and other industry sources. Ten years later, The Negotiated Rulemaking Act passed by Congress furthered efforts to seek assistance on this issue from the private sector. A panel of officials from the AFL-CIO, Army Corp of Engineers, National Erectors Association, Associated General Contractors of America (AGC), and Associated Builders and Contractors (ABC) working together with OSHA for three years fine-tuned the program that former Assistant Secretary of Labor Charles Jeffries said would produce a new set of OSHA regulations that would prevent

30 fatalities and 1142 injuries each year—at a savings to employers of nearly $40 million.

Finally enacted on July 17, 2001, only time will tell if Mr. Jeffries' predictions are correct, but it certainly seems like the process of government and industry working together is a step in the right direction.

What to do when an OSHA inspector appears at the jobsite. If a project manager is unable to meet with the OSHA inspector when notified that an inspector is onsite, the project superintendent should be familiar with the proper procedures to follow when such an inspection occurs. These entail the following:

1. Once the inspector makes their presence known to the job superintendent, the superintendent should verify the inspector's credentials and determine the nature of the inspection. Is it routine or initiated by receipt of a complaint? The inspector should be asked if they intend to tour the entire site or just selected portions.

2. The superintendent should call their home office and advise their project manager or Director of Safety that an OSHA inspector is onsite in case either one wishes to accompany the inspector on their tour.

3. The superintendent should contact the foremen of all major trades and request that they accompany him or her on the inspector's tour.

4. The superintendent or project manager (PM) should bring with them a pad and paper, and if a camera is available, bring that along in case a photo is needed to dispute a citation.

5. If any OSHA-related documents are requested, bring these, too, but no more documents than specifically requested.

6. Do not volunteer information, but if a violation is discovered and the inspector inquires about the amount of time necessary to correct the situation, the PM or superintendent should respond but without acknowledging responsibility for the violation.

7. Avoid making any statement that may be construed as admission of a violation of any OSHA rule or regulation.

8. If citations are issued during the walkthrough, don't voice an objection at the time of the inspection. Wait until officially notified of the citation, then respond properly.

9. A closing conference with the OSHA inspector should be requested to insure that everyone understands the findings of the inspection.

10. Citations will be issued by OSHA's area director, and when received they should be posted on the jobsite for three days. Any corrective work must be completed within the time frame indicated in the citation.

Safety Pays—In More Ways Than One

An awareness of a shortage of skilled construction workers has been apparent for years. These shortages became more acute in many parts of the country in the 1990s and continue today, likely growing in the foreseeable future.

Loss of a skilled worker, either on a temporary or permanent basis resulting from an accident or job-related injury will be sorely felt by those contractors dependant upon productive crews to get the job done. This further shows the need for effective administration of a safety program.

The positive effects of a good safety record

A significant side benefit of a good safety record is lower costs, primarily in the field, via increased productivity, and reduced insurance premiums.

Excessive insurance costs can add several percentage points to a contractor's overhead, or conversely, shave several points off corporate overhead—two or three percentage points, one way or another, may be the determining factor in winning a bid. Research and dissemination of information developed by the insurance industry has also aided the contractor in their search to improve safety. Insurance companies have a major stake in halting runaway insurance premium costs, and must accumulate, tabulate, and distribute reams of statistical data to their contractor clients regarding the types, and occurrence rate, of various job-related injuries. Some insurance companies conduct studies on topics from slip-resistant materials to the development of exercise regimens that aid workers in avoiding or lessening the severity of job-related injuries.

Soaring worker compensation insurance costs can increase corporate overhead significantly. The following case illustrates the point. Several years ago, the author's company contracted with a New England structural steel fabricator and erector that had a poor worker compensation insurance rating of 1.4, which meant its accident rate was 40-percent higher than the average for their trade. One of their subsequent projects was with a general contractor with a very strong safety program, whose subcontract agreement included a requirement to abide by very stringent safety rules and regulations. One of these regulations required all ironworkers working above six feet off the ground to wear safety belts and harness. This regulation was strictly enforced.

Although many of the strong-willed ironworkers initially balked at this rule, the general contractor firmly stated, "Abide by the safety rules or leave the job." The management of this structural steel subcontracting firm began to recognize the merit of these safety rules and adopted them into their own company safety policy. An adjustment to worker's compensation insurance rates is made after a review of the previous three years' experience, and it took this steel subcontractor several years to work off their prior poor safety record, but they remained committed to do so.

Several years later, their ironworkers had worked 225,000 accident- and injury-free hours, and as a result the company saved hundreds of thousands of dollars in insurance premiums.

Workers' compensation insurance

Worker compensation laws are governed by each individual state. Therefore, provisions vary somewhat from state to state, but three factors remain more or less constant in calculating insurance premiums.

- *Experience Modification Rate (EMR).* This is a "multiplier" that is determined and based upon previous insurance experience of the company policy holder and is used to forecast future benefit payments to employees who have filed claims.

- *Manual Rate.* This is the insurance premium based upon the type of work performed. Various craft trades are classified into "families", referred to as classification codes which are assigned four digit numbers. Each classification code has a corresponding premium rate based upon worker accident claim experience for that trade.

- *Payroll Units.* Payroll units are determined by dividing the employer's total annual direct labor costs by 100.

Calculating workers' compensation insurance premiums. Workers' compensation insurance premiums (WCIPs) are determined by utilizing the following formula:

$$\text{WCIP} = \text{EMR} \times \text{Manual rate} \times \text{Payroll units}$$

The hypothetical examples shown in Table 13.2 reflect how a good experience modifier can reduce workers' compensation insurance rates.

Safety *does* pay! In Table 13.3, a good safety record in this hypothetical situation resulted in a savings of $107,893.00 for that year.

In December 1998, CNA Commercial Insurance in Chicago, Illinois published a chart reflecting the amount of sales and/or new income that must be generated to pay for accident costs. CNA indicates that although direct costs of an accident are paid for by the insurance company, most indirect costs are not, and these indirect costs are anywhere from three to five times the amount covered by the insurance policy.

Owners get involved in contractor safety. For years, owners have been concerned about jobsite safety. No one wants an accident or fatality on their construction site for reasons both humanitarian and PR. An accident

TABLE 13.2   A Good Experience Modifier Can Reduce Workers' Compensation Insurance Rates

Enterprise Construction Company—Insurance Cost Breakdown

Class	Payroll	Developed Rate	Safe Premium*	Normal Premium
Clerical	$100,000	0.38	$380.00	$380.00
Sales	100,000	0.90	900.00	900.00
Carpenters	1,000,000	14.50	145,000.00	145,000.00
Executives	200,000	6.80	13,600.00	13,600.00
Raw total			$159,880.00	$159,880.00
Experience modifier			0.77	1.00
			$123,107.00	$ 159,880.00
State assessment (17.5%)			+21,554.00	+ 27,979.00
*Premium discount (24.5)			−30,385	− 0
Premium before dividend			$114,276.00	$- - -
*Dividend (30%)			−34,310.00	− 0
Total insurance costs			$79,966.00	$187,859.00

*Premiums derived from good safety experience.

scene on the local or national news is what corporations want to avoid, at all costs.

Many owners began requesting copies of contractor safety programs in their bid documents, and now many owners are getting more specific in their scrutiny of bidder's safety records. Intel Corporation with $1.95 billion in ongoing construction projects in 2005 required that bidders have an experience modifier rate of 1.0 or lower. One Intel contractor says, "They force you to put your money where your mouth is." Intel views this strict policy on safety as serving a long-term workforce mobility problem and enhancing the quality of life for workers.

Developing the Company Safety Program

The structure of a company safety program encompasses the following components:

TABLE 13.3   Chart Reflecting the Amount of Sales and/or New Income that Must Be Generated to Pay for Accident Costs

Accident Costs	1% Profit Margin	2% Profit Margin	3% Profit Margin	4% Profit Margin	5% Profit Margin
$1000	$100,000	$50,000	$33,000	$25,000	$20,000
$5000	$500,000	$250,000	$167,000	$125,000	$100,000
$10,000	$1,000,000	$500,000	$333,000	$250,000	$200,000
$25,000	$10,000,000	$5,000,000	$3,333,000	$2,500,000	$2,000,000

- A statement of company policy
- The objective of an accident prevention program
- The appointment, duties, and responsibilities of a Safety Director or Safety Coordinator
- The stated responsibilities of field supervisors in administering the plan, and their relationship with the Safety Director/Safety Coordinator
- Procedures for reporting job-related injuries and illnesses
- The working rules and regulations of the safety program
- A Hazard Communication (HazCom) program, as required by OSHA
- Procedures for dealing with safety violations and violators

The statement of company policy

The statement of company policy simply outlines the reason for implementing the safety program. It can be as succinct as the following:

> The Enterprise Construction Company recognizes that accident prevention is a problem of organization and education which can and must be administered to avoid pain and suffering to our employees and also reduce lost time and operating costs incurred by our company. Accordingly, I state and pledge my full support and commitment to the following:
>
> - That the company intends to fully comply with ALL safety laws and ordinances
> - That the safety of our employees and the public is paramount
> - That safety will take precedence over expediency and shortcuts
> - That every attempt will be made to reduce the possibility of accident occurrence
>
> Sincerely,
John Hancock—President

The objective of the accident prevention program. The objective of the accident prevention program is almost a repeat of the statement of company policy, except that it goes into greater detail.

Objective of the Accident Prevention Program

> The Enterprise Construction Company recognizes that accident prevention is a problem requiring both organization and education in order for it to be overcome. To be successful, our policy must be administered vigorously and intelligently to reduce lost time, avoid much pain and suffering to our employees, and last, but not least, to reduce insurance and operating costs. We rely primarily upon our supervisory personnel in both the field and office to furnish the sincere and constant cooperation required to administer this program.

Effectiveness of the Accident Prevention Program will depend upon the participation and cooperation of management, the supervision of employees, and a coordinated effort to carry out the following basic procedures:

1. Planning all work to minimize losses due to personnel injury and property damage.
2. Maintaining a system that facilitates the prompt detection and correction of unsafe practices and conditions.
3. Making available and enforcing, the use of personal protective equipment, and physical and mechanical guards.
4. Maintaining an effective system of tool and equipment inspection and maintenance.
5. Establishing an educational program that instructs all participants in the basics of accident control and prevention by instituting:
   a. New employee orientation training
   b. Periodic safety meetings
   c. Use and distribution of safety bulletins and related materials
   d. Instruction in the proper and prompt reporting of all accidents and a system that institutes an immediate investigation into such events to determine the cause of the accident and take steps to prevent any recurrence.

Safety equipment. The use of personal protective equipment must meet or exceed minimum OSHA standards, and all state and local requirements are *mandatory*. Such equipment will be furnished by the company.

Inspection of tools and equipment. All equipment and tools must meet or exceed minimum OSHA regulations. Equipment and tool inspection programs must also meet or exceed minimum OSHA requirements, including the maintenance of required records and other documentation.

General Safety Reference. The reference material for this safety program is contained in the U.S. Department of Labor's Occupational Safety and Health Administration (OSHA) manual CFR 1926.

The safety director/safety coordinator

The duties of the safety director/coordinator include the following:

1. Coordinate and monitor the accident prevention program:
   a. Oversee accident investigations. All accident investigations involving serious injuries or those that could have resulted in serious accidents must be investigated by the safety director.
   b. Oversee the proper use of safety equipment.
   c. Perform frequent and unannounced jobsite safety inspections.
   d. Attend and participate in regular safety meetings.
2. Continually review job safety reports and the preparation and dissemination of monthly summaries of safety violations, field inspections, and general program administration items.

3. Immediately document critical conditions and the steps to be taken (and by whom) to correct these conditions.

4. Maintain communication with insurance carriers regarding accident prevention problems.

5. Review and take action, as required, on all safety program violators.

The responsibilities of field supervisors and their relationship with the safety director. The field supervisor (superintendent) who is present daily at the jobsite is the first line of defense in any accident program, and these supervisors not only need formal training in accident prevention, but they need to develop the means of communicating the accident and safety program to everyone on the site—both their own employees and the subcontractor employees.

The safety manual must include provisions for Weekly Tool Box meetings, the content of which can be furnished by most insurance carriers. These simple and concise Tool Box meetings generally focus on only one or two safety-related items each week and when conducted in small groups can be very effective in getting their message across.

Procedures in the field relating to injury and illness reporting. The part of the safety program in which injuries and illness in the field are handled must contain clear and concise information on the reporting of all job-related accidents and injuries. It must be stressed that the field supervisors are not to use their judgement as to which accidents and illnesses should be reported. *All* accidents and illnesses must be reported.

Two types of accident and illness reporting forms are required for this:

- One that's required by the company's Safety Director
- One that complies with OSHA reporting requirements

The field supervisors should have both types of report forms in their safety packet, along with all of the other forms given to them, and must understand how and when these forms are to be used.

Working rules and regulations of the safety program. The rules and regulations of the safety program are its "nuts and bolts," and outline the specific items of protective personal equipment designed for general use and for specialized use—for example, goggles/face shields employed for all metal-cutting tasks, ear protection used when operating certain pieces of equipment, and so forth. When powder-actuated tools are employed, a separate training program outlining the proper use of those tools is required.

Details relating to "red tagging" (taking defective equipment out of use) should be included in this section of the safety plan, and it's wise to include detailed procedures for the proper use of electrical extension cords and electrically operated tools since these items are included in the top OSHA violations lists.

A hazard communication program

The Hazard Communication (HazCom) program is another OSHA requirement that is often not given the attention it deserves, which is why it ranks among the Top 25 Hit List Violations imposed by OSHA.

HazCom was initiated when OSHA became aware of products used on construction sites, or incorporated into projects, which contain hazardous materials. The improper handling, storage, and use of these hazardous products caused a significant number of injuries and job-related illnesses—hence the creation of the Hazard Communications program.

The material safety data sheet.   Manufacturers of materials deemed to contain hazardous substances are required, by law, to prepare a material safety data sheet (MSDS) for the product, which describes its hazardous nature, proper handling and storage instructions, and in case of contact or ingestion by a worker, the necessary first aid/medical procedures to follow.

Prior to shipment to the site of any hazardous materials or substances, the supplier of these materials is required to transmit an MSDS to the office for dissemination to the field. When the product does arrive at the construction site, the provisions of the corresponding MSDS are to be followed with respect to handling and storage instructions. In case of spillage (if liquid), or contact with the skin or eyes, or if the product is ingested, the instructions in the MSDS will include first-aid procedures and any follow-up medical treatment required. The field supervisor must keep an accurate up-to-date orderly file of all MSDS documents and insure that all such products delivered are treated in full compliance with these safety data sheets. One of many documents that an OSHA inspector will want to review is the field office's file of MSDS sheets, so keep them in a convenient place and in an orderly manner.

Dealing with safety rule violators—the need for a model policy.   One of the safety director's more important responsibilities is the monitoring of safety violations. Each project manager and their field supervisors must be instructed in the prompt and proper reporting of safety violations. If workers perceive that the company is lax in dealing with violations of the safety program, the entire program can become a toothless tiger.

A disciplinary process for violators of safety rules and regulations is an integral part of any safety program and should be reviewed from time to time with long-term workers, and should become a mandatory part of the orientation safety program for new workers. In this "quick to sue" era, highlighting restraints on dismissals without sufficient documentation, it's important that all employees review the company's disciplinary policy in the presence of the safety director, who should respond to all questions. After this review, each employee should sign an acknowledgment sheet indicating that they have received and read, and fully understand the provisions of the company's policy as it relates to violations of safety rules and regulations. This will forestall any complaints from a worker who has committed a serious violation and then decided to hide it under the guise of "I didn't know that was against company policy."

Rules regarding the use of unauthorized drugs or controlled substances on the site must be approached carefully for fear of wrongful accusations. The oft-told story concerning the firing of an employee who appeared to be "spaced out" on drugs, but who was then later found to be suffering from side effects of a doctor's prescribed medication needs no repeating. But the company who allowed that unlucky supervisor to indiscriminately fire that worker is probably still paying off the cost of the lawsuit.

The following is a typical disciplinary procedure section from a safety program:

> Compliance with OSHA and company safety rules and regulations is a condition of employment at (*company name*). All employees working for (*company name*) will be trained in, and must familiarize themselves with both OSHA and company safety rules and regulations before beginning work on a construction project. A copy of the company's safe work rules and procedures will be provided by your supervisor.
>
> Management personnel at all levels—including project managers, field supervisors, and foremen—are responsible for taking action when a violation is observed. If a violation occurs, they must take action immediately to correct the violation and enforce this disciplinary policy. Employees who fail to follow safety rules and regulations established to protect them and their fellow employees endanger themselves and others.

Note: We've all been guilty of walking through the jobsite during an inspection, or meeting with the architect and observing a safety violation, but rather than break away and advise the violator, we continue on our way ignoring the violation. This is not lost on the work crews in and about the area. When a violation is observed, STOP and bring it to the attention of the violator with a stern warning, followed by a written notification of that violation.

The following procedures should be instituted and distributed to all employees so they fully understand the actions the company will take when violations are observed.

First warning: The first time an employee is observed violating any safety rule, the employee shall be given a "first warning". The first warning will be an oral one and will be so noted in that employee's personnel file.

Second warning: The second time the employee is observed violating any safety rule, the employee shall be given a "second warning". The second warning will be an oral warning accompanied by a written safety violation notice. A copy of the written safety violation will be given to the employee, the employee's union steward (if applicable) and the company's safety director. A copy of the notice will be placed in the employee's personnel file. The employee will be required to meet with the safety director for counseling.

Third warning: The third time an employee is observed violating any safety rule, the employee shall be given a "third warning". The third warning will be a written safety violation notice, a copy of which will be given to the employee, the employee's union steward (if applicable), and the safety director. A copy of the notice shall be placed in the employee's personnel file. A meeting with the employee, their immediate supervisor, the safety director, and a representative of top management will be held to determine why the employee failed to comply with the company's safety program—for the third time. Top management must determine what action will be taken at this time. One option open to management, which is included in the disciplinary section of the safety program is suspension from work. For example, the policy might state that "Employees who accumulate three warnings in a 12-month period may be suspended from work, without pay, for up to one week."

Fourth warning: The fourth time an employee is observed violating any safety rule, the employee shall be given a "fourth warning. A fourth warning will be a written safety violation notice. Employees who do not follow safety rules, especially after being warned several times previously, are a threat to themselves and their co-workers. Therefore, employees who receive a fourth warning, may, at management's discretion, be terminated from employment or be subject to other disciplinary action deemed appropriate by management.

The actions previously listed must be taken whenever a safety violation is observed.

Consistency and fairness are key factors in administering the safety program. It is critical that the administration of the safety program be consistent and fair. Employees will not take the disciplinary policy seriously if they know that safety violations have occurred and no oral or written warnings were issued.

Each project manager, supervisor, and lead man ought to have copies of safety violation forms in their briefcase or file cabinet in the job trailer.

The carrot and stick approach to safety.   The disciplinary portion of the safety program certainly qualifies for the "stick" approach. Some companies, however, supplement their enforcement program by offering incentives to maintain safe working conditions. These incentives can be based upon achieving either short-term or long-term goals, or a combination of both.

A short-term incentive can take the form of presenting a small gift such as a Swiss army knife to an employee who exhibited some on-the-spot safety measure. Safety stickers, to be worn on one's hard hat, are a highly visible short-term safety award.

Long-term incentives, for continued safety consciousness or accident-free records can take the form of cash or merchandise bonuses, extended vacation time with pay, or a prepaid weekend for two at a nearby hotel.

Fluor Daniel, the Virginia-based subsidiary of international constructor Fluor Corporation, developed a Hazard Elimination Program several years ago that has been responsible, in large part, for achieving millions of work hours each year without a time-lost accident. Their Hazard Elimination Program works on a "point" basis. Whenever an employee spots a safety violation, takes corrective action, and reports the violation and action to their supervisor, the employee safety committee then evaluates the hazard as either a Type I, Type II, or Type III hazard. Type I designates a minor hazard, while Type III constitutes a violation that could cause imminent danger. Type II lies between a minor and major infraction.

Based upon the category of the violation, points are awarded to the individual, totaled monthly, and entered into the program's tally sheet. Those employees who achieve high scores receive company recognition and awards ranging from gifts to paid time-off. The program has proven itself to be highly effective. In fact, one Missouri construction project exceeded five million work hours without a disabling injury.

Remember, a solid safety program, administered vigorously, reduces corporate overhead, increases employee morale, creates positive public relations, and just makes good business sense. The attached forms, shown in Figures 13.7 through 13.12, may come in handy while formulating your company's safety plan.

SAFETY VIOLATION NOTICE
COMPANY

To: _____   Date _____   Time _____

_____   Job #

_____   Project

ATTN: _____   Location

Company _____

The company identified above is not following safe practices as follows:

_____

_____

_____

_____

_____

_____

_____

_____

Heffner & Weber requires that all practical effort be taken to eliminate or minimize exposure of workers to unsafe and/or unhealthy conditions on our job sites. Subcontractors and visitors are obligated to cooperate with this policy.

FAILURE OF A WORKER TO COOPERATE FULLY IN CARRYING OUT OUR SAFETY PROGRAM IS CAUSE FOR DISCIPLINARY ACTION INCLUDING REMOVAL FROM SITE, PROBATION, SUSPENSION AND / OR DISMISSAL.

Very Truly Yours,   Received By

_____   _____

Project Superintendent   For _____

                                          Date _____   Time _____

FIGURE 13.7   Safety Violation Notice—Company violation.

SAFETY VIOLATION NOTICE
INDIVIDUAL

To: _____   Date _____   Time _____

_____   Job #

_____   Project

ATTN: _____   Location

Individual _____   ☐ Heffner & Weber Employee

Employer _____   ☐ Subcontractor Employee

Craft _____   ☐ Site Visitor

The individual identified above is not following safe practices as follows:

_____

_____

_____

_____

_____

Heffner & Weber requires that all practical effort be taken to eliminate or minimize exposure of workers to unsafe and/or unhealthy conditions on our job sites. Subcontractors and visitors are obligated to cooperate with this policy.

FAILURE OF A WORKER TO COOPERATE FULLY IN CARRYING OUT OUR SAFETY PROGRAM IS CAUSE FOR DISCIPLINARY ACTION INCLUDING REMOVAL FROM SITE, PROBATION, SUSPENSION AND / OR DISMISSAL.

Very Truly Yours,   Received By

_____   _____

Project Superintendent   For _____

                                    Date _____   Time _____

FIGURE 13.8 Safety Violation Notice—Individual violation.

ACCIDENT / INCIDENT REPORT
WITNESS STATEMENT

JOB NUMBER _____   DATE _____   SHEET _____

JOB NAME _____   SUPERINTENDENT _____

A. Witness Name And address

Name _____   DATE _____   SHEET _____

Address _____   Home Phone _____

City _____   Work Phone _____

State ZIP   Employer

B. Witness Description Of Accident / Incident

Line	Time	Witness Statement
1		
2		
3		
4		
5		
6		
7		
8		
9		
10		
11		
12		
13		
14		
15		
16		
17		
18		

Witness
Signature _____   Date _____

Project
Superintendent _____   Date _____

FIGURE 13.9   Accident Report—Witness Statement.

ACCIDENT / INCIDENT REPORT
MINOR

JOB NUMBER _____ DATE _____ SHEET _____

JOB NAME _____ SUPERINTENDENT _____

1. **WAS ANYONE INJURED?** ☐ **YES - ACCIDENT REPORT** ☐ **NO - INCIDENT REPORT**

2. Name Of Employee Injured _____ Employer* _____

   Describe Injury. _____

   Was Medical Attention Required? Describe. _____

   Treated By _____ Location _____

3. Name Of Witness(es) _____ Employer _____

   _____ _____

   _____ _____

4. Describe Work Being Performed At The Time Of The Accident / Incident And Applicable Safety Rules

   _____

   _____

5. Were Safety Rules Violated? ☐ NO ☐ YES (If YES Describe Violation)

   _____

   _____

6. Describe What Happened In Sequence.

   _____

   _____

   _____

7. What Caused The Accident / Incident

   _____

   _____

8. Describe What Was Done Or Can Be Done To Prevent This From Taking Place Again.

   _____

   _____

9. Project Superintendent - Describe Observations And Immediate Response Actions Taken

   _____

   _____

FIGURE 13.10 Accident/Incident Report—Minor.

ACCIDENT / INCIDENT REPORT
MINOR CONT'D

10. Continuation from above sections. Enter Question Number and Line number being continued.

CONT'D	Time	Continuation

Signature _____ Date _____
Project Superintendent   * If Subcontractor Attach Subcontractor Accident / Incident Report

FIGURE 13.10   (*Continued.*)

ACCIDENT / INCIDENT REPORT
Sheet 1

JOB NUMBER _____   DATE _____   SHEET _____

JOB NAME _____   SUPERINTENDENT _____

1. **WAS ANYONE INJURED?**   ☐ YES - ACCIDENT REPORT   ☐ NO - INCIDENT REPORT

2. Enter Data for all injured parties. Use second sheet if additional space is needed.

Name	Employer	Injury	Treatment

3. **WERE THERE WITNESSES?** ☐ NO   ☐ YES - Enter below. Use second sheet if additional space is needed.

Name	Employer	Job Function	Location

4. **PROJECT SUPERINTENDENT - DESCRIBE OBSERVATIONS AND IMMEDIATE RESPONSE ACTIONS TAKEN.**

Line	Time	Observation / Action
1		
2		
3		
4		
5		
6		
7		
8		
9		
10		
11		
12		

FIGURE 13.11   Accident/Incident Report.

ACCIDENT / INCIDENT REPORT
Sheet 1 Cont'd

5. DESCRIBE RELATED WORK BEING PERFORMED AT THE TIME OF THE ACCIDENT / INCIDENT AND
   APPLICABLE SAFETY RULES

Line	Time	Work In Process / Safety Rules
1		
2		
3		
4		
5		
6		
7		
8		
9		
10		
11		
12		

6. WERE SAFETY RULES VIOLATED?   ☐ NO   ☐ YES - Enter on next page.

7. DESCRIBE WHAT HAPPENED IN SEQUENCE.

Line	Time	Sequence Of Events
1		
2		
3		
4		
5		
6		
7		
8		
9		
10		
11		
12		

FIGURE 13.11   (*Continued.*)

ACCIDENT / INCIDENT REPORT
Sheet 2 Cont'd

JOB NUMBER _____   DATE _____   SHEET _____

JOB NAME _____   SUPERINTENDENT _____

8. Continuation from above sections. Enter Question Number and Line number being continued.

CONT'D	Time	Continuation

FIGURE 13.11   (*Continued.*)

FIGURE 13.12 Emergency Number List including local utility company listings.

EMERGENCY NUMBER LIST

Project _____ Plat # _____ Tax # _____
Location _____ Street Address _____
Project Superintendent _____ City _____ State _____ ZIP _____
Job # _____

	Type Of Emergency Service	Name Of Emergency Provider	Emergency Telephone Number	Limits On Hours Of Operation	Limits On Capabilities	Non-Emergency Telephone	Point Of Contact	Address Of Emergency Provider	Fees / Pre-Use Agreements Or Contracts
1	FIRE								
2	POLICE								
3	MEDICAL								
4									
5	Ambulance								
6	Hospital								
7									
8	Water Break								
9	Gas Leak								
10	Sewer Leak								
11	Elect - OH								
12	Elect - UG								
13	Tele - OH								
14	Tele - UG								
15	Cable - OH								
16	Cable - UG								
17	HazMat								
18	Environ								
19									
20									

EMERGENCY NUMBER / ADMINSTRATIVE NUMBER

324

14

Design-Build

Design-build is neither a new concept nor a complex one. This project delivery system is similar to that of the Master Builder approach practiced by the Egyptians 4500 years ago during the construction of their pyramids, and in the United States during the early part of the twentieth century when contractors self-performed most of their work and included design services if the client so desired.

Today, the design-build team of architect, contractor, and engineer provide the project owner with a single contract that encompasses design and construction. In 1986, approximately 3 percent of all construction project delivery systems were utilizing design-build. By 1998, this figure had grown to 27 percent, and according to the Design-Build Institute of America (DBIA), it's projected that 45 percent of all projects will be constructed via design-build during the first decade of the twenty-first century. So far, this projection appears not far off the mark given how public and private sectors have both experienced the benefits of design-build.

General contractors and subcontractors need to become more familiar with design-build, not only because it represents a substantial market, but also because it is a system that allows the contractor/subcontractor to enter a new market—one in which projects can be negotiated, where "low bid" is not the overriding factor, and where profit margins are slightly higher.

What Is Design-Build?

Under a design-build approach, a team consisting of a general contractor, architect, and engineer is assembled to provide an owner with a one-source point of contact for the design and construction of their proposed project. Design-build companies can provide all of these services with their in-house staff of engineers and construction professionals, or they can subcontract for the services they don't normally provide.

Why has design-build garnered so much attention recently?

Numerous surveys have revealed that design-build decreases costs, as well as the overall design and construction time. It is 23-percent faster than a CM-at-risk project and 33-percent faster than design-bid-build projects according to a Pennsylvania State University College of Engineering study. A Construction Industry Institute (CII) report revealed that design-build produced a median of 9000 square feet of construction per month, as opposed to 4500 square feet using design-bid-build.

The more rapid cycle of design and construction allows owners to occupy their facility more quickly and reduce the time for which costly construction financing is required. As far as project cost savings, the CII survey revealed that design-bid-build experienced the highest cost escalation at 4.84 percent while initial costs via design-build escalated only 2.37 percent—presumably because of the impact of change orders associated with design-bid-build projects.

Benefits accruing the design-build team.   Zweig White, a management consulting firm headquartered in Chicago, conducts annual surveys of the design, engineering, and construction industries. Their *2005 Design-Build Survey of Design and Construction Firms* asked the question to 98 design and construction firm executives, "Are design-build projects more or less profitable than those projects completed using traditional project delivery methods?" Eighty-four percent of respondents said "Yes." The breakdown of this survey in Figure 14.1 includes the demographics of the survey respondents.

The survey also asked respondents to provide reasons why they thought design-build was more profitable. Their answers are shown in Figure 14.2.

The advantages to the owner and design-builder of one-source responsibility. Combining design and construction in one contract provides the owner with a single source, thereby relieving them of considerable management and coordination responsibility. Since design and budget can be dual-tracked, conformance to the budget as the project proceeds through design development can be carefully monitored, while redesign—with its associated costs and delays—can be avoided entirely or mitigated.

Single-source responsibility avoids the finger-pointing that often arises between the client's design consultants and the contractor if

Breakdown of the survey sample

Sample size

98 design and construction firms completed and returned a valid questionnaire.

Firm type

Integrated design/build 43%
Design services/consulting 27%
Construction 31%

Year founded

Prior to 1945. 29%
1945 - 1959 . . . . 16%
1960 - 1969 . . . . 11%
1970 - 1979 . . . . 9%
1980 - 1989 . . . . 15%
1990 - 1999 . . . . 14%
2000 to present 3%
Unspecified. 2%

Region of headquarters

New England 6%
Middle Atlantic. 10%
South Atlantic 18%
North Central. 28%
South Central 6%
Mountain. 12%
Pacific. 17%
Unspecified. 2%

Staff size

Minimum. 2
Lower Quartile 30
Median 150
Mean. 1,392
Upper Quartile 650
Maximum 35,000

Staff size *(breakdown)*

1 - 49. . . . . 33%
50 - 99 . . . . 11%
100 - 249 . . . . 14%
250 - 499 . . . . 13%
500 - 999 . . . . 5%
1,000 + ... 20%
Unspecified. 3%

2004 gross revenue

Minimum $720,000
Lower Quartile $10,000,000
Median $50,000,000
Mean $321,818,352
Upper Quartile $220,000,000
Maximum $5,000,000,000

Note: due to rounding, percentages for some questions do not total 100.

FIGURE 14.1 A breakdown of the 2005 Zweig White survey on the design, engineering, and construction industries. (With permission from Zweig White Information Services, LLC, Natick, MA.)

errors and omissions or coordination issues surface. This is because the design-build team "owns" those costs to correct—not the owner.

By combining design and construction, the adversarial relationships that often exist between architect, contractor, and owner are lessened substantially and usually disappear entirely. Experience has shown that the number and type of change orders are substantially reduced, along with disputes and claims that often occur during the design-bid-build process.

In Great Britain, design-build is referred to as a "package project," an appropriate name because the owner no longer buys a service—design or construction—but instead buys a "package"—a product. Studies in England conducted by the University of Reading's Design-Build Forum revealed that the DB delivery system produced a 12-percent improvement in the speed of construction, and a 30-percent increase in overall project delivery. These British researchers also found that design-build resulted in a 13-percent reduction in square-foot costs, and projects were more likely to be completed within a 5-percent range of their original budget.

Design/build profitability

Issues   Are design/build projects in general more or less profitable than projects completed using traditional project delivery methods? Why?

Background   Design, consulting, and contracting firms are only going to pursue the design/build market if it is profitable for them to do so. Do firms believe design/build is a viable, *profitable* way to bring a project to completion?

Survey Findings   ■ The vast majority of firm leaders (84%) believe design/build projects are more profitable than traditional projects. One of the most common reasons firms cite for this belief is that design/build projects allow the builder to have greater control.

■ Firms that consider design/build projects to be less profitable than traditional projects do so for reasons including higher costs and increased risk.

In your opinion, are design/build projects in general more or less profitable than projects completed using "traditional" project delivery methods?

More profitable. 84%

Less profitable 12%

Unspecified 4%

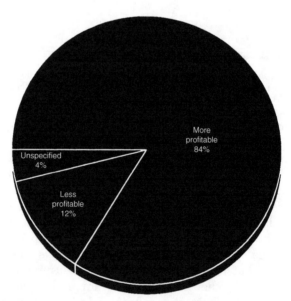

FIGURE 14.2 Responses by participants as to why design-build was thought more profitable. (With permission from Zweig White Information Services, LLC, Natick, MA.)

Design/build profitability *(continued)*

Why are design/build projects in general *more* profitable than projects completed using "traditional" project delivery methods?

- Ability to directly manage all costs and to plan and design to cost
- Able to control costs more
- Assuming owner/regulators don't hamstring design/build on project submittal requirements
- Assumption of greater risk if managed properly should produce greater reward
- Because of "team effort," fewer documents are used
- Best value selection
- Better control (2)
- Better coordination between A/E and contractor makes project more efficient and therefore more profitable
- Better defined scope and more efficiencies provide more value to owner and more room under budget for higher contractor fees
- Better risk control when true "team" relationship exists
- Budget control
- But not enough yet— market still learning the value
- Can control schedule
- Client is buying value
- Constant communication with owner prevents out-of-budget changes
- Control
- Control added early in the process
- Control is in hands of those who can keep overall project on track
- Control materials
- Control of cost
- Control schedule
- Control scope
- Control specifications
- Cost savings
- Cost-control ability, but more risk
- Costs are controlled continuously

- Design firms paid for ideas, not hours
- Design/build firm has opportunity to meet scope more efficiently with right design partners
- Faster delivery speed
- Flexibility in design solutions with cost savings and still maintain quality
- For both design/build team and owner provides increased time savings, reduced claims, improved community relations
- For integrated firms— control; client satisfaction = repeat business = profit
- For our firm, we are entering the construction side of the equation, which we didn't do before
- Full control of the project
- Higher design and construction fee
- Higher fee
- Higher profits for partners by adjusting project costs to accommodate profit
- Higher risk
- If estimated properly
- If managed properly, more profit can be made
- If properly managed
- If you learn to identify, quantify, and manage risk
- It is not (usually) purely fee driven
- Less competition (3)
- Less litigation
- Less time
- Longer-term customer relationship
- More control (2)
- More control during project cycle
- More control equals more cost/profit control
- More control over schedule
- More control over scope
- More cushion/contingency to be spread out on project
- More input into the means and methods
- Need to reduce risk by defining the project
- Negotiated
- No "bad" projects

FIGURE 14.2 *(Continued.)*

The downside of design-build (with counter-arguments in italics). Critics of design-build cite the following problems associated with general contractors, design consultants, and owners in a design-build mode, many of which represent *institutional* barriers that are often difficult to overcome.

- The cultural differences between the architect and the general contractor must be reconciled, since each may have had a different agenda

Design/build profitability *(continued)*

Why are design/build projects in general *more* profitable than projects completed using "traditional" project delivery methods? *(continued)*	Why are design/build projects in general *less* profitable than projects using "traditional" project delivery methods?
■ Not negotiating with public clients ■ Opportunity to negotiate fee ■ Revenue based on risk assumed, not hours spent ■ Reward matches risk ■ Risk, but you manage it ■ Risk/reward ■ Scope coverage ■ Solving client problems rather than building buildings ■ Team work ■ Teaming mentality ■ The client is willing to pay for the additional value of single point of responsibility ■ Typically are negotiated work ■ Value	■ Due to excessive marketing and proposed preparation costs ■ Emphasis shifts more to project satisfaction, rather than client relationship— decreasing long-term continuing work ■ Have not overcome losses where team lost bids ■ Less opportunity for change orders ■ More risk ■ Owner says, "Should have covered that in your price." ■ Poor ability to quantify costs so early in project ■ The higher risk ■ They are awarded on a fee basis, which is not very sophisticated ■ Thought is that design/build is responsible for all cost growth, regardless of reason ■ Too many unforeseen risks

FIGURE 14.2   *(Continued.)*

based upon their experience with more conventional design-bid-build project delivery systems. The contractual relationship between the contractor and designer is distinctly different in the design-build process, requiring them to rethink ways to work together harmoniously. *The ability to develop a design-build capability, with its potentially higher profit levels, and an opportunity to negotiate contracts rather than hard bid is a strong incentive to set aside those cultural and institutional differences.*

■ The owner may be placed in the role of traffic cop, having to insure that their interests are being addressed. In the more conventional design-bid-build program, the architect acts as the owner's agent and is charged with overseeing the contractor's obligations and responsibilities to some degree. *If the owner engages a CM, they will have their traffic cop, or the owner can hire an experienced consultant to act as their representative.*

■ The cost savings generally touted as one of the advantages of design-build are not always there. If none accrue, an owner may be left with the feeling that they have been "sold a bill of goods." *If the design-build team has done a first-rate job of extracting the owner's program, even if no cost savings accrue, the project will probably have been completed more quickly and with little or no change orders and disputes.*

- To some team members—whether they be owners, general contractors, or architect/engineers—at least initially, design-build is a venture into uncharted waters with all the dangers that such ventures possess. *This is true, but with the demonstrated advantages offered by design-build, a carefully orchestrated venture with a small pilot-type project may produce confidence, knowledge, and experience leading to more such projects.*

- The checks and balances present in the conventional design-bid-build process are not fully present in the design-build delivery system. *Once again, employing a CM as the owner's agent or hiring an experienced construction professional will provide the client with checks and balances.*

- Both surety and insurance companies take a much more cautious role in the design-build process because their experience to date has been somewhat limited and the liability issues so clearly defined in the design-bid-build process become intertwined. *Surety concerns can be lessened when the contractor is the lead team in a design-build venture since they will generally have had prior bonding experience. Insurance issues are somewhat different in a design-build mode and more and more insurance companies are becoming sophisticated in their approach to this process.*

Developing a Design-Build Capability

Several ways exist in which a general contractor can develop a design-build capacity: hire an architect to provide in-house design capability, create a joint venture with an architect, or purchase designs just like any other subcontracted service.

Creating in-house design capability

An experienced architect comes at a cost and may seek some form of equity incentive if they were earlier successfully employed by an established firm. Thus, overhead will increase sharply and some losses should be anticipated until such time as additional projects are brought in-house and the added overhead has been absorbed. However, an experienced architect may have developed a list of loyal clients that could form the core of this new venture and bring new business into the construction firm.

The joint venture

The joint venture (JV) is a one-time business entity created for a specific project; when the project is completed, the JV dissolves. This is one vehicle to combine both design and construction and can be formed with either architect or builder as the lead member. Licensing laws in some states may limit the ability of an architect to obtain a builder's license or a contractor to practice architecture, so an investigation of any state laws regarding such ventures needs to be made.

The JV must specify the rights, responsibilities, and obligations of each member of the joint venture, which can be accomplished by creating a teaming agreement. The Associated General Contractors of America (AGC) publishes a number of design-build–related contracts, and their AGC Document No.499—Standard Form of Teaming Agreement For Design-Build Project (see Figure 14.3) is an excellent example of a teaming agreement.

The limited liability corporation—the LLC

The LLC is another business entity that can be employed to form a design-build team. The LLC offers the liability protection of a corporation but is usually created for a single purpose or specific venture. Like other business entities, the LLC requires the services of an attorney to set it up. Apart from its limited liability features, there are other benefits of the LLC:

- Unlike a regular corporation, no formal meetings are required and therefore no minutes are necessary.
- No corporate resolutions are required.
- The distribution of profits can be individually tailored.
- All business profits, losses, and expenses flow through the corporation to the individual members of the company, thus avoiding the double taxation of paying corporate and personal taxes on the money earned.

The disadvantages of an LLC are

- The LLC is dissolved when a member dies or undergoes bankruptcy, whereas a corporation can theoretically live forever.
- Because of the nature of an LLC, lending institutions are reluctant to provide funds without personal guarantees from its officers.
- Clients may be reluctant to do business with an LLC because they recognize its single-subject nature.

An architect- or contractor-led design-build team?

When considering a design-build team, the question of who is best suited to assume leadership may arise: builder, architect, or engineer? There are several answers to this question, but Question No.1 may well be

THE ASSOCIATED GENERAL CONTRACTORS OF AMERICA

AGC DOCUMENT NO. 499
STANDARD FORM OF TEAMING AGREEMENT
FOR DESIGN-BUILD PROJECT

This Agreement is made this _____ day of _____ in the year _____ , ◆

by and between

TEAM LEADER _____ ◆
(Name and Address)

and **TEAM MEMBER** _____ ◆
(Name and Address)

and **TEAM MEMBER** (if applicable) _____ ◆
(Name and Address)

and **TEAM MEMBER** (if applicable) _____ ◆
(Name and Address)

the parties collectively referred to as the **TEAM** for services in connection with the following **PROJECT**

_____ ◆
(Name, Location and Brief Description)

for **OWNER** _____ ◆
(Name and Address)

FIGURE 14.3   AGC document no.499—standard form of teaming agreement for design-build project contract. (*All materials displayed or reproduced are with the express written permission of the Associated General Contractors of America under copyright license No.0115.*)

ARTICLE 1

TEAM RELATIONSHIP AND RESPONSIBILITIES

1.1 This Agreement shall define the respective responsibilities of the Team Members for the preparation of responses to the Owner's request for qualifications and request for proposals for the Project. Each Team Member agrees to proceed with this Agreement on the basis of mutual trust, good faith and fair dealing and to use its best efforts in the preparation of the statement of qualifications and proposal for the Project, as required by the Owner, and any contract arising from the proposal.

1.2   The Team Leader, _____ ◆

_____ shall provide overall direction and leadership for the Team and be the conduit for all communication with the Owner. In addition the Team Leader shall provide expertise in the areas of (a) construction management and construction; (b) the procurement of equipment, materials and supplies; (c) the coordination and tracking of equipment and materials shipping and receiving; (d) construction scheduling, budgeting and materials tracking; and (e) administrative support. The Team Leader's representative shall be: _____ ◆

1.3   The principal design professional is Team Member,

_____ , ◆ who shall perform the following design and engineering services required for the Project: _____ ◆

_____
_____
_____
_____
_____ .

In addition this Team Member shall coordinate the design activities of the remaining design professionals, if any. This Team Member's representative shall be: _____ ◆

_____
_____ .

1.4   Team Member, _____ ◆

_____ , shall provide expertise in the following areas:

_____ ◆
_____
_____
_____
_____
_____
_____
_____ .

This Team Member's representative shall be: _____ ◆

_____
_____ .

1.5   Team Member, _____ ◆

_____ , shall provide expertise in the following areas:

_____ ◆
_____
_____
_____
_____
_____
_____ .

This Team Member's representative shall be: _____ ◆

_____
_____ .

1.6   Each Team Member shall be responsible for its own costs and expenses incurred in the preparation of materials for the statement of qualifications and the proposal and in the negotiation of any contracts arising from the proposal, except as specifically described herein:

_____ ◆
_____
_____
_____
_____
_____
_____
_____
_____
_____
_____
_____ .

Any stipends provided by the Owner to the Team shall be shared on the following basis:

_____ ◆
_____
_____
_____
_____
_____
_____ .

1.7   EXCLUSIVITY No Team Member shall participate in Owner's selection process except as a member of the Team, or participate in the submission of a competing statement of qualifications or proposal, except as otherwise mutually agreed by all Team Members.

2

AGC DOCUMENT NO. 499 • STANDARD FORM OF TEAMING AGREEMENT FOR DESIGN-BUILD PROJECT
© 2001, The Associated General Contractors of America

FIGURE 14.3   *(Continued.)*

"Which firm through their marketing efforts, or through personal contact was approached by a prospective client to discuss considering design-build for their next project?"

Contractor as lead.   The predominance of contractors as team leaders has evolved for many practical reasons:

ARTICLE 2

STATEMENT OF QUALIFICATIONS AND PROPOSAL

2.1   The Team Members shall use their best efforts to prepare a statement of qualifications in response to the request of the Owner. Each Team Member shall submit to the Team Leader appropriate data and information concerning its area or areas of professional expertise. Each Team Member shall make available appropriate and qualified personnel to work on its portion of the statement of qualifications in the time frame proscribed, and shall provide reasonable assistance to the Team Leader in preparation of the statement of qualifications.

2.2   The Team Leader shall integrate the information provided by the Team Members, prepare the statement of qualifications and submit it to the Owner. The Team Leader has responsibility for the form and content of the statement of qualifications and agrees to consult with each Team Member, before submission to the Owner, on all matters concerning such Team Member's area of professional expertise. The Team Leader shall represent accurately the qualifications and professional expertise of each Team Member as stated in the submitted materials.

2.3   If requested by the Owner, the Team Members shall prepare and submit a proposal for the Project to the Owner. Each Team Member shall support the Team Leader with a level of effort and personnel, licensed as required by law, sufficient to complete and submit the proposal in the time frame allowed by the Owner. A clear and concise statement of the division of responsibilities between the Team Members will be prepared by the Team Leader. The Team Leader shall make all final determinations as to the form and content of the proposal. The Team Leader shall use its best efforts, after the Team has qualified for the Project, to obtain the contract award, and each Team Member shall assist in such efforts as the Team Leader may reasonably request.

ARTICLE 3

CONFIDENTIAL INFORMATION

3.1   The Team Members may receive from one another Confidential Information, including proprietary information, as is necessary to prepare the statement of qualifications and the proposal. Confidential Information shall be designated as such in writing by the Team Member supplying such information. If required by the Team Member supplying such Confidential Information, a Team Member receiving such information shall execute an appropriate confidentiality agreement. A Team Member receiving Confidential Information shall not use such information or disclose it to third

parties except as is consistent with the terms of any executed confidentiality agreement and for the purposes of preparing the statement of qualifications, the proposal and in performing any contract awarded to the Team as a result of the proposal, or as required by law. Unless otherwise provided by the terms of an executed confidentiality agreement, if a contract is not awarded to the Team or upon the termination or completion of an contract awarded to the Team, each Team Member will return any Confidential Information supplied to it.

ARTICLE 4

OWNERSHIP OF DOCUMENTS

4.1   Each Team Member shall retain ownership of property rights, including copyrights, to all documents, drawings, specifications, electronic data and information prepared, provided or procured in furtherance of this Agreement or any contract awarded as a result of a successful proposal. In the event the Owner chooses to award a contract to the Team Leader on the condition that a Team Member not be involved in the Project, that Team Member shall transfer in writing to the Team Leader, upon the payment of an amount to be negotiated by the parties in good faith, ownership of the property rights, except copyright, of all documents, drawings, specifications, electronic data and information prepared, provided or procured by the Team Member pursuant to this Agreement and shall grant to the Team Leader a license for this Project alone, in accordance with Paragraph 4.2.

4.2   The Team Leader may use, reproduce and make derivative works from such documents in the performance of any contract. The Team Leader's use of such documents shall be at the Team Leader's sole risk, except that the Team Member shall be obligated to indemnify the Team Leader for any claims of royalty, patent or copyright infringement arising out of the selection of any patented or copyrighted materials, methods or systems by the Team Member.

ARTICLE 5

POST AWARD CONSIDERATIONS

5.1   Following notice from the Owner that the Team has been awarded a contract, the Team Leader shall prepare and submit to the Team Members a proposal for a Project-specific agreement of association among them. (Such agreement may take the form of a design-builder/subcontractor agreement, a joint venture agreement, a limited partnership agreement or an operating agreement for a limited liability company.) The Team Members shall negotiate in good faith such Project-specific agreement of association so

3

FIGURE 14.3   *(Continued.)*

- Relationships with lending institutions have created significant lines of credit that are necessary in that business.
- Long-term dealings with surety provide easy access to bonds.
- A detailed estimating capability is available from a large database of current costs.

that a written agreement may be executed by the Team Members on a schedule as determined by the Team Leader or by the Owner, if required by the request for proposal. The Team Leader shall use its best efforts, with the cooperation of all Team Members, to negotiate and achieve a written contract with the Owner for the Project.

ARTICLE 6

OTHER PROVISIONS   ◆

This Agreement is entered into as of the date set forth above.

WITNESS: TEAM LEADER: _____ ◆

_____ ◆   BY: _____ ◆

                                            PRINT NAME: _____ ◆

                                            PRINT TITLE: _____ ◆

WITNESS: TEAM MEMBER: _____ ◆

_____ ◆   BY: _____ ◆

                                            PRINT NAME: _____ ◆

                                            PRINT TITLE: _____ ◆

WITNESS: TEAM MEMBER: _____ ◆

_____ ◆   BY: _____ ◆

                                            PRINT NAME: _____ ◆

                                            PRINT TITLE: _____ ◆

WITNESS: TEAM MEMBER: _____ ◆

_____ ◆   BY: _____ ◆

                                            PRINT NAME: _____ ◆

                                            PRINT TITLE: _____ ◆

12/01

4

AGC DOCUMENT NO. 499 • STANDARD FORM OF TEAMING AGREEMENT FOR DESIGN-BUILD PROJECT
© 2001, The Associated General Contractors of America

FIGURE 14.3   (*Continued.*)

- Relationships with subcontractors and the legal contracts that evolved to establish and control these relationships is already in place.

- A cadre of experienced field supervisors and project managers are on hand.

- Accounting procedures to handle receivables, payables and to monitor cash flow and project budgets have been firmly established.

- Schedule preparation and the experienced personnel to create, monitor, and update them are already in place.

Contractor as prime.   A builder can engage an architect and form a design-build venture by merely subcontracting work to them in much the same manner as they subcontract work to the trades. By negotiating the services to be provided, and the interaction between contractor and subcontractor (in this case, the architect and their design consultants), the method of payment for services and other obligations and responsibilities similar to a Teaming Agreement can be developed. This prime/subcontractor relationship can effectively operate as a design-build team.

The architect as team leader.   Architects often have very close relationships with clients, and the institutional concept of the architect as the defender of the client is hard to dismiss, adding much to their possible role as team leader. There are other rationales behind the idea of the architect acting as team leader. These include the following:

- A proven track record of successfully designing the type of project currently under consideration
- Extensive experience in providing construction services
- A nucleus of experienced field supervisors
- Experience in a specific design-market niche, such as pharmaceuticals, medical devices, or recreational facilities

It was not until 1978 that the American Institute of Architects lifted their ban on design-build. But by 1985 they had issued their first edition of design-build documents. As of 1996, a complete new series of design-build documents for architects and others was available from AIA.

The architect as prime contractor; builder as subcontractor.   One only has to look at high-end residential construction to see an operation where the architect is prime, and the builder is the subcontractor. The same process can be applied to nonresidential design-build work.

The AIA, recognizing the need for a contract between architect and contractor, developed their Document A491-Standard Form of Agreement Between Design/Builder and Contractor, which in 2004 superseded AIA A141. These contracts were two part agreements:

- Part1—The contractor is to provide the preliminary evaluation of the owner's program, advise on the selection of materials and constructability issues, prepare the schedule and preliminary estimates, and develop a fixed-price or guaranteed maximum price (GMP) depending upon the form of contract.

- Part 2—This part of AIA A491 is basically a standard contract for construction requiring that the contractor provide all labor, materials, and equipment to complete the work outlined in the plans and specifications.

Developing a Design-Build Program

One of the first requisites to consider in developing a design-build strategy is to review the company's accumulated data base of historical project costs for that segment of the design-build market the company wishes to enter. What is the contractor's experience? Is it in warehousing, commercial office buildings, medical facilities?

What type of project(s) has the company successfully completed in the past five or ten years that would lend itself to a design-build marketing approach?

Other issues to consider:

- Is there a database of costs for these projects, including square-foot costs, unit costs, extractable component costs for structures, finishes, and MEP work?

- Does the company have deep experience in conceptual estimating and have they utilized these skills in negotiating work?

- Has the company worked successfully with a cadre of subcontractors experienced in critical trades—structures and MEP—that they could call upon as members of a design-build team?

A competent staff and a strong database of costs for the selected sector of the industry will arm the general contractor with the tools necessary to let the world know that they have the necessary skills to be considered a candidate for the design-build project under consideration. Having worked well with an architectural firm or two in a specific field of design may also become the genesis of a design-build venture.

Developing design-build proposals

Several executives from successful design-build firms say that a preliminary proposal to a prospective owner should not be free. There will be lots of owners on "fishing expeditions" without any real interest in pursuing a new design-build project, and these experienced design-builders strongly recommend setting a fee for any proposal. The fee doesn't have to be substantial and maybe not even cover all the costs, but it should be large enough to weed out the interested from the merely curious.

After an initial conference with a prospective client, a typical phased proposal will be prepared, listing the various stages of project development and their associated costs. For projects in the $10 million range, an initial fee of $25,000 to $50,000 might be proposed, depending upon the complexity and type of project under consideration. If accepted, this initial proposal would provide the owner with the following:

- A floor plan and a typical floor plan if multistoried
- Wall sections and elevations—front and typical
- Definition of the structural system to be employed
- Finish schedules, door schedules
- One-line electrical drawing
- Riser diagrams for HVAC, plumbing, and fire protection systems
- Outline specifications for all components included the previous entry
- A narrative describing the design
- Additional information that more closely defines the project—for instance, catalog cuts and manufacturer's product brochures

The total cost of the project will be included as either a lump-sum or guaranteed maximum price contract format.

If this initial phase is acceptable to the client, the design-builder will move on to the next phase—at an incrementally increased cost. This next phase—call it Phase B—will provide the client with a 50-percent complete design, but will include more defined plans of all the construction components and systems via detailed specifications. Phase C will afford the client 100-percent complete design documents—for an additional fee. Phase D is actually the contract signing and the authorization to proceed with building.

At any step along with way, the client can decide to terminate their agreement with the design-builder and upon receipt of their total fee, the design-builder will turn over all documents developed to that point to the client for their unconditional use.

Another approach

Other design-builders, following similar steps outlined earlier, may limit their proposal to three steps:

- Step 1—Provide the client with a schematic design and preliminary estimate for a relatively nominal fee. If acceptable, go to the next step.
- Step 2—Develop enough scope to provide a guaranteed maximum price within a variance of plus or minus 5 percent.
- Step 3—If acceptable, go to the contract and develop construction documents while executing competitive pricing as the design proceeds,

thereby creating the fast-track process that will reduce interest on capital and enable the client to occupy the facility more rapidly. As the last buy-outs occur, generally the MEPs, the owner will be presented with a fairly good estimate of what the final GMP will be. Each-month-costs to date and projected costs to complete are prepared by the design-builder and presented to the owner.

With this type of proposal, a small fee will be negotiated with the client for the first stage, while a much larger fee is assessed to proceed to Step 2. The final step or stage is actually the contract signing with both GMP and the design-builder, in which the fee structure is included.

How Owners Select Design-Builders

Most design-build projects are awarded on the basis of either a direct competition cost/design evaluation process or the low bid, which is very similar to design-bid-build.

Owners generally look for the following qualifications when considering a design-build firm:

- The builder's financial and bonding capabilities
- The design-build team's experience in the type of construction being considered
- A track record of successful design and technical engineering competence
- The experience of both the designer and contractor's key personnel and staff
- The firm's overall experience in the design-build process
- The organization and management of the design-build entity
- The design-build team's quality control program and administration
- The design-build team's record of on-budget performance
- The design-build team's record of on-time delivery performance

The selection process

The most common selection process for awarding design-build projects can take any one of the following paths:

- *Direct Selection:* Similar to a negotiated contract process, an owner may select a design-builder on the basis of reputation and previous experience, or on the recommendation by others. This method of engagement is generally limited to private sector work.

- *Competitive Negotiation:* A short list of design-builders may have submitted bids, so the owner will review each bid for compliance with the program, preliminary design considerations, fees, the time frame for design and construction, and the experience of the design and construction personnel. The owner may then elect to negotiate the project with one of the bidders, a practice common in the private sector.

- *Cost/Design Evaluations:* Also referred to as the "best value" selection, this method is used by many owners. A series of technical criteria is submitted to a short-listed group of bidders, and upon the receipt of responses, an evaluation process to review each proposal is established. Both cost and qualitative considerations are considered in the selection process.

- *Cost Competition:* Much like the design-bid-build process, this selection method is based upon a bidder's response to a fairly complete and tightly prescribed set of requirements provided by the owner in their Request for Proposal (RFP). The owner, assuming that all bidders are submitting proposals of equal scope, then makes a selection based upon the "low bid."

A typical contract between owner and design-builder is one prepared by The Associated General Contractors of America—AGC Document No.400—Standard Form of Design-Build Agreement and General Conditions between Owner and Design-Builder. This contract, shown in Figure 14.4, is included at the end of this chapter.

A typical design phase agreement between a design-build entity and an owner. The American Institute of Architect's (AIA) Document A491—Standard Form of Agreement between Owner and Design-Builder—is divided into two parts. Part 1 deals with the preliminary design, budget, and schedules, and the submission and acceptance of the design-builder's proposal prior to the execution of Part 2—the agreement for construction and other services required to complete the design-build project. Whether this specific contract form is used or not, this two-phased approach is nearly standard practice in both private and public sector work.

The Basic Services clause in AIA A491 sets forth the elements to be included in the proposal in order to obtain acceptance by the owner and thus proceed to Part 2, the completion of design and the start of construction. These first-phase elements include.

- A description of the preliminary design documents. Preliminary design documents can include a typical floor plan, an elevation, wall sections, rough rendering, and a site plan outline specifications.

- A statement regarding the proposed contract sum and the form of contract being proposed—for instance, lump-sum, cost-plus-a-fee, or GMP.

THE ASSOCIATED GENERAL CONTRACTORS OF AMERICA

AGC DOCUMENT NO. 400
PRELIMINARY DESIGN-BUILD AGREEMENT
BETWEEN OWNER AND DESIGN-BUILDER

This standard form agreement was developed with the advice and cooperation of the AGC Private Industry Advisory Council, a number of Fortune 500 owners' design and construction managers who have been meeting with AGC contractors to discuss issues of mutual concern. AGC gratefully acknowledges the contributions of these owners' staff who participated in this effort to produce a basic agreement for construction.

This Agreement is intended to be used in conjunction with AGC Document No. 410 or 415.

TABLE OF ARTICLES

This Agreement has important legal and insurance consequences. Consultation with an attorney and an insurance consultant is encouraged with respect to its completion or modification.

AGC DOCUMENT NO. 400 • PRELIMINARY DESIGN-BUILD AGREEMENT BETWEEN OWNER AND DESIGN-BUILDER
© 1999, The Associated General Contractors of America

FIGURE 14.4   AGC document no.400—Standard form of design-build agreement and general conditions between owner and design-builder contract. (*All materials displayed or reproduced are with the express written permission of the Associated General Contractors of America under copyright license No.0115.*)

- A proposed schedule for both the completion of the design and the start and completion of construction.
- A statement containing any deviations from the owner's stated program.

As stated previously, Phase 2 is basically a contract for construction.

AGC DOCUMENT NO. 400
PRELIMINARY DESIGN-BUILD AGREEMENT
BETWEEN OWNER AND DESIGN-BUILDER

ARTICLE 1

AGREEMENT

This Agreement is made this _____ day of _____ , ◆

in the year _____ , by and between the   ◆

OWNER   ◆
(Name and Address)

and the
DESIGN-BUILDER   ◆
(Name and Address)

for preliminary services in connection with the following
PROJECT   ◆
(Name, location and brief description)

Notice to the parties shall be given at the above addresses.

2

FIGURE 14.4   *(Continued.)*

Formalizing the design-build team in two phases.   The team of contractor, architect, engineer, and possibly some key subcontractors will form the basis of the design-build team either in a contractor- or architect-led team.

Because a design-build venture with an owner may be a two-part affair, as described previously, so must the relationship between the design-build team follow this two-part, or two-phased, relationship.

ARTICLE 2

TEAM RELATIONSHIP

2.1   The Owner and the Design-Builder agree to proceed on the basis of trust, good faith and fair dealing, and shall take all actions reasonably necessary to perform this Agreement in an economical and timely manner.

ARTICLE 3

DESIGN-BUILDER'S RESPONSIBILITIES

3.1   The Design-Builder shall exercise reasonable skill and judgment in the performance of its services. Architectural and engineering services shall be procured from licensed, independent design professionals retained by the Design-Builder or furnished by licensed employees of the Design-Builder, or as permitted by the law of the state where the Project is located. The person or entity providing architectural and engineering services shall be referred to as the Architect/Engineer. If the Architect/Engineer is an independent design professional, the architectural and engineering services shall be procured pursuant to a separate agreement between the Design-Builder and the Architect/Engineer. The standard of care for architectural and engineering services performed under this Agreement shall be the care and skill ordinarily used by members of the architectural and engineering professions practicing under similar conditions at the same time and locality. The Architect/Engineer for the Project is_____
_____ . ◆

3.2   The Design-Builder is responsible for the following Preliminary Design-Build Services:

3.2.1   OWNER'S PROGRAM If requested by the Owner as an Additional Service, the Design-Builder shall assist the Owner in the development and preparation of the Owner's Program, which is an initial description of the Owner's objectives. The Owner's Program may include budget and time criteria, space requirements and relationships, flexibility and expandability requirements, special equipment and systems, and site requirements.

3.2.2   PRELIMINARY EVALUATION The Design-Builder shall review the Owner's Program to ascertain the requirements of the Project and shall verify such requirements with the Owner. The Design-Builder's review shall also provide to the Owner a preliminary evaluation of the site with regard to access, traffic, drainage, parking, building placement and other considerations affecting the building, the environment and energy use, as well as information regarding applicable governmental laws, regulations, and requirements. The Design-Builder shall review the Owner's existing test reports but will not undertake any independent testing nor be required to furnish types of information derived from such

testing in its preliminary evaluation. The Design-Builder shall also propose alternative architectural, civil, structural, mechanical, electrical and other systems for review by the Owner, in order to determine the most desirable method of achieving the Owner's requirements in terms of cost, technology, quality and speed of delivery. Based upon its review and verification of the Owner's Program and other relevant information, the Design-Builder shall provide a Preliminary Evaluation of the Project's feasibility for the Owner's acceptance. The Design-Builder's Preliminary Evaluation shall specifically identify any deviations from the Owner's Program.

3.2.3   PRELIMINARY SCHEDULE The Design-Builder shall provide a preliminary schedule for the Owner's written approval. The schedule shall show the activities of the Owner and the Design-Builder necessary to meet the Owner's completion requirements.

3.2.4   PRELIMINARY ESTIMATE The Design-Builder shall prepare for the Owner's written approval a preliminary estimate utilizing area, volume, or similar conceptual estimating techniques. The level of detail for the estimate shall reflect the Owner's Program and any additional available information. If the preliminary estimate exceeds the Owner's budget, the Design-Builder shall make written recommendations to the Owner.

3.2.5   SCHEMATIC DESIGN DOCUMENTS The Design-Builder shall submit for the Owner's written approval Schematic Design Documents based on the agreed upon Preliminary Evaluation. Schematic Design Documents shall include drawings, outline specifications and other conceptual documents illustrating the Project's basic elements, scale and their relationship to the Worksite. One set of these Documents shall be furnished to the Owner. When the Design-Builder submits the Schematic Design Documents, the Design-Builder shall identify in writing all material changes and deviations from the Design-Builder's preliminary evaluation, schedule and estimate. The Design-Builder shall update the Preliminary Schedule and preliminary estimate based on the Schematic Design Documents.

3.2.6   ADDITIONAL SERVICES The Design-Builder shall provide the following Additional Services: ◆

3

FIGURE 14.4   (*Continued.*)

The teaming agreement

One of the first agreements to be created when a design-build venture is being considered is the teaming agreement (see Figure 14.3) between the contractor and designer to designate the rights and responsibilities of each participant during the proposal stage. If the design-build proposal

ARTICLE 4

OWNERSHIP OF DOCUMENTS

4.1 Upon the making of payment as required by this Agreement, the Owner shall receive ownership of the property rights, except for copyrights, of all documents, drawings, specifications, electronic data and information prepared, provided or procured by the Design-Builder, its Architect/Engineer, Subcontractors and consultants, and distributed to the Owner for this Project ("Design-Build Documents"). The Owner shall not have the right to use, reproduce and make derivative works from the Design-Build Documents for other projects without the written authorization of the Design-Builder, who shall not unreasonably withhold consent. The Owner's use of the Design-Build Documents on other projects or without the Design-Builder's written authorization or involvement is at the Owner's sole risk, and the Owner shall defend, indemnify and hold harmless the Design-Builder, its Architect/Engineer, Subcontractors and consultants, and the agents, officers, directors and employees of each of them from and against any and all claims, damages, losses, costs and expenses, including but not limited to attorney's fees, costs and expenses incurred in connection with any dispute resolution process, arising out of or resulting from such use of the Design-Build Documents. The Design-Builder shall obtain from its Architect/Engineer, Subcontractors and consultants property rights and rights of use that correspond to the rights given by the Design-Builder to the Owner in this Agreement.

ARTICLE 5

OWNER'S RESPONSIBILITIES

5.1 The Owner shall provide to the Design-Builder all relevant information for the Project, including the Owner's Program, unless the Owner's Program is developed and prepared with the assistance of the Design-Builder as an Additional Service. The Owner shall timely review and approve schedules, estimates, Schematic Design Documents, and other documents provided under this Agreement.

5.2 OWNER'S ELECTION TO PROCEED If the Owner elects to proceed with the Project beyond the Preliminary Design-Build Services provided in this Agreement, the Owner and the Design-Builder shall enter into an additional agreement for the completion of the design and the construction of the Project. If the Owner elects not to proceed with the Project, the Owner shall have no further obligation to the Design-Builder other than its indemnity obligation pursuant to Paragraph 4.1 and the payment of compensation as set forth in this Agreement.

ARTICLE 6

CONTRACT TIME

6.1 The Design-Builder's Services provided under this Agreement shall commence on or about _____ ◆ _____ ,and shall be completed on or about_____ . ◆

ARTICLE 7

COMPENSATION

7.1 The Owner shall compensate the Design-Builder monthly for Preliminary Design-Build Services and any Additional Services performed under the Agreement on the following basis: ◆

(State whether a stipulated sum, actual cost or other basis. If a stipulated sum, state what portion of the sum shall be payable each month.)

4

AGC DOCUMENT NO. 400 • PRELIMINARY DESIGN-BUILD AGREEMENT BETWEEN OWNER AND DESIGN-BUILDER
© 1999, The Associated General Contractors of America

FIGURE 14.4 *(Continued.)*

is ultimately accepted by the owner and a contract for design and budget (Part 1 or Part A) is awarded, the design-build team will use the teaming agreement as a guide in the preparation of a contract between each member of the team. Part of the teaming agreement will be devoted to establishing the organizational structure between contractor and designer, defining the division of responsibility between team members and dealing with the issues during the proposal stage, which, hopefully,

7.2 Reimbursable expenses under this Agreement shall include: ◆

(List those reimbursable expenses that are not included above.)

This Agreement is entered into as of the date entered in Article 1.

ATTEST:_____ ◆   OWNER:_____ ◆

                                           BY_____ ◆

                                           PRINT NAME _____ ◆

                                           PRINT TITLE _____ ◆

ATTEST:_____ ◆   DESIGN-BUILDER: _____ ◆

                                           BY_____ ◆

                                           PRINT NAME _____ ◆

                                           PRINT TITLE _____ ◆

3/00
5

AGC DOCUMENT NO. 400 • PRELIMINARY DESIGN-BUILD AGREEMENT BETWEEN OWNER AND DESIGN-BUILDER
© 1999, The Associated General Contractors of America

FIGURE 14.4 *(Continued.)*

will terminate when an award for construction (Part 2 or Part B) is executed.

The teaming agreement will concern itself with the following issues:

- *Organizational Structure*
  - Joint venture

- Limited Liability Corporation (LLC)
- Prime-subcontractor agreement where the general contractor is usually, but not always, the "prime," and the design team the "subcontractor(s)"
- *Division of Responsibility among Team Members*
  - During the proposal stage
  - During the construction stage
  - During post-construction
- The Proposal Preparation Stage
  - Which party will prepare the technical and design work necessary to comply with the owner's Request For Proposal (RFP)? Will the architect be the lead in the design work, or will the various design disciplines report to the general contractor as the Team Captain?
  - How will the costs for the proposal preparation and presentation be shared?
  - If confidential information, such as financial statements, must be shared by the team members, how will this confidential and/or proprietary information be protected?
  - Exclusivity and noncompeting clauses need to be addressed so that one party or the other cannot withdraw from the process and team up with another group competing for the same project.
  - The establishment of penalties if one party or the other withdraws from the proposal process and the remaining members have to seek the services of other participants or abandon the project.
- *Pre-Client Contract Award Issues*
  - Agreement on the design-build team contract format
  - Agreement on the scope of participation and corresponding remuneration for services
  - Establish construction and design fees to be incorporated into the design-build contract with the owner
  - Establish a contingency amount for both design and construction
  - Set procedures for the coordination and tracking of design with the construction budget, and how adjustments are to be made in order to meet the budget
  - If there is a savings clause in the contract with the owner, how are savings to be shared among the team members?
  - Indemnity, bonding and insurance considerations, participation, and responsibilities are to be defined.
  - Warranty issues—including design errors, omissions, and construction matters—are to be clarified.

The owner's responsibility to the design-build team. A successful project will depend upon dealing with a knowledgeable owner who has developed a detailed, well thought-out, and defined building program. Just as the owner will investigate the design-build team's capabilities, so must the design-build team investigate the owner's ability to provide the management structure to support such a venture. Some of the questions that ought to be asked, or areas to be observed, are:

- Does the owner's management team fully support a design-build approach to the project?

- Are the owner's representatives involved in the project technically capable of dealing with design and construction professionals?

- Will these owner representatives have the power to act, or will they only be conduits to decision makers? And will those decision be made promptly?

- Will any outside consultants join the management team, and if so, what will be the extent of their participation (for instance, permitting, zoning, legal, or technical)?

- What financial controls will have to be dealt with for requisition approval, and payment and changes to the budget?

- Does the owner require hands-on involvement in the design and construction phases in order to feel comfortable with the process? Will their representatives require notification of all meetings in case they wish to attend?

Other owner responsibilities to be considered.   Owners in the design-build process, generally retain responsibilities relating to physical site conditions—not to be confused with sitework—and in that respect furnish the following information and/or services.

- Property surveys, metes and bounds, topographical studies, the location of existing utilities

- Geotechnical surveys of subsurface conditions, and test borings for foundation-bearing design determinations

- Temporary and permanent easements, zoning compliance, notification of any encumbrances affecting land use; any rules, regulations, or ordinances that would affect the type of construction being considered

- A legal description of the property

- Any existing records or drawings of previous structures on the site

- Any environmental or hazardous conditions studies pertaining to the site, such as impact studies

- Other documents to sufficiently establish the size of the project; the anticipated quality levels; a description of the structural, MEP systems and types of materials to be used in the project.

Contract Provisions Unique to the Design-Build Process

The two-part contractual relationship between the design-builder and the owner includes a termination clause that can be affected in one of two ways:

> By either party, upon seven (7) days notice if the other party fails to perform "substantially in accordance with the terms through *no fault of the party initiating the termination.*" A "no fault" provision. *or* By the owner without cause, upon at least seven (7) days written notice to the design-builder.

In the event that the cause for termination was not the fault of the design-builder, they will be compensated for all services performed to the date of termination along with reimbursable and termination costs attributed to that termination, including a reasonable amount for overhead and profit.

The Part 1 agreement generally provides for an initial payment upon the execution of the agreement. During the life of the Part 1 agreement, the design-builder may receive reimbursement of expenses on the basis of "multiples" of the amounts actually expended, if a list of reimbursable expenses was included in that agreement. This allows the design-builder to recapture some costs during the design phase, including overhead and profit.

Special provisions of design-build contracts

The standard of care provision.   To ensure that the designers will use the same degree of professionalism in preparing a design for a design-build project as they would for a design-bid-build project, a standard of care provision, similar to the one listed next, is often included in the contract with the owner.

> *Standard of Care*: The standard of care for all design services performed under this agreement shall be the care and skill ordinarily used by a member of the architectural and/or engineering profession practicing under similar conditions at the same time and locality. Notwithstanding the previous, in the event that the contract documents specify that portions of the work are to be performed in accordance with a specific performance standard, the design services shall be performed so as to achieve such standards.

Contingency clauses.   Contingency clauses are common provisions in design-build contracts—their purpose being to somewhat cushion the design-builder against such events as errors or omissions that may have occurred during the design, review, and approval stages. These contingency clauses must be very explicit and leave no doubt in anyone's mind why they are there and for what purpose they may be tapped. One such clause in a GMP contract could read as follows:

> The GMP includes a contingency in the amount of $_____ which is available for the design-builder's exclusive use for costs that are incurred in performing work that was not included in a specific line item, or which constitutes the basis of a change order under the agreement. By way of example, and not as a limitation, such costs include trade buy-out differentials, overtime, acceleration, costs in correcting defective or damaged or nonconforming work, nonnegligent design errors and omissions, and subcontractor defaults. The contingency is not available to the owner for any reason, including changes in the scope or any other items which would enable the contractor to increase the GMP under the agreement. The design-builder shall notify the owner of all anticipated charges against the contingency.

When this contingency issue is discussed with the owner, it should be stressed that the owner should also include a contingency in their project proforma to be reserved strictly for their use and be employed for unanticipated costs such as unforeseen subsurface conditions, severe weather, the contractor's compensable delays, and so on.

Contract provisions relating to defining the owner's program

When the design-build proposal is being formulated with the owner to define their program, any deviations to that program need to be defined and discussed with the owner. This can best be accomplished by developing an Exclusions List (or Deviations List), prefaced by a contract clause similar to the following:

> The design-builder's proposal shall specifically identify any deviations from the owner's program, the identification of which shall be set forth on a separate exhibit in the proposal identified as either Exclusions or Deviation List. In case of an inconsistency, conflict, or ambiguity between the owner's program and the design-builder's proposal, the inconsistency, conflict, or ambiguity shall be resolved in accordance with the following order of precedence:
>
> 1. Deviation or Exclusion List
> 2. Owner's program
> 3. Design-builder's proposal (excluding the Deviation or Exclusion List)

Some owners may include an "intent" clause in the contract to insure that the design-build team fully understands their program, particularly when the design documents are not fully complete at contract signing.

> The intent of the contract documents is to include all of the work required to complete the project, except those portions specifically excluded (*and set forth in the Exclusion/Deviation List mentioned earlier*). It is acknowledged

that as of the date of the contract, the plans and/or specifications are not complete but define the scope and nature of the work and are sufficient to establish the contract sum. No adjustment shall be made in the contract sum if, as a prudent contractor, the contractor should have been aware of the anticipated work as may be required to produce a first-class (office building or whatever the project is)."

The role of the subcontractor in the design-build process

A design-build team will be strongly advised to assemble a select group of qualified subcontractors and suppliers during the proposal stage and actual design stage.

Contractors, architects, and engineers recognize the value that experienced, quality subcontractors can bring to the table during the initial stage of a project's development. The subcontractor's intimate knowledge of cost, constructability, what works, and what doesn't work is invaluable to both builder and designer whether developing conceptual estimates, discussing design considerations, proposing value engineering options, or establishing schedules.

Subcontractors who can provide these services are sought out by design-build companies—and if they prove themselves, they can develop long-term relationships with the design-build team. Subcontractor involvement with the contractor during the design-build exercise can take several forms, some of which may be different from their relationship in a design-bid-build project. Key subcontractors assembled in the early stages of the project require some type of commitment in order to entice them to expend their time and money in the project. The subcontractor may agree to fully participate in the process in exchange for an award of a subcontract agreement, if the design-build venture is successful. But some "what if" scenarios must be explored, such as the following:

- If the project proceeds, will these subcontractors be awarded contracts for their particular scope of work via a negotiated contract?

- If the project is aborted along the way, will the subcontractors and vendors expect to be compensated for their involvement to date, and if so, what will be the order of magnitude of their compensation, and who will pay them (another item to be included in the teaming agreement)?

- If the project proceeds to contract, but along the way the owner requires that competitive bids be obtained for all major items of work, will these subcontractors agree to that arrangement?

This meeting of the minds needs to be worked out early in the game. An owner may initially elect to accept a lump-sum construction contract, but as the project proceeds through the design stage, may elect to switch to a

GMP contract and therefore wish to receive competitive bids on all items of work. The project manager should be prepared for this turn of events.

Tracking design development and the estimate.   At each stage of the schematic or design development (DD) drawings, the estimate must be compared with the scope of work being defined in these preliminary documents. Now is not the time to scrimp on the costs of reproducibles, even if insufficient sums were not included in the estimate. Copies of the DD drawings should be distributed to all interested parties. Subcontractors should be sent copies of these drawings with a request to review and comment on the scope of the work being developed. Periodic meetings ought to be held to receive any comments that surface during these reviews, and if scope increases are discovered, they need to be dealt with promptly and professionally. At these meetings, all parties will be asked the same questions:

- Is the owner's program being met?
- Does it appear that the emerging design is compatible with the budget?
- If the answer to the preceding question is "Yes," no further questions need to be asked. If the answer is "No," then three more questions will be necessary: Why not? Where is the scope increase? And what do we need to do to get back on budget?

Each design development meeting must produce meeting minutes documenting agreement or disagreement on key issues, whether further development is required to finalize the design of certain systems or components, or agreement that certain developing components or systems must be changed in order to meet budget requirements but still remain within the owner's program parameters. If action items are noted in these meeting minutes, the party(s) expected to respond and the proposed response date should be included.

As outline specifications are being fleshed out during design development, these too require scrutiny to insure that the proper product(s) designated, as well as the materials specified, are in the "acceptable" price range. This process will continue until the final set of plans and specifications has been completed and accepted by all parties as contract documents.

The project manager assigned to the design-build project must keep meticulous and detailed notes during the various exchanges between designers, subcontractors, and vendors. Misunderstandings will occur, and when subcontract agreements or vendor purchase orders are being finalized, it is not uncommon to hear, "I never agreed to that" or "You do recall, I took exception to that" or "Don't you recall that I said my price included X and not Y which is part of the plans or specifications."

TABLE 14.1   Design-Build Teams Assume Risks Different from Those Associated with the More Conventional Design-Bid-Build Process

Risk Category	Traditional Design-Bid-Build	Design-Build
Geotechnical services	Owner	Design-builder
Design criteria	Owner	Design-builder
Design defects	Owner	Design-builder
Constructability	Owner	Design-builder

Detailed notes taken during this period of intense "give and take" may prove invaluable if such disagreements occur—and they will.

Several design-builders have stated that costs to assemble and monitor the design development of a design-build project are considerable and draw heavily on human resources within the company. A project manager may be required to work full time on one design-build project from initial proposal through design development and well into the construction process. So allocation of the company's manpower needs to be reviewed when considering embarking on their first design-build project.

Risk allocation and design-build. The design-build team will assume risks considerably different from those associated with the more conventional design-bid-build process as noted in the following. Not only will the design-builder assume risks associated with the design of the structure, but may also assume responsibility for site conditions if they are requested by an owner to provide geotechnical services. Some design-builders exclude this service and request that the owner provide basic geotechnical surveys for evaluation. (See Table 14.1.)

Design-Build in the Public Sector

In July 2004, the Commonwealth of Massachusetts became the 46th state to adopt design-build as an acceptable project delivery system. One universal benefit from design-build, as reported from states surveyed by the American Association of State Highway and Transportation Officials (AASHTO) in 2002, is more rapid delivery time. In Florida, a review of 11 completed design-build projects revealed a 36-percent decrease in design and construction time, while in North Carolina, their department of transportation, NCDOT, reported that the speed and innovations provided by design-build can shorten some highway projects by three years.

Public agency contracting methods. Various methods were developed to create a public-private partnership that would allow private corporate innovation and participation while preserving the public interest. In response, these states began to develop a series of bid selection criteria for that purpose, which included the following:

- *Direct selection:* A competitive process where the design-builder is selected based on definable, objective criteria, prior experience, complete scope of work, terms, and price.
- *Best value:* An award based on the combination of price and qualitative evaluations.
- *Equivalent design/low bid:* A best value selection where technical submissions are followed by a critique of the proposal and respondents are afforded an opportunity to change their design and adjust their bid accordingly.
- *Fixed-price design:* The agency's Request for Proposal (RFP) contains the maximum cost of the project, with the award based on the best qualitative design proposal.
- *Adjusted low bid:* On selection of the qualified low bidder, the price may be adjusted by further negotiations.

The two-part RFP.   At the federal level, Federal Acquisition Regulation (FAR) 48FAR, Chapter 1, Part 15 is representative of the way in which a two-part design-build proposal is offered, a process which many states have adopted.

- *Part 1 or Part A:* This portion of the RFP is devoted to establishing the bidder's qualifications, which will be evaluated before shortlisting and proceeding on to the next phase. This questionnaire invites responses to:
  - Verify the bidder's technical competence and experience in the type of project being considered.
  - Document past performance of the proposed design-build team— both contractor and design consultants.
  - Detail the capacity of the team to meet the criteria included in the RFP.
  - Answer other factors that may be appropriate to the specific situation or project at hand.
- *Part II or Part B:* This phase of the RFP requires bidders to:
  - Provide a technical proposal to meet the goals established by the agency.
  - Provide cost and pricing information commensurate with the technical data submitted.

As more and more private owners and public agencies seek design-build as their preferred project delivery system, a construction company should consider the many options open to them to enter this dynamic and profitable aspect of the business.

Sustainability and Green Buildings

In June 1993, President Bill Clinton established the President's Council on Sustainable Development, which was given the mission to develop and implement bold new approaches for integrating economic, social, and environmental policies to guide this country to a more environment-friendly approach in the coming new century. In 1996, the Council issued their report *Sustainable America* that essentially started this country down the road to a new way of looking at the impact we all have on nature's fragile and intricate framework. As a result, the word *sustainability* entered the lexicon of architectural, engineering, and construction communities.

Paul Hawkins, in his book *The Ecology of Commerce*, provides a precise definition of the term: "Sustainability is an economic state where the demands placed upon the environment by people and commerce can be met without reducing the capacity of the environment to provide for future generations."

Green building construction is based upon designs that are more environmentally sensitive, and that preserve our physical resources. The two topics—sustainability and green buildings—have since become mainstream, and both public and private owners recognize the savings that can result from incorporating many of these environmentally friendly schemes into their building programs.

Advocates of green buildings can no longer be viewed as "tree huggers" as more communities and corporations seize upon new opportunities to affect savings, protect the environment, and create more public awareness of the growing need to preserve our planet. The process of building factories, office buildings, and homes has had a major impact on our ecosystem in the past, but it is a process that can be mitigated and turned around.

The Impact of Construction on the Environment

Commercial and institutional buildings have a dramatic impact on our environment.

- Buildings in the United States consume 36 percent of our total energy use and 65 percent of all electrical consumption.
- Our buildings are responsible for 30 percent of all greenhouse gas emissions.
- Buildings consume 30 percent of our raw materials.
- Buildings produce 30 percent of our total waste output—approximately 136 million *tons* annually.
- Buildings consume 12 percent of all potable water.

The U.S. Department of Energy reports that there are 4.6 million commercial buildings in the United States, occupying more than 67 billion square feet of space, and these buildings consume one-sixth of the world's fresh water supply, half of the virgin wood harvested, and two-fifths of our materials and energy reserves.

This provides impetus to the sustainable movement: we must preserve what we have because our resources are not limitless.

As contractors, we have a great deal of control over our environment—from recycling asphalt, concrete, and rebars to reducing harmful emissions by properly maintaining equipment that uses fossil fuels.

What Do We Mean by Sustainability?

Sustainability, as explained earlier, is the quest to *sustain* economic growth while maintaining long-term environmental health. When applied to construction, sustainability means creating designs that seek to balance the short-term goals of a project with the long-term goals of efficient operating systems that protect the environment and nature's resources. Sustainable buildings represent a holistic approach to construction that combines the advantages of modern technology with proven construction practices—using nature to enhance the building's efficiency rather than fight it.

Using fenestration to let natural light into the building employs the latest technology of inert gas–filled insulated glass panels with low emissions coatings and thermal break frames to reduce interior space lighting requirements and diminish building heating and cooling loads, as well as their related energy costs. Oriented strand board (OSB) and medium density fiberboard (MDF) are two perfect examples of sustainability—using waste and recycled wood products to create new products that, in some cases, are more durable and more maintenance-free than the virgin wood from which they came.

Whole-building design

Whole-building design is a process wherein the building's structure, envelope, interior components, mechanical and electrical systems, and even its site orientation are viewed holistically. The whole-building concept considers site, energy, materials, indoor air quality, acoustics, natural resources, and their interrelationship with each other.

In this process, new and proven technologies are discussed, weighed, debated, and incorporated or discarded.

The benefits of whole-building design are directed toward the following goals:

- Reducing energy costs
- Reducing both capital and maintenance costs
- Reducing the environmental impact of the building to the site and environs
- Increasing occupant comfort, health, and safety
- Increasing employee productivity

The history of green building construction in this country is proof that all of these requirements can be met, and at little or no initial cost to the project. The cost-effectiveness of these green buildings, over the somewhat long term, is just beginning to be documented, and it validates their reason for being. But let's discuss the term *sustainability* in today's vernacular a little closer.

LEED versus sustainability. Sustainability is the process involved in designing and building an environmentally friendly structure, while LEED (Leadership in Energy and Environmental Design) is a trademark-protected rating system developed by the United States Green Building Council (USGBC), a program of standards and certification for accreditation purposes. LEED addresses a variety of types of construction, all with one purpose in mind: to define high-performance buildings that are environmentally responsible, healthy, and profitable. The LEED program encompasses the following:

- *LEED-NC:* New Construction
- *LEED-EB:* Existing Buildings
- *LEED-CI:* Commercial Interiors
- *LEED-C&S:* Core and Shell

- *LEED-H:* Homes
- *LEED-ND:* Neighborhood Development

The rating systems were developed by the USGBC committees and allow for four progressive levels of certification:

- Certified (the lowest level)
- Silver
- Gold
- Platinum (the highest level)

Six credit areas exist in each category, with points awarded for the degree of compliance.

- Sustainable sites
- Energy and atmosphere
- Water efficiency
- Indoor environmental quality
- Materials and resources
- Innovation in design

Within each credit area, a number of points can be earned, the total of which determines the level of certification achieved. For example, the total number of award points is 69.

- Basic certification requires 26–32 points.
- Silver certification requires 33–38 points.
- Gold certification requires 39–51 points.
- Platinum certification requires 52 points or more.

The Basic certification level must meet 40 percent of the LEED system; Silver must meet 50 percent, Gold 60 percent, and Platinum 80 percent.

LEED approaches construction of a green building by first focusing on the site. The total number of points available for a *sustainable site,* for example, is 14, with one point offered when the criteria for each element has been achieved. These 14 areas are concerned with

- Erosion and sedimentation control
- Site selection
- Urban redevelopment
- Brownfield redevelopment
- Alternative transportation—public transportation access
- Alternative transportation—bicycle-friendly

- Alternative transportation—alternative fuel refueling stations
- Alternative transportation—parking reductions
- Reducing site disturbance—protecting and restoring open spaces
- Reducing site disturbance—maximizing open space
- Stormwater management—flow reduction
- Stormwater management—flow treatment
- Landscape and exterior design to reduce heat islands, nonroof surfaces
- Landscape and exterior design to reduce heat islands, roof surfaces

Government takes the LEED. According to a study released by USGBS in February 2005, 41 cities in the United States have adopted some type of LEED certification program for construction or major renovation work in their public facilities. Bidders on these designated projects will have to show proficiency in delivering LEED-certified buildings in order to be qualified.

Of the 41 nationwide municipal participants, the following shows a few program specifics:

- *Atlanta, GA:* All city-funded projects larger than 5000 square feet (465 square meters) or costing at least $2 million must meet a LEED Silver Rating Level.
- *Austin, TX:* LEED certification is required on all public projects larger than 5000 gross square feet (465 square meters).
- *Berkeley, CA:* Municipal buildings larger than 5000 square feet (465 square meters) were required to be LEED-certified in 2004; in 2006, buildings of this size must achieve Silver certification.
- *Dallas, TX:* All city buildings larger than 10,000 square feet (929 square meters) are required to have at least LEED Silver certification.
- *Boston, MA:* This city established LEED Silver as the goal for all city-owned buildings.
- *Chicago, IL:* All new city-funded construction and major renovation projects will require LEED Silver certification at a minimum.
- *Kansas City, MO:* All new city buildings must be designed to meet LEED Silver certification at a minimum. The city is participating in a LEED-EB (existing buildings) pilot program for their city hall.
- *San Francisco, CA:* All municipal new construction, additions, and major renovation projects larger than 5000 square feet (465 square meters) must achieve LEED Silver certification.
- *Scottsdale, AZ:* In March 2005, the city passed Resolution 6644 requiring all new public buildings to be certified as LEED Gold.

In Canada, the number of sustainable buildings is also growing:

- *Calgary:* The city's sustainable building policy requires all new or significant renovations larger than 500 square meters (5380 square feet) to achieve LEED Silver certification or higher.

- *Vancouver:* All new civic buildings larger than 500 square meters (5380 square feet) have adopted green building standards LEED-British Columbia (LEED-BC). New public buildings must achieve LEED Gold certification at a minimum.

Green buildings in the private sector

Private developers have also recognized the value of green buildings both in terms of costs and in public relations.

The Swiss Reinsurance Tower in London reported 50-percent less energy consumption than in conventional buildings. Closer to home, the Conde Nast Building in Manhattan uses 35- to 40-percent less energy than standard construction design requires, and the Solaire, a 27-story, 293-unit, Gold-rated apartment building further downtown in Battery Park City is 35-percent more energy efficient than required by code, resulting in 67-percent lower power demands. During construction, 93 percent of recoverable materials were diverted from the local land-fills. The $1 billion 1 Bryant Park building in Manhattan, when completed in 2008, will have glass double-wall technology that actually dissipates the sun's heat, will have under-floor ducts, carbon dioxide sensors to insure the flow of fresh air, and a rainwater and waste water collection system that is estimated to save 10.3 million gallons of water annually.

Out West, Toyota embraced green building technology with their new $87 million sales campus in Torrance. This 624,000-square-foot facility has 53,000 square feet of solar panels that generate 536 kilowatts and is projected to pay for itself in seven years. Motion sensors control all of the building's lighting, and ceramic floor tiles are made from recycled glass and recycled concrete.

Pennsylvania in the LEED.   Pennsylvania's Department of Environmental Protection has been at the forefront of green construction with five LEED-registered projects on stream as of 2005. The state's first LEED Gold-level green building was built in Cambria, and this 40,000-square-foot project came in at $90.00 per square foot, slightly under comparable costs for conventional buildings. The building has triple-pane high-performance windows that ultimately reduced their heating and cooling loads savings by $20,000 in initial costs and continues to reduce operating costs. The DEP reports that their LEED Silver-level buildings cost virtually the same as conventional construction.

Even the Pentagon is interested in savings. Hensel-Phelps Construction Company, while working on a Pentagon renovation project, discovered a wheat-straw board product that was suitable to use as backer boards in electrical closets. This simple substitution of product saved the government $30,000.

Some Design-Build/Sustainable Building Guidelines

There are eight simple principles of sustainable design that Tony Loyd and Donald Caskey, senior vice presidents and principals of Orange County, California–based Carter & Burgess set as guidelines to design, construction, and operation:

- A multidiscipline, integrated approach is the key to success.

- Simple is better than complex.

- The overriding framework in these types of projects should reflect a respect for nature so that it is not depleted or harmed.

- Life-cycle costs are more significant than first costs (in the age old battle between capital versus expenses).

- Minimize energy use in the selection of building materials, mechanical systems, and appliances.

- Since maintenance of the structure is important, plan accordingly.

- Build with local materials whenever possible to reduce transportation costs. Local materials may be better suited to that environment.

- Consider passive strategies whenever possible, such as building orientation, overhangs and sun shades, thermal mass, and natural lighting.

Are green buildings more expensive than conventional construction?

A study of the cost and benefits of green buildings was conducted by the State of California after Governor Gray Davis issued Executive Order D-16-00 in August 2000 that funded the research. The complete study, titled *A Report to California's Sustainable Building Task Force—October 2003*, is available on the Internet at www.usgbc.org/Docs/News/News477.pdf.

This rather detailed examination showed that while green buildings may cost more than conventionally designed buildings, the premium for sustainability is much lower than generally perceived. Figure 15.1 reveals the premium costs associated with the four certification levels established by USGBC. Figure 15.2 contains the average premium for green offices and schools.

Level of Green Standard	Average Green Cost Premium
Level 1 – Certified	0.66%
Level 2 – Silver	2.11%
Level 3 – Gold	1.82%
Level 4 – Platinum	6.50%
Average of 33 Buildings	1.84%

Source: USGBC, Capital E Analysis

FIGURE 15.1   The premium costs associated with the four certification levels established by USGBC.

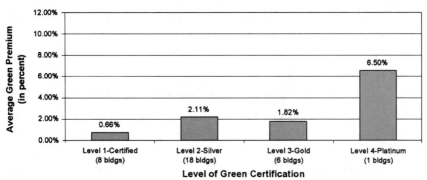

Average Green Premium vs. Level of Green Certification (for Offices and Schools)

Source: USGBC, Capital E Analysis

Year of Completion	Average Green Cost Premium
1997-1998	2.20%
1999-2000	2.49%
2001-2002	1.40%
2003-2004	2.21%
Avg. of 18 Silver buildings	2.11%

FIGURE 15.2   The average premium for green offices and schools.

While the total number of buildings surveyed in this Task Force study is not large, only 33, it does reveal that, in general, green buildings have an average premium cost of just about 1.84 percent.

Green costs are coming down every year as additional architects and engineers, equipment manufacturers, and builders become more familiar with the concept and gain greater experience in its development.

This California study indicated that minimal increases in upfront costs of about 2 percent would, on average, result in life-cycle cost savings of about 20 percent. For example, an initial investment of $100,000 to incorporate green building features into a $5 million project would result in a savings of $1 million in today's dollars over the life of the building, according to the findings in the report.

The financial benefits of green buildings, as pointed out in the survey, includes lower energy costs, lower waste disposal costs, lower water costs, lower environmental and emissions costs, lower operating and maintenance costs, and the increased productivity and health of the workers occupying these types of buildings.

The energy costs and water savings were rather easy to predict, but the productivity and health gains were much less precise and much harder to predict. Figure 15.3 shows the percent of reduction in energy costs for certified, silver, and gold certifications.

	Certified	Silver	Gold	Average
Energy Efficiency (above standard code)	18%	30%	37%	28%
On-Site Renewable Energy	0%	0%	4%	2%
Green Power	10%	0%	7%	6%
Total	28%	30%	48%	36%

Source: USGBC, Capital E Analysis

As discussed above, green buildings use an average of 30% less purchased energy than conventional buildings. In addition, green buildings are more likely to purchase "green power" for electricity generated from renewable energy sources. Green power purchases can take two forms:

- Customers can purchase green power directly from their utility or from a local green power provider. In this case customers are paying for electricity generated from renewable energy sources, typically by a local provider in the state or utility jurisdiction. About 40% of US electricity customers have this option.

- Customers can purchase green certificates, or green tags. In purchasing green certificates, a customer is buying ownership of the reduced emissions (and by implication the environmental and health benefits) associated with renewable power, even though the green generating facility is frequently not in the customer's vicinity. All electricity consumers have this option.

FIGURE 15.3 The percent of reduction in energy costs for certified, silver, and gold certifications.

Let's take a look at some of the positive effects attributable to green building construction in the California study, effects that will obviously vary from state-to-state, but which nonetheless represent an order of magnitude that can be adjusted accordingly.

Energy Use.   These buildings were 25- to 30-percent more energy efficient when compared to ASHRAE 90.1-1999. Interactions between lighting versus heating and cooling and between fresh air and humidity control are analyzed simultaneously allowing designers to prepare a holistic approach to energy-consuming equipment and building performance. Except for isolated areas in this country, air conditioning is the overriding requirement, particularly in buildings with high occupancy rates; therefore, special attention needs to be paid to this building component. Innovative approaches to satisfying cooling loads include

- Incorporating more efficient lights, task lighting, sensors to cut unnecessary lighting, and using daylight, which will not only reduce power consumption but also reduce cooling loads

- Increasing ventilation effectiveness, which will help cut cooling loads during peak periods through improved system optimization

- Using under-floor air distribution systems; use of an under-floor plenum to deliver space conditioning typically cuts fan and cooling loads

- Commissioning in a systematic approach to insure that systems as designed are installed and are operating as planned.

- Using heat island reduction measures; increased roof reflectivity will lower building temperatures and reduce cooling loads. Albedo is the unit used in measuring the reflectivity of solar energy striking a roof—the higher the Albedo number, the higher the reflectivity.

- Generating energy onsite via photovoltaics, which in some climates can generate 20 percent of total consumption

Projected savings

The California study showed that the reduction in energy costs will provide the following energy savings over 20 years using the present value-cost analysis:

Thirty-percent reduced consumption at an electricity price of $0.11 per kWh is about $0.44/ft^2/yr $\times$ 20 years = $5.48 per ft^2.

The additional value of peak demand reduction from green buildings was estimated at $0.025/ft^2/yr $\times$ 20 years = $0.31 per ft^2.

Together, the total 20-year present value of energy savings from a typical green building is $5.79 per ft^2.

Water conservation.   Green building water conservation is divided into four sectors:

- Potable water is used more efficiently through better designs and new technologies.

- Gray water—nonfecal wastewater from bathroom sinks, tubs and showers, washing machines, and drinking fountains—is captured and used for lawn and planting irrigation.

- Onsite stormwater is captured for use onsite or to recharge groundwater tables.

- Recycled or reclaimed water is made available for other uses.

The California studies showed that, taken all together, these measures can reduce water consumption in the building to levels 30-percent lower than code requirements and can reduce exterior water demands by as much as 50 percent. In areas where water supplies are being overloaded, reclaimed water projects are taking on added importance. The Bay Area of California expects fully 50 percent of their new water supply to come from reclaiming. These reclaiming projects typically cost about $600–$1100 per acre/foot based on estimates from the East Bay Municipal Utility District.

Waste reduction.   We are known as the disposable generation: use something a few times and discard it. In fact, packaging costs often exceed the value of the item these days, and that outer package always seems 500-percent larger than the product itself.

Reducing waste is a national concern and a nationwide problem. Not only are trucking and removal costs higher due in no small part to increases in gasoline and diesel fuel, but many states are simply running out of room and have no place to dump their waste. California estimates that their total annual waste, as of 1998, amounted to 33 million tons, 21 million of which is generated by nonresidential buildings. An updated study would most likely show a much higher figure.

Green building attempts to reduce waste and focus on recycling and reuse—two things which can begin during the construction process and continue on throughout the lifetime of the building.

Steps that can be taken during construction to start this process include

- Reusing and minimizing construction and demolition debris, and diverting some of it from landfills to recycling facilities. Good examples of this are recycling cast-in-place concrete to remove rebars and then converting the concrete to aggregate. The recycling of masonry materials for use as a base course under paving has proven to be an effective use of construction debris.

- Source reduction—using materials that are more durable and easier to repair/maintain.

- Using reclaimed materials, such as (mentioned earlier) aggregate for the base course under paving, or employing ground glass as a reflective material in asphalt paving.

- Using materials that can function in a dual role (for example, exposed structural systems and ductwork, and staining concrete floor slabs).

- Incorporating an existing structure into a new building program.

During the life of the building, the following can be done:

- Develop an indoor recycling program.

- Design for deconstruction.

- Design for flexibility via the use of movable walls, modular furniture, movable task lighting, and other reusable building components.

Construction and demolition diversion rates reached as high as 97 percent with some California projects and are typically 50 to 75 percent in green buildings.

Other revealing but not so apparent benefits of green buildings

The obvious effects of green building design and construction come readily to mind, but there are other subtle and compelling reasons to support this movement.

Recycling creates jobs.   One interesting sidebar to this question of disposal or recycling is how it affects employment. The total impact from diversion of waste material is nearly twice as much as the impact for disposal.

A study conducted by University of California, Berkeley, revealed that one additional ton of waste disposed of in a landfill generated $289 of total output in the state economy. One additional ton of waste diverted as a recyclable generated an average of $564. Only 2.46 jobs were created for every 1000 tons of waste disposed, but 4.73 jobs were created for waste diverted as recyclable.

Productivity gains from improvements in the working environment.   The quality of the working environment affects both health and productivity. A healthy environment leads to a reduction in sick days that impacts productivity. The pie chart in Figure 15.4 reflects the results of a further study by California that lists the financial benefits of LEED certified and silver buildings, and worker productivity and health are far and away the best beneficiaries to have with a green program.

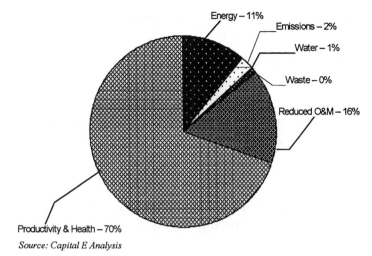

Source: *Capital E Analysis*

FIGURE 15.4 The results of a study listing the financial benefits of LEED certified and silver buildings.

Lastly, Figure 15.5, the conclusion of the California Build Task Force report is that building green is quite cost-effective, thus making lots of financial sense.

The sustainable approach to construction. The process of designing and constructing a structure to green standards involves not only the building itself but the site on which it will be located. This includes access to the site as well.

The following goals and objectives can be viewed as a primer for this form of sustainability.

Category	20-year NPV
Energy Value	$5.79
Emissions Value	$1.18
Water Value	$0.51
Waste Value (construction only) - 1 year	$0.03
Commissioning O&M Value	$8.47
Productivity and Health Value (Certified and Silver)	$36.89
Productivity and Health Value (Gold and Platinum)	$55.33
Less Green Cost Premium	($4.00)
Total 20-year NPV (Certified and Silver)	**$48.87**
Total 20-year NPV (Gold and Platinum)	**$67.31**

Source: *Capital E Analysis*

FIGURE 15.5 The conclusion of the California Build Task Force report.

The Site. *Sitework goal*: Meet or exceed standards for erosion and sedimentation control by doing the following:

- Prevent loss of soil during construction due to stormwater run-off and wind erosion.
- Prevent siltation of existing storm sewers and streams.
- Protect topsoil stockpiles for reuse, or modify soils to meet topsoil acceptable standards.

Site utilities goal: Reduce stormwater run-off and reuse by accomplishing the following:

- Minimize or totally eliminate stormwater run-off by carefully planning infiltration swales and basins to reduce impermeable surfaces instead of installing detention ponds.
- Retain or recharge existing water tables by minimizing disturbances, saving trees and natural vegetation, and supporting and enhancing natural landforms and drainages.
- Store roof run-off for future use as gray water or reclaimed water.
- Install onsite a small footprint, state-of-the-art treatment plant to recycle water for irrigation purposes.

Open space and landscaping goal: Protect and restore existing vegetation by doing the following:

- Protect trees, which enhances the value of the site and lowers cooling loads. Indigenous landscaping supports wildlife and biodiversity and does not require the level of irrigation necessary for new ground cover. It also eliminates the need for chemical treatment.
- Minimize pesticide use by installing weed cloth, mulches, and dense plantings.

Circulation and transportation goal: Improve circulation and decrease the need for private transportation by doing the following

- Tie development or building to transit nodes and emphasize alternatives such as organized car pooling, water taxies (if available), buses, car sharing.

The Building. *During construction goal*: Reduce waste by doing the following:

- Divert at least 75 percent of construction, demolition, and land clearing from disposal as landfill.

- Deconstruct all existing structures with substantial recoverable materials and dispose of them to recyclers.

- Adjust new site contours to provide for a balanced site. Modify non-topsoil soils to acceptable topsoil requirements.

The Holistic Approach—Again.   Energy-efficient building components are all-encompassing. Energy-efficient heating and cooling systems, and building envelope products like double/triple-glazed windows come easily to mind, as do advanced, programmable control systems. And what about foundation insulation, roof insulation, and Albedo (roof reflectivity) values? Energy-efficient plumbing fixtures and lighting fixtures with built-in power management systems improve every year. Office equipment that goes into "sleep" mode when not used not only reduces electrical costs but also lowers the heat load.

Passive solar design, the technology of heating, cooling, and lighting by converting sunlight into a power source can work effectively with other energy-efficient materials and products. Photovoltaics can supplement or replace power from local utility companies.

NREL and Oberlin College's pilot program.   The National Renewable Energy Laboratory (NREL) was established in 1974 and is the principal laboratory for the Department of Energy's Office of Energy Efficiency and Renewable Energy. Their mission is to develop renewable and energy-efficient technologies.

Oberlin College in Oberlin, Ohio wanted to design and construct a building to serve as a model and teaching aid for students in their environmental studies program. To that end, they built the 13,600-square-foot (1260-square-meter) Adam Joseph Lewis Center on campus, containing classrooms, offices, and an atrium.

The goal of the project was to construct a building that was not only energy efficient, but also one that was able to export energy to the local grid system. In order to do so, they would install passive solar designs, use natural ventilation wherever possible, design an enhanced thermal building envelope and use geothermal heat pumps for heating and cooling. The building's roof would incorporate an integrated photovoltaic (PV) system to allow for solar generation of electricity for the building.

After the building was completed in 2000, the NREL began to monitor the structure to evaluate its energy performance. Their findings would serve three purposes:

- Evaluate the performance of the building and several of its subsystems.

- Provide suggestions to improve the initial performance.

- Document lessons learned to improve the design of future low-energy buildings.

This study, while sophisticated in its analysis of the performance of mechanical and electrical systems, dramatically describes steps that can be taken to reduce energy demands.

NREL's study of the Oberlin College building ended in 2003, but it stated that more work was required to fulfill the original goal of the project as being one of net energy exported, but the strides taken in this venture further the case for energy self-sufficiency.

Some of the lessons learned by NREL are generic in nature and apply to any sustainable building project:

- PV systems must be engineered to minimize transformer balance and system losses. These losses can represent a significant portion of the overall system production.

- PV systems may not significantly reduce the building demand. In this case, any small demand reduction due to PV is from load diversity.

- During summer months, on average, large PV systems in commercial buildings can export electricity from 8:00 A.M. to 6:00 P.M. From a utility perspective, this building was a net positive during daylight hours in the summertime and provided power when it was most needed by the grid.

- Control design must be completely integrated using the full capabilities of the equipment in the building, such as CO_2 sensors, motion sensors, and thermostats. A balance must be achieved between human operations and automation.

- Dark ceilings must be avoided to take full advantage of daylighting and uplighting.

- Daylighting sensors are needed in all daylit areas. It is not sufficient to rely on manual controls

- Daylighting must be designed into all occupied areas. The daylighting design should consider additional heating and cooling loads imposed upon the building. Overglazed areas such as the atrium in this building provided abundant daylighting but resulted in additional heating and cooling loads.

- Specifications for heat pumps must work with appropriate groundwater temperatures.

- Electric boilers can be employed as a back-up source if they are used sparingly and do not cause excessive demand charges on the building. Controls and staging are essential for the integration of limited-use systems such as these.

The Greening of Existing Buildings

The LEED Certification—EB was established to deal with the upgrading of existing buildings to green standards. JohnsonDiversey is a manufacturer of cleaning and hygiene products located in Sturtevant, Wisconsin. It is housed in a 277,440-square-foot building, built in 1997, 70 percent of which is office space, while 30 percent is devoted to research laboratories. It is a breakaway company from SC Johnson Inc. in Racine, Wisconsin, the well known producer of Johnson's Wax. JohnsonDiversey's legacy for innovation extends back to the parent company in Racine, which was among one of the first corporations in America to recognize the importance of good architecture and its positive effect on the working lives of their employees. The Frank Lloyd Wright–designed SC Johnson headquarters in Racine was not only a monument to progressive corporate policy when built in 1936 but remains so today with its famous lily-pad columns in the building's main room. SC Johnson Wax voluntarily eliminated CFCs, that ozone depleting refrigerant, from their aerosol product line in the 1970s, and led the development of more environmentally compatible propellant products.

In 2004, the Sturtevant facility at JohnsonDiversey earned its LEED certification as an existing building for a structure containing 80,000 square feet (7435 square meters), about 30 percent of its 278,000-square-foot (25,836 square meters) building.

Their LEED certification included the following modification/remediation measures:

- Native prairie plants and restored wetlands were developed on more than half of the 57-acre site.
- Stormwater collection for turf grass reduced potable water use by 2 to 4 million gallons per year.
- Low-flow fixtures reduced water use.
- More than 50 percent of solid waste was recycled.
- Ninety percent of interior space receives reflected light.
- Personal environment controls are installed at each work station.
- Rapidly renewable, locally available materials such as maple wood are used throughout.

Several innovative programs exist at the site, some of which do not involve substantial cash outlays, and one that is as simple as encouraging alternative transportation choices.

Of the 580 parking spaces provided, 10 percent (or 58) are reserved for hybrid vehicles, while 16 car/vanpool spaces are allotted to encourage car pooling. The personal environment modules (PEMs) installed in 93 percent of the total building office area allow for the individual control of temperature, air flow, lighting, and acoustics at each designated workstation.

JohnsonDiversey converted water usage from 2.5 gallons per minute (gpm) to 0.5 gpm by installing aerators on all lavatory faucet fixtures, and additionally reduced usage from 2.5 gallons per minute to 1.8 gpm by the installation of aerators on all shower fixtures. In combination with flush valve replacement diaphragms rated at 1.6 gallons per flush (gpf) for toilets to 0.5 gpf for urinals, they have reduced water use performance to very low levels.

They have reduced waste disposal through a vigorous recycling program and employee awareness, and have distributed a recycling card to each employee providing information on what is to be recycled, where to take recyclables, and who to contact for questions. Twenty-four recycling areas for cans, plastics, and glass are situated throughout the building, where the recyclables are collected and emptied into large containers on the loading dock.

Table 15.1 shows their annual waste generation profile.

For all construction projects within the building, they require that staff or contractors recycle and/or salvage at least 30 percent by weight any construction, demolition, or land-clearing waste.

Items like toxic materials source reduction were addressed by inventorying such things as existing light fixtures and bulbs. They now purchase 32W T-8 Alto lamps from Phillips that have a mercury content of 18.6 parts per million (ppm), which is considerably under the limit of the 25.0 ppm code.

The Green Building Rating System for Existing Buildings was issued in October 2004 and is referred to as LEED-EB. USGBC launched Version 2.2 of the Green Building rating systems in late 2005, reflecting experience gleaned from comments made regarding the previous iteration. A direct dialog with ASHRAE resulted in new calculations in order to achieve some performance goals. New application guides for health-care facilities, schools, and laboratories are also in the works.

TABLE 15.1   The Annual Waste Generation Profile of the SC Johnson Headquarters

Garage	208,000 lbs
Waste-recycled	74,800 lbs
Paper	116,480 lbs
Commingle (cans, glass, plastic)	5200 lbs
Total waste stream	404,480 lbs
Total recycled	196,480 lbs
Percent recycled	49 percent

As the Green Building movement spreads across the private and public sector, new opportunities await those design and construction firms that become intimate with the requirements of sustainable structures.

Thus, as you can see, it is easy to be green—and also profitable.

Interoperability and Building Information Modeling (BIM)

Interoperability, the ability to share electronic information seamlessly among all participants on a construction project, and building information modeling (BIM), the computer-assisted design process whereby 3-D and 4-D images are developed, will forever change the way we design and construct buildings.

Looking at the Last Several Decades

It was the federal government's Telecommunications Act of 1996, which permitted local telephone companies to compete for customers with long-distance carriers, that was at the forefront of an information explosion as each of these new telecommunication companies sought to build their own fiber-optic infrastructure to connect Internet users around the world.

One such fiber-optic company, Global Crossing, gambled that these local, national, and international phone companies would have a huge demand for transmission lines, and banking on an explosion in the new digital technology, began laying fiber-optic cables to bind the globe together. Though Global Crossings itself is no more, the fiber-optic networks they installed now connect the world, and are only beginning to reach their potential.

The fiber-optic infrastructure and "open protocols" allow digital devices to "talk" to each other, resulting in a global communications network. With the advent of HTML (Hypertext Markup Language), URLs (uniform resource locators) that locate and display web pages universally, and HTTP (Hypertext Transfer Protocol) to move these documents around, the need for interoperability became apparent as new software and program developers began to introduce their own proprietary language. By the end of the 1990s, 2-D design had progressed to 3-D, affording architects, engineers, builders, and owners the opportunity to view a virtual model of the construction product.

Construction software company Autodesk®, with their product Revit®, produced a CAD system that was data-based so that a Bill of Quantity list of materials incorporated in the design would be produced as the project progressed. Figure 16-1 is a composite of three components produced by the Autodesk® design process: a plan view (lower right), a take-off of materials (upper left), and a 3-D model of the building (upper right). Any change in design is then reflected in a change of the list of building materials and the 3-D model. The global fiber-optic network permitted various segments of design to be outsourced anywhere in the world, with the potential to not only reduce costs but to speed up the entire design process. The end of the workday in New York is now the beginning of the workday half way around the world.

Eventually, off-shore engineering companies in India began to advertise their services over the Internet. As one company said: "We offer a top-flight engineering service. Why pay $53,240 for a CAD drafter in Los Angeles, when we can supply fully qualified people at about $12,000 per year?" Another site stated: "They have a 35-hour workweek in Europe. Here in India we have a 35-hour workday."

Contractors Slow to Embrace Technology

Many contractors were, and still are, reluctant to embrace these new technologies with their ability to store and retrieve information electronically. Builders cited many reasons for hanging on to paper documents. The following lists a few of them:

- The cost of hardware and software is still too expensive.
- They still don't have full confidence that information won't be lost through computer "crashes," or from a temporary loss of power.
- The old say: "We've always done it that way. It works, so why change?"
- Contractors routinely communicate with subcontractors and vendors who don't use computers for anything other than payroll and accounting functions.
- Local, county, and state offices frequently require some paper format and documentation for filing.
- Requirements for original seals/signatures on documents filed with various government agencies are still out there.

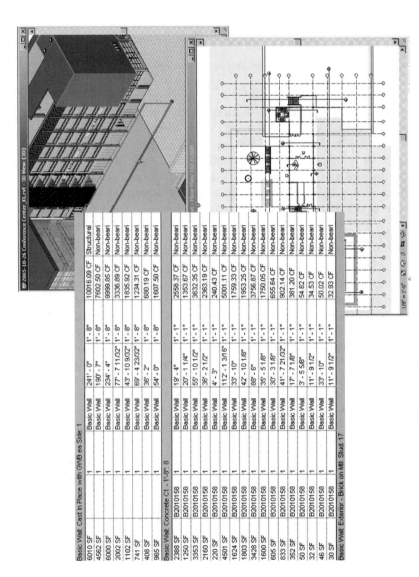

FIGURE 16.1 Autodesk® Revit® software showing plan view, bill of quantities and 3D image. (*By permission: Autodesk® Revit® Structures U.S.A.*)

- The use of electronic media on the construction site by employees not accustomed to the medium is inefficient and therefore prone to inaccuracies.

- Belief that paper records are more official and are legally more acceptable than stored electronic data.

- There is no real incentive to work electronically.

The Construction Finance Management Association (CFMA) reported in a recent study that EXCEL is the most widely used software application in today's businesses, and that predominant construction industry software includes AccuBid, Bidmaster Plus, McCormick Estimating, and Precision Collection. The most common forms of project management software is Primavera Enterprise and Expedition and Prolog Manager. Only 25 percent of the construction firms surveyed by CFMA used collaborative software such as Buzzsaw, Constructware, or Meridian Project Talk. The scheduling software used is primarily Suretrak and Primavera.

Interoperabilty—what is it and why is it so important?

One definition of *interoperability* is the ability to exchange and manage electronic information seamlessly, and the ability to comprehend and integrate this information across multiple software systems. Another definition is "an open standard for building data exchanges." Interoperability simply means that your system can "talk" to mine, and we can all "talk" to the designers, contractors, subcontractors, vendors, and owners' representatives in the same electronic language. There is little interoperability in the AECO (architect, engineer, contractor, owner) community today, but many organizations, recognizing its importance, are aggressively attacking the problem—a problem not confined to the design and construction communities.

One German automobile manufacturer was alerted to the problem of interoperability after receiving a fair amount of customer warranty complaints about various system component failures in the electronics installed in their high-priced models. Apparently, there was no central protocol in place governing or controlling the "language" of computer chips supplied by each of those disparate component vendors, and when all of these parts from a variety of suppliers were installed, they could not "talk" to each other, which manifested itself, in the eyes of the customer, as a system failure. It took some time to uncover the cause and correct the problem, but in the meantime there were a lot of very unhappy customers.

Recently, several trade and private organizations have begun to recognize the missed opportunities and tremendous cost of not fully embracing interoperability and the resultant seamless integration of the entire project's database—from design to construction to commissioning and continuing, on through the building's entire life cycle.

The NIST report. In 2002, the National Institute of Standards and Technology (NIST) concluded their study to quantify the cost for inefficient interoperability in commercial, institutional, and industrial facilities for both new and "in place" construction. According to NIST, this inability to seamlessly exchange and manage electronic information in the construction industry adds an astounding $6.18 per square foot to project costs in addition to operations and maintenance costs of $0.23 per square foot. In total, inefficient interoperability cost the construction industry, per this report (see Fig. 16-2), a whopping $15.8 billion in 2002. The manufacturing sector has dealt with this problem with considerable success, but it should be kept in mind that, on the whole, they deal with a flow of similar products in a controlled environment, and they also enjoy economies of scale. The construction industry is mainly a one-off product and even when a similar product is built, say a motel chain project or a fast food restaurant, various zoning and building regulations and site conditions frequently impact the structure's basic design.

In the September 2004 issue of *Architectural Record* magazine, Mr. Ken Sanders, FAIA, author of the mid-'90s book *The Digital Architect*,

Stakeholder Group	Planning, Design, and Engineering, Phase	Construction Phase	Operations and Maintenance Phase	Total
Architects and Engineers	1,007.2	147.0	15.7	1,169.8
General Contractors	485.9	1,265.3	50.4	1,801.6
Specialty Fabricators and Suppliers	442.4	1,762.2	—	2,204.6
Owners and Operators	722.8	898.0	9,027.2	10,648.0
Total	**2,658.3**	**4,072.4**	**9,093.3**	**15,824.0**

Source: RTI estimates. Sums may not add to totals due to independent rounding.

Cost Category	Avoidance Costs	Mitigation Costs	Delay Costs
Architects and Engineers	485.3	684.5	—
General Contractors	1,095.40	693.3	13.0
Specialty Fabricators and Suppliers	1,908.40	296.1	—
Owners and Operators	3,120.00	6,028.20	1,499.80
Total	**6,609.10**	**7,702.00**	**1,512.80**

Source: RTI estimates. Sums may not add to totals due to independent rounding.

FIGURE 16.2   NIST Study of cost of interoperability. (*National Institute of Standards and Testing, Gaithersburg, MD*).

in his article entitled "Why Building Information Isn't Working . . . Yet," compared the difference in technology use between the automobile and aerospace industries and the construction industry, stating ". . . most importantly, cars and planes are the products of an integrated design-build process. The designer and builder are one and the same. This is rarely the case with building design and construction."

Why contractors are slow to embrace information management. The NIST study uncovered many reasons why the construction industry suffers from inefficiency in information management, often operating in isolation and not effectively communicating with other internal and external partners in the design and construction process.

- Collaboration software is not integrated with a contractor's other systems. Some builders use collaborative software, but it is generally not integrated with other systems—it is used in a stand-alone application, defeating the purpose of the software.

- Many parties work together on only one project, so there is little incentive to invest in long-term collaborative software, each project frequently being unique, having different participants, scope, workforce, and teams, and operating in a different location.

- Life-cycle management processes are fragmented and not integrated across the project's life cycle.

- There are inefficiencies and communication problems when participants to the project from all parts of the life cycle have various versions of the same software or different software.

- A lack of data standards inhibits the transfer of data between different phases in the life of a project and their associated systems and applications.

- Internal business processes are fragmented and inhibit interoperability. NIST found that in some firms: *an estimated 40 percent of engineering time is dedicated to locating and validating information gathered from disparate sources.*

- Many firms use automated and paper-based systems to manage data and information, while hard-copy construction documents are routinely used on the jobsite.

- Many smaller construction firms do not employ, or have only limited use of, technology in managing their business processes and information.

The federal government push for interoperability and BIM. On January 24, 2005, the General Services Administration sent out a Request For Information (RFI) to the capital facilities industry (design consultants, general contractors, subcontractors, and vendors) with the following statement:

Interoperability problems in the capital facilities industry stem from the highly fragmented nature of the industry's continued paper-based business practices, a lack of standardization, and inconsistent technology adoption among stakeholders. Based on interviews and surveys, it is found that $15.8 billion in annual interoperability costs were quantified for the capital facilities industry in 2002. Of these costs, two-thirds are borne by owners and operators, which incur most of these costs during on-going facility operation and maintenance (O&M).

The United States General Services Administration (GSA)/Public Buildings Service (PBS) is seeking information from industry partners on Industry Foundation Classes (IFC)-Based Integrated and Interoperable Building Information Model (BIM) technology as part of its effort to improve project deliveries in the capital construction program. The GSA/PBS currently has an active pipeline of more than 200 major capital construction projects conclusively exceeding a value of $11 billion.

The GSA, in this RFI, proclaimed an opportunity for firms in the design, construction, and facility management and real property industries to submit suggestions on the use of IFC-BIM technology. This information will be used by the government to establish potential sources in the marketplace that have knowledge and experience in the use and practice of this state-of-the-art technology. Look for some future government projects to require bidders to have interoperable software, and the ability to provide BIM modeling as part of their qualifications package.

The industry movement toward interoperability.   The International Alliance for Interoperabililty (IAI) has held discussions with industry leaders in 19 countries to define a single building information framework. Using heating and cooling as an example, IAI asked ASHRAE in the U.S. and their counterparts in CIBSE in England, and DIN in Germany, to get together and define a process for calculating a building's HVAC requirements. They wanted to develop a generic model for systems development in order to provide a seamless flow of information for mechanical systems across all national boundaries.

This is a process termed Industry Foundation Classes (IFCs) that must be repeated by other design and construction teams to develop the specific nongraphic common language required for interoperability. Each IFC thereby becomes a dictionary for project component information to be shared by owners, architects, engineers, general contractors, and specialty contractors.

Just like how the HTML and HTTP protocols allowed the transmission of web pages to become a universal event, a new technology is needed to generate the cross-referencing and dissemination of design and construction information on a global basis.

The current state of affairs.   FIATECH is a nonprofit research and development consortium based in Austin, Texas that focuses on developing and delivering technologies to the construction industry to improve the design, engineering, and construction of capital projects. Recently, they have been working on several approaches to advancing the interoperability of construction software.

Extensible Markup Language (XML) is a simple, flexible text format originally designed to meet the needs of large-scale publishing, but which now plays a major role in exchanging data over the Internet. AecXML was chartered in 1999 to promote and facilitate interoperability among software applications for information exchange in architecture, engineering, and construction. AgcXML., a program sponsored by the Associated General Contractors of America, and planned for delivery in 2006, will create an XML schema (plan) to deal with the following common construction documents:

- Requests for Information (RFIs)
- Submittals
- Purchase orders
- Contracts—both AGC forms and other industry standard forms
- Pay applications
- Change order requests (CORs) and change order approvals
- Punch lists
- Daily reports
- Addendum notifications
- Meeting minutes
- Requests for Proposals (RFPs) and pricing

The Open Building Information Xchange (oBIX) is a movement backed by facility managers and industry sources to use the programming of XML for seamless Internet- and intranet-based communications between building systems in order to run a building on standard protocols and techniques, thus creating a format by which buildings, facility managers, and owners can interface with the Internet.

The civil engineering profession has developed LAND XML to tackle this problem, while the steel industry has created an interoperability protocol called CIS/2.

The steel industry becomes a leader

In 2004, the American Institute of Steel Construction, Inc. (AISC) issued a white paper entitled "Interoperability and the Construction Process" in which they explained their efforts, and that of the steel industry, toward achieving interoperability. AISC initiated the CIMSteel Integration Standards Version 2 (CIS/2), enabling designers and specialty steel contractors to exchange data. CIS/2 is compatible with other software programs such as Bentley, RAM, GT Strudl, Robot and ISS drafting software, Tekla and Design Data detailing software, and Fabtrol shop fabrication software.

ASCI, in their report says "The neutral file format allows stand-alone programs—such as structural analysis and design, detailing and manufacturing information systems, as well as CNC driven fabrication equipment—to communicate with each other by translating a program's native format into a neutral format to allow data interchange across multiple platforms."

Dealing with the coordination problem. A structural engineer can now design a steel structure in the BIM (3-D) mode, and concurrently and instantaneously transmit the design to the architect and MEP design consultants so they can begin to incorporate their work into this "skeleton" framework. If a general contractor is on board at the time, a copy of the 3-D design can be forwarded to them, and possibly onto their steel fabricator. The design consultants will then be able to "talk" to each other and to the general contractor and subcontractors in a paperless fashion. Suggested changes offered by any member of the team can be distributed, reviewed, and addressed immediately, and changes can be effected and distributed so that steel shop drawings are then produced quickly and e-mailed to the engineer of record for approval, bypassing the old paper trails and thereby dramatically speeding up the process.

Interference problems are thus highlighted early in the design process and corrections made before the design is completed rather than uncovering these difficulties during construction.

All of this is accomplished without having to handle rolls and rolls of design and shop drawings, without the time-consuming tasks of packaging and repackaging, and the delivery charges incurred back and forth. The potential savings as a result of fewer reproducibles and reduced handling and shipping costs may be minute on small projects, but on larger ones it could mean tens of thousands of dollars.

Designs that really work, that eliminate the need for RFIs to resolve questions, are thus addressed and resolved *during* design development, not *after* the construction contract award, resulting in a set of drawings that are *really* coordinated—one of the goals of 3-D modeling.

Just think how many more projects each participant could manage, and how much more time could be spent focusing on the project at hand, and not getting bogged down in paper pushing and generating of RFIs, RFCs, and hundreds of transmittals, when this process becomes commonplace.

Interoperability and BIM as envisioned by the steel industry.   According to Tom Faraone, Senior Regional Engineer for AISC Marketing, LLC, an organization affiliated with the American Institute of Steel Construction, the steel industry is already using bar codes to speed up product fabrication and delivery, and is working on other ways to utilize these devices more effectively. There is an increased interest in radio frequency identification devices (RFID), a micro radio transmitter providing fabrication data, that can be affixed to each structural steel member as it enters the fabricator's shop. Upon leaving the shop it could then convey to a computer-operated crane its precise position within the structural framework.

The goal of AISC is to develop a system in conjunction with its members that will accelerate the entire design-fabricate-deliver-erection process of a structural steel building. If time is money, then it surely applies to this industry as well.

The New York Times, in an article dated April 13, 2005, reported on a project in Boston called the Charles Street Jail, which consisted of the redevelopment of this historic building into a four-star hotel. The developer budgeted the project at $50 million in 2003 but was devastated by the sharp increase in structural steel that occurred at the time. An eight-month redesign was required to reduce the updated cost of $74 million down to a more manageable $64 million. Although the consumer price index (CPI) showed an inflation rate in the 2 to 3 percent range, this wasn't so in the building business, where some estimating services pegged inflation in the industry at 12 percent for the year 2004.

The final design of the Charles Street Jail required the architect to delete the planned basement, reduce the floor to floor height and add a 15th floor. Mr. Richard Friedman, CEO of Carpenter & Company, the developer, summed it up in four words: "It's been a nightmare."

A more rapid design and review cycle can become an effective guard against the forces of inflation, and AISC says their CSI/2 system can produce a 50-percent savings in scheduling.

Case study—the Lansing community college project, Lansing, Michigan.   The interoperable process, by maximizing efficiencies between designer and fabricator, allowed the Lansing Community College Health and Human Services Career and Administration Building project to lower their costs to add a 4th floor by $315,000 or $2.35/sf , according to AISC. The electronic transfer of information between the designers and fabricator permitted the building team to rapidly review alternative design schemes, make changes, and get them reviewed and approved, resulting in the elimination of 700 members and a savings of 190 tons of steel. Without this interoperability process, changes of this nature would have required multiple manual re-entries of data, long delays in the revision, review, and approval of shop drawings, and, almost certainly, a justifiable delay in completion, the cost of which might have completely negated all or much of the savings that would accrue with the design change.

Larry Kruth, Engineering and Safety Manager at Douglas Steel Fabricating Corporation, the contractor that fabricated and erected the project's structure, is sold on interoperability. On an unrelated project, Larry said that the design engineer had specified several large rolled sections, W44 × 265s, which were only available at a mill in Luxembourg. Larry quickly notified the engineer, suggesting a switch to a W40, available in this country. The design change was made quickly electronically, and the project's progress continued—seamlessly.

The Denver art museum project—another example of 3-D/interoperability success. The addition to the Denver Art Museum was a 147,000 square-foot structure consisting of 16,500 pieces of steel with a total combined weight of 2,750 tons. There were 3100 pieces of primary steel sections, 50,000 bolts, and 28,500 pounds of field and shop welds. The intricate connection information was passed from the design team to the detailer using simple sketches of each individual connection. Each sequence of steel fabrication was detailed in a 3-D model, and two-dimensional shop drawing details were created. Then, as each sequence of shop drawings were completed, the detailer provided the design/construction team with 3-D electronic models in addition to the hard copy drawings so that the architect could verify and check the geometric control and coordination with other architectural elements. The end result of this design/fabricate/erect process was

- 3-D graphic aids were freely shared by designer/contractor/subcontractor to facilitate each one's own work and thus improve the overall product.

- Minimal shop issues were encountered due to the level of coordination during the 3-D design.

- Minimal field issues were encountered and erection proceeded without any major field adjustments or fixes.

- The fast track approach of overlapping design, fabrication, and erection resulted in a faster start and more rapid completion.

- The preliminary interactive work by all members of the team during design smoothed out the fabrication and erection process resulting in the completion of erected steel three months ahead of schedule.

In the AISC white paper, they quote Mr. David I. Ruby, P.E., a principle in the firm of Ruby & Associates, who described the current process of steel design:

> The architect would present a defined building concept to the structural engineer who would design the structure utilizing a structural analysis program, prepare design drawings, and submit (them) to the fabricator. The fabricator would take the drawings, (and) have a material specialist prepare a full take-off by hand to determine the material required for the structure. The fabricator would review all of the material from the engineer—page by page, sheet by sheet, floor by floor. They'd take a yellow crayon and mark off every beam, and another person would re-check with a red crayon indicating it was checked again so the fabricator knew that the shop bill accounts for all the materials. Manually, this process took a week or more. And we're not talking just 40 hours of labor, but two or three people putting in 40 hours or more to pull that all together. With interoperability, this process takes just a few hours. We can now send files at noon, and by 3 o'clock the fabricator has the bill of materials to order.

Mr. Ruby goes on to say:

> You always want to purchase at the best cost, and the best cost comes from purchasing mill material which is normally rolled and/or stocked between 40 and 60 feet long. So you have to multiply it. That means if you need three 18-foot beams, you don't order exact pieces—you order one 55-footer and cut it to length in the shop. All of these calculations used to be done by hand.

What Is Building Information Modeling All About?

Building Information Modeling (BIM) is the transition from 2-D to 3-D design and is sometimes used synonymously with Virtual Building Model (VBM) or Virtual Design and Construction (VDC), each of which refers to the ability to produce a three-dimensional view of a construction project as building components are designed, modified, or deleted.

Figure 16-3 displays the sequencing from 2-D to 3-D and 4-D computer-assisted design.

Coordination and interference issues addressed

Most project specifications include a requirement for the general contractor to prepare "coordination" drawings, such as:

> The general contractor is fully responsible for coordinating the work . . . Coordination space requirements and installation of mechanical and electrical work are indicated *diagrammatically* on the drawings. Prepare coordination drawings for all areas where close coordination is required for installation of products and materials . . .

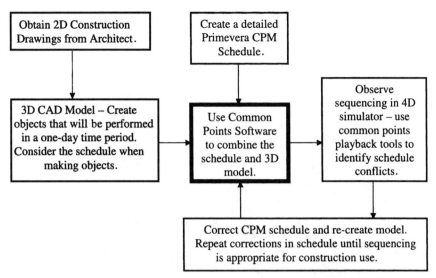

FIGURE 16.3   Sequencing from 2-D to 3-D and 4-D computer-assisted design.

This is where interference problems occur. Piping and ductwork often must be rerouted or resized to avoid interference with structural members or other systems. Ceiling heights may need to be lowered to ensure that all above-ceiling MEP work is fully concealed, and when lots of things don't fit, added costs and delays add to the frustration of all parties.

Building information models create a digital database that can be shared by all parties on the construction project and can be distributed from the architect to the engineer, and to the contractor and the contractor's subcontractors and vendors—all through file sharing. The two-dimensional plan can be displayed in 3-D fashion as each layer of design is added (Fig. 16-4), allowing both architect and engineer to comment on coordination issues, and suggest and make changes with a certainty that all other affected components are adjusted accordingly. Just as important, all members of the construction team are instantly apprised and buy into these changes.

When a "negotiated" project is underway, these changes, when a data-based design system like Autodesk's Revit® is employed, allows the general contractor to review the revised quality take-off (Fig. 16-5) and verify or adjust their estimate.

Some recipients of BIM information will be read-only, while others can review and recommend changes which, if implemented, will be reflected in all parts of the design affected by that change or changes.

This means that the time normally spent manually checking all of the drawings by design consultants, and by the contractor and their

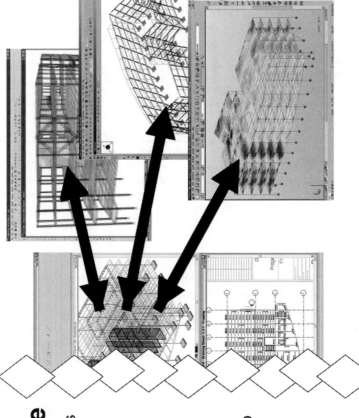

FIGURE 16.4 A 3 dimensional display as layers are added to the structural design. (*By permission: Autodesk® Revit® Structures, U.S.A.*)

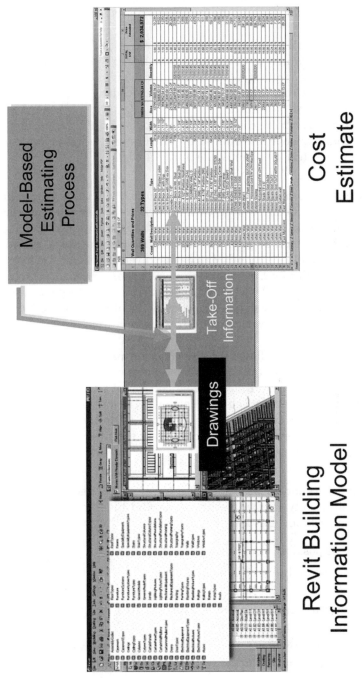

Model-Based Estimating Process

Revit Building Information Model

Drawings

Take-Off Information

Cost Estimate

FIGURE 16.5 Leveraging BIM in quantity take-off process. (*By permission: Autodesk® Revit® Structures, U.S.A.*)

389

subcontractors, will be reduced considerably or totally eliminated, giving all parties additional time for project management, quality control, and scheduling matters.

3-D modeling brings all of these interference problems to the fore during the design stage, and not during the hectic construction process as the design passes from structural engineer to architect to MEP design consultants. Thus, conflicts will be immediately identified and resolved through a collaborative effort.

This is certainly a more cost-effective way of dealing with interference problems than squeezing a duct size to fit under a beam or punching through a structural member while in the field.

4-D modeling. 4-D modeling adds a *time* factor to a 3-D model, allowing display of the design's progression during the construction cycle. CPM schedules can thus be transformed into living breathing presentations where the actual progression of construction over a period of time, such as a week or a month, can be displayed against an as-planned schedule. Schedules become more than just paper presentations when 4-D modeling is used. At weekly project meetings, the general contractor can now visually display specific parts of the "planned schedule" and graphically show the "as built" field condition at that point in time.

These presentations allow all parties to view problems, seek acceptable recovery methods, and look at the results of their efforts in next week's 4-D presentation. Delay claims can be either strengthened or defended against by using selected sequences of a 4-D presentation of *actual* versus *planned* events. And imagine a 4-D Two-Week Look Ahead schedule that's presented at one subcontractor meeting and then viewed at the next to see if everyone's goals have been achieved.

BIM—its promises and its problems. As a project management tool, the ability to effectively coordinate drawings and highlight any systems interference problems has a profound impact on the project by:

- Reducing or possibly eliminating Requests For Information (RFIs)

- Reducing or possibly eliminating Architects Supplementary Instructions (ASIs)

- Drastically reducing changes orders related to coordination/component conflict (interference) problems

- Reducing the potential for cost overruns by allowing more control over the factors that generate or create change orders

- Reducing delays in design and construction schedules

As a single source for building information, a data-based CAD BIM system presents many advantages:

- Plans, elevations, wall sections, and schedules are always consistent—change one, and all related work is changed.

- The coordination across different disciplines eliminates the problems previously associated with ensuring that everything fits in its allotted space—horizontally and vertically.

- Schedules for finishes, doors, windows, and hardware are easily generated and updated as changes occur in plan and elevation design.

- The ability to generate quantities of materials during design facilitates the procurement and, particularly in a design-build or negotiated project mode, tracking design with the budget.

- The data created by BIM continues to have a useful life during both commissioning and the continuing operation and maintenance of the building.

BIM can impact quality.   Because changes to one system or one item are reflected back through the data base to related systems, we may have finally gotten rid of that typical problem where a window size may have changed, but no corresponding change was made to the exterior masonry opening. That 3070 door in Room 507 changed to a 3468 did not update the door schedule. With BIM, these problems that created confusion and ate up man hours may no longer exist or, at least, will be dramatically reduced. And because it is a data base system and one change is recognized and adjusted throughout the design automatically, architects and engineers may find that they have a little more time to review and tweak their designs. The contractor relieved somewhat, or completely, from the task of issuing RFIs to questions relating to coordination or missing data can spend more time on processes, schedules, and quality.

Owners, tired of the finger pointing that happens whenever errors and omissions type change orders occur, will have one less argument to resolve—and one less cost to pay.

Cause and effect.   A recent survey of general contractors in the southeastern United States engaged in traditional design-bid-build projects revealed that 78 percent of the respondents reported the following frequency of problems:

- Problems with specifications—100 percent
- Unrealistic schedules—84 percent
- Physical interference problems—75 percent
- Tolerance problems—73 percent

This same survey revealed that 75 percent of responding general contractors attributed constructability problems to their inability to provide input during design. It's likely that these same problems affect contractors throughout the country, and by employing 3-D and 4-D database modeling, many of these problems affecting the entire industry can possibly be avoided.

Both developers and contractors look with dismay at the high cost of construction today. The $15.8 annual cost attributed to the inadequacy of today's interoperability (as reported by NIST) is too large an amount to be ignored. The federal government and the private sector of design professionals and contractors need to embrace this interoperability process as an effective way to deal with those rising costs.

BIM, with its promise of 3-D and 4-D modeling, may become more prevalent in the industry and make design and construction even more cost-effective.

Index